REVIEWING PHYSICS

The Physical Setting

with Sample Examinations

THIRD EDITION

Judah Landa
Former member New York State Regents Physics Committee
Former coordinator of Physics
Midwood High School, Brooklyn, New York

David R. Kiefer
Former Assistant Principal for Physical Science
Midwood High School, Brooklyn, New York
Science Supervisor, Indian Hills High School
Oakland, New Jersey

Amsco School Publications, Inc.,
a division of Perfection Learning®

The publisher wishes to thank the following Teachers who acted as reviewers for the third edition of
Reviewing Physics: The Physical Setting

Jack A. DePalma
Physics Teacher
Brooklyn Technical High School
Brooklyn, New York

Gregory Guido
Physics Teacher
Northport High School
Northport, New York

Tom Lynch
Science Department Supervisor
Oyster Bay High School
Oyster Bay, New York

Tom Meyer, Ph.D.
Science Teacher
Passaic High School
Passaic, New Jersey

Cover: Megan J. Shupe
Text and Composition: Northeastern Graphic, Inc.
Art: Hadel Studio

Physics in Your Life and Science, *Technology, and Society* features written by

Christine Caputo

Please visit our Web sites at: *www.amscopub.com* and *www.perfectionlearning.com*

When ordering this book, please specify:
either **1351101** *or* **REVIEWING PHYSICS: THE PHYSICAL SETTING, THIRD EDITION**

ISBN: 978-1-56765-914-6

Note to the Teacher

The newly revised, third edition books of this series—*Reviewing Biology: The Living Environment, Reviewing Earth Science: The Physical Setting, Reviewing Chemistry: The Physical Setting*, and *Reviewing Physics: the Physical Setting*— offer an innovative format that reviews the National Science Standards-based New York State Core Curriculum. The presentation is aligned with the New York City Scope and Sequence. The physics book also offers in the Enrichment sections material that is covered in a college-preparatory science course. The series is specifically geared to the needs of students who want to refresh their memory and review the material in preparation for final and Regents exams. Each book is readily correlated with the standard textbooks for this level.

The text begins with the *Introduction to Physics*, a review of measurement and mathematical skills that are useful to physics students. The remainder of the material in *Reviewing Physics: The Physical Setting, Third Edition* is divided into five chapters, each of which is subdivided into major topic sections followed by questions that are grouped in Part A and Part B-1. Each chapter contains a reading passage, Physics in Your Life, which discusses current topics in physics. In addition, each chapter contains a set of Chapter Review Questions, which include Part A, B-1, B-2, and C questions. Diagrams that aid in reviewing and testing the material often accompany questions. The more than 1,000 questions found in the text can be used for topic review throughout the year, as well as for exams and homework assignments. The tests at the back of the book can be used for final exams or practice for the final exam. The *Enrichment*, the final part of each Chapter, presents topics that are no longer part of the Core Curriculum.

The book is abundantly illustrated with clearly labeled drawings and diagrams that illuminate and reinforce the subject matter. Important science terms are **bold-faced** and defined in the text. Other terms that may be unfamiliar to students are *italicized* for emphasis. In addition, the large work-text format and open design make *Reviewing Physics: The Physical Setting, Third Edition* easy for students to read.

A section called *Laboratory Skills* follows the five chapters. This special section reviews the skills that all students should master in the course of completing one year of physics instruction at this level. *Reviewing Physics: The Physical Setting, Third Edition* also contains the 2006 edition of *Reference Tables and Charts*. There is a full Glossary, where students can find concise definitions of significant scientific terms. The extensive Index should be used by students to locate fuller text discussion of these and other physics terms.

Also included in this edition of *Reviewing Physics: The Physical Setting, Third Edition* are seven new *Science, Technology, and Society* features that explore current issues in physical science, technology, and society. Reading comprehension, constructed-response, and research questions presented at the end of each feature encourage students to evaluate the issues and to make their own decisions about the impact of science and technology on society, the environment, and their lives.

Contents

Introduction to Physics

Measurement

Science is built on observation and measurement. Measurement involves the comparison of an unknown quantity with known, standard units. Measurements are always inexact because they are subject to error.

Errors in Measurement

Factors that affect the accuracy of measurements include flaws in the method used to obtain the measurement, fluctuations in the environment, limitations of the instruments used, and human error. Measurement errors fall into two basic groups—systematic and random.

Systematic errors tend to be in one direction, either too high or too low. For example, if a thermometer reads 5°C instead of 0°C in an ice and water mixture, then all of its readings will be too high by 5°C. Systematic error can be reduced by checking the method used and adjusting, or calibrating, all instruments.

Random errors tend to produce readings that fluctuate; some readings are too high, some are too low. Such fluctuations are always present in measurements. For example, repeated measurements of temperature of a liquid might produce readings of 56.4°C, 56.5°C, and 56.2°C. Random error can be reduced by taking the average of a large number of measurements and by controlling environmental fluctuations.

Precision

Precision refers to the smallest decimal place obtained in a measurement. Units are always specified when giving the precision of a measurement. (See Table I-1.)

Table I-1	
Measurement	**Precision**
143 m	1 m (the units place in meters)
4.8 g	0.1 g (the tenths place in grams)
24.962 s	0.001 s (the thousandths place in seconds)

The scale markings on an instrument determine the possible precision of measurements made with that instrument. A measurement may be *estimated* to one-tenth of the smallest interval printed on the scale. Thus, the last recorded digit in a number obtained by a measurement is usually an estimate based on a reading between the smallest intervals marked on the scale. For example, the smallest printed interval on a 10-cm ruler is 1 mm, or 0.1 cm (Figure I-1). The possible

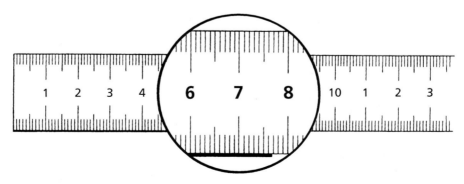

Figure I-1. Measuring the length of a line.

Table I-2.

Rule	Example	Number of Significant Figures
1. Zeros located at the end of a number and to the right of a decimal point are significant. They indicate the precision possible with the instrument used.	3.0 g	2
	12.3000 km	6
	1.000 s	4
	5.20 N	3
2. Zeros located between significant digits are significant.	30.9 V	3
	402.06007 mm	8
	1.030 ml	4 (rules 1 & 2)
3. Leading zeros are not significant. They may be included for clarity of format or to "hold place," but they are not the result of a measurement.	.042 kg	2
	0.042 J	2
	0.00000009 m	1
	0.160 A	3 (rules 1 & 3)
	0.106 W	3 (rules 2 & 3)
	0.016 m/s	2
	0.0010100 s	5 (rules 1, 2, & 3)
4. Zeros located at the end of a number and to the left of a decimal point are significant.	40. °C	2
	3000. K	4
	250,600. g	6 (rules 2 & 4)
5. Zeros located at the end of a number are not significant if they are not followed by a decimal point.	40 °C	1
	3000 K	1
	250,600 m	4 (rules 2 & 5)

precision when using this ruler is one-tenth of 1 mm, or 0.01 cm. Such a ruler could be used to obtain a reading of 7.68 cm, with the last digit (8) as an estimate. A reading of 7.683 cm would be beyond the possible precision of a measurement made with this ruler.

Significant Figures

Significant figures are those digits that are obtained properly and directly from an instrument, including the final, estimated digit. In determining the significant figures in a measurement, keep in mind that any digit from 1 to 9 is always significant. The only digit that may not be significant is 0 since zeros are sometimes used as "place holders." Table I-2 gives rules for determining which zeros in a measurement are significant and which are not significant.

Accuracy

The **accuracy** of a measurement refers to the agreement, or closeness, of its value to the true or accepted value. Accuracy may be expressed in terms of **absolute error** and *percent error*. In both cases, the absolute value (indicated by vertical lines) of the difference between the measured and accepted values is used to obtain a positive answer.

Absolute error
$= |$ measured value $-$ accepted value $|$

Percent error
$= \dfrac{|\text{measured value} - \text{accepted value}|}{\text{accepted value}} \times 100\%$

Sample Problem

1. The average of several measurements of the mass of an object is 48.60 g. Find the absolute error and percent error of this measurement if the actual mass of the object is 48.75 g.

Solution:

$$\text{Absolute error} = |48.60\text{ g} - 48.75\text{ g}| = 0.15\text{ g}$$

$$\text{Percent error} = \dfrac{|48.60\text{ g} - 48.75\text{ g}|}{48.75\text{ g}} \times 100\%$$

$$= 0.31\%$$

Rounding in Calculations

A chain is as strong as its weakest link. Similarly, a calculated answer can be only as precise as the least precise measurement involved in the calculation. As a result, calculated answers often must be rounded. If the digit to be dropped is less than 5, the digit to the left of it remains unchanged. For example, if 27.23 is rounded to three significant figures, it becomes 27.2. If the digit to be dropped is 5 or more, the digit to the left of it is increased by 1; for example, 27.46 rounded to three significant figures becomes 27.5.

In addition and subtraction, the answer must be rounded to the same precision as the *least* precise number in the calculation. For example:

Addition:

$$
\begin{array}{r}
6.12 \text{ g} \\
18.3 \text{ g} \\
+ 0.044 \text{ g} \\
\hline
24.464 \text{ g} = 24.5 \text{ g}
\end{array}
$$

In this calculation, 18.3 g is the least precise number—to the tenths place.

Subtraction:

$$
\begin{array}{r}
48.3639 \text{ m} \\
- 13.21 \text{ m} \\
\hline
35.1539 \text{ m} = 35.15 \text{ m}
\end{array}
$$

Here, 13.21 m is the least precise number—to the hundredths place.

In multiplication and division, the answer must be rounded to contain the same number of significant figures as the measurement with the *least* number of significant figures. For example:

Multiplication:

$$
\begin{array}{r}
9.78 \text{ m} \\
\times 1.4 \text{ m} \\
\hline
13.692 \text{ m}^2 = 14 \text{ m}^2
\end{array}
$$

In this calculation, 1.4 m has the least number of significant digits—2.

Division:

$$
\frac{180 \text{ g}}{5020.00 \text{ mL}} = 0.0358565 \text{ g/mL} = 0.036 \text{ g/mL}
$$

Here, 180 g has the least number of significant digits—2.

Sample Problem

2. Find the perimeter and area of the rectangle in Figure I-2.

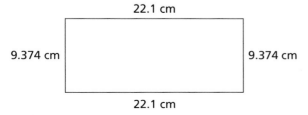

Figure I-2.

Solution:

Perimeter: The perimeter of a rectangle is equal to the sum of the lengths of the sides.

$$
\begin{array}{rll}
22.1 & \text{cm} & \text{(tenths)} \\
9.374 & \text{cm} & \text{(thousandths)} \\
22.1 & \text{cm} & \\
+ 9.374 & \text{cm} & \\
\hline
62.948 & \text{cm} = 62.9 \text{ cm} \\
& \text{(rounded to tenths place)}
\end{array}
$$

Area: The area of a rectangle is equal to its length multiplied by its width.

$$
\begin{array}{rll}
9.374 & \text{cm} & \text{(4 significant figures)} \\
\times 22.1 & \text{cm} & \text{(3 significant figures)} \\
\hline
207.1654 & \text{cm}^2 = 207 \text{ cm}^2 \\
& \text{(3 significant figures)}
\end{array}
$$

USING MEASUREMENTS

After making careful measurements, you may be asked to use those measurements to solve a problem or to draw a graph. Using measurements is often easier if you express the numbers in scientific notation.

Scientific Notation

In **scientific notation**, numbers are expressed in the form $A \times 10^n$, where the coefficient A is any number equal to or greater than 1 but less than 10, and the exponent n is an integer. Scientific notation is used to indicate the number of significant figures in a measurement and to make mathematical operations with very large and very small numbers easier. All of the digits in the coefficient are significant. The value of n is determined by counting the number of decimal places that the decimal point in the original number must be moved to form A. If the decimal point of the original number is moved to the left, n is positive; if the decimal point is moved to the right, n is negative. For example:

93,000,000 m becomes 9.3×10^7 m
(The decimal point is moved 7 places to the left.)

0.0002040 g becomes 2.040×10^{-4} g
(The decimal point is moved 4 places to the right.)

If 93,000,000 m represented a measurement that is precise to the thousands place, only the first five digits are significant. This cannot be expressed in decimal form, but in scientific notation it is possible to indicate *any* desired number of significant figures. The number can be written as 9.3000×10^7 m.

When adding or subtracting numbers in scientific notation, the value of n for each number in the calculation must be made identical before the coefficients are added or subtracted. The resulting number must be adjusted for correct precision and form. For example:

Addition:

$$(3.2 \times 10^4) + (4.9 \times 10^3)$$
$$= (3.2 \times 10^4) + (0.49 \times 10^4)$$
$$= 3.69 \times 10^4 = 3.7 \times 10^4$$

Subtraction:

$$(1.254 \times 10^{-1}) - (8.5 \times 10^{-2})$$
$$= (1.254 \times 10^{-1}) - (0.85 \times 10^{-1})$$
$$= 0.404 \times 10^{-1}$$
$$= 0.40 \times 10^{-1} = 4.0 \times 10^{-2}$$

When multiplying numbers in scientific notation, multiply the coefficients and add the exponents. The product of the coefficients should then be rounded off to the correct number of significant figures. For example:

$$(8.12 \times 10^2) \times (2.13 \times 10^5)$$
$$= 17.2956 \times 10^7$$
$$= 17.3 \times 10^7 = 1.73 \times 10^8$$

When dividing numbers in scientific notation, divide the coefficient of the numerator by the coefficient of the denominator and subtract the exponent of the denominator from the exponent of the numerator. For example:

$$\frac{6 \times 10^9}{8.75 \times 10^{12}} = 0.6857143 \times 10^{-3}$$
$$= 0.7 \times 10^{-3} = 7 \times 10^{-4}$$

Units and Equations

In mathematical calculations, units are treated like algebraic quantities. Units should appear in every step of a calculation. They must appear in the substitution step and the answer. For example:

$$2 \text{ m}^2 + 3 \text{ m}^2 = 5 \text{ m}^2$$

$$1.0 \text{ km} \times 3.0 \text{ km} \times 4.0 \text{ km} = 12 \text{ km}^3$$

$$8.0 \text{ kg} \times (3.0 \text{ m/s})^2 = 8.0 \text{ kg} \times 9.0 \text{ m}^2/\text{s}^2$$
$$= 72 \text{ kg} \cdot \text{m}^2/\text{s}^2$$

$$\sqrt{\frac{64 \text{ m}}{32 \text{ m/s}^2}} = \sqrt{2.0 \text{ s}^2} = 1.4 \text{ s}$$

In a formula, the units on either side of the equal sign must be equivalent. In some cases, it is necessary to convert some units to others before this equivalence can be demonstrated. When this is the case, three steps should be used:

1. Obtain needed relationships between units.
2. Substitute units with their equivalences.
3. Simplify.

Sample Problem

3. Given that d is the distance in meters, t is the time in seconds, v is the speed in meters/second, and a is the acceleration in meters/second^2, demonstrate that the formula $d = vt + \frac{1}{2}at^2$ is balanced as far as the units are concerned.

Solution:

$$d = vt + \tfrac{1}{2}at^2$$

$$\text{meters} = \left(\frac{\text{meters}}{\text{seconds}}\right)(\text{seconds})$$
$$+ \left(\frac{\text{meters}}{\text{seconds}^2}\right)(\text{seconds})^2$$

$$\text{meters} = \text{meters} + \text{meters}$$

$$\text{meters} = \text{meters}$$

Drawing Graphs

When drawing a graph, begin by giving a title to the graph. The next step is to label both axes with the appropriate units. By convention, the dependent variable is placed on the y-axis and the independent variable is placed on the x-axis. For example, when drawing a graph representing the change in distance of a moving object over time, distance (dependent variable) is placed on the y-axis. Time (the independent variable) is placed on the x-axis.

After labeling the axes, you must choose a scale for each axis with appropriate scale divisions. "Appropriate" means that most, but not necessarily all, of the grid is used and that the scale divisions allow for simple estimation between the lines. The next step is to plot the data points accurately.

The plotted data points frequently form a pattern. When you are asked to draw the **best-fit line** based on that pattern, use a transparent ruler or straightedge. A best-fit line is not achieved by connecting the data points but by drawing a straight line between them. If some or all of the data points do not lie on the line, they should be balanced on both sides of the line (Figure I-3). When the best-fit line is a curve, draw it as accurately as possible (Figure I-4).

slope of a curve at a particular point, draw the tangent to the curve at that point and calculate the slope of the tangent line.

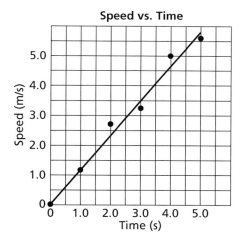

Figure I-3. Best-fit straight line.

To determine the slope of a straight line, pick any two points on the line and use the slope formula $m = \dfrac{\Delta y}{\Delta x}$ to determine the value of the slope. Be sure to include the units. To determine the

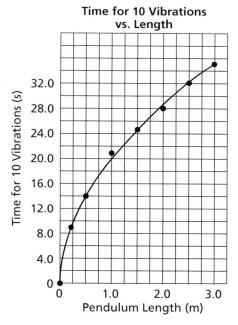

Figure I-4. Best-fit curve.

CHAPTER 1

Mechanics

VECTORS

The branch of physics known as **mechanics** deals with motion and forces. A **force** is a push or a pull. Within this branch are **kinematics**, the study of motion; **statics**, the study of forces on stationary objects; and **dynamics**, the study of the relationship between forces and motion.

Units of Measurement

Physicists use SI (International System) units, an extension of the metric system, for standards of measurement. Table 1-1 lists the seven fundamental SI units. Three of these physical quantities are basic to the study of mechanics: length, time, and mass. The fundamental units for these quantities are the *meter* (m), the *second* (s), and the *kilogram* (kg), respectively. Units that consist of combinations of the fundamental units are called **derived units**. For example, the unit of force, the newton, is a derived unit that is equivalent to a kilogram meter per second squared ($kg \cdot m/s^2$).

Table 1-1
Fundamental SI Units

Quantity	Fundamental Units	Symbol
Length	Meter	m
Mass	Kilogram	kg
Time	Second	s
Electric current	Ampere	A
Temperature	Kelvin	K
Amount	Mole	mol
Luminous intensity	Candela	cd

Scalar and Vector Quantities

Most of the quantities that we measure in daily life are **scalar quantities**, such as 9.3 seconds, 16°C, and 15 meters. A scalar quantity can be expressed by a number and an appropriate unit. It has magnitude but no direction.

Other physical quantities, called **vector quantities**, have a specific direction as well as magnitude. For example, we may need to specify that a car has traveled 50 km due south, that a plane is located 75 m southeast of the runway, or that a force of 20 newtons (N) acts to the right rather than to the left. (Force and newtons will be discussed later in this chapter.) Arrows called **vectors** represent such quantities. The length of the arrow represents the magnitude of the vector quantity (based on a specified scale), and the direction of the arrow is the direction of the vector quantity. For example, Figure 1-1 represents a displacement of 5 meters (using the given conversion scale) that is directed 40° from the horizontal.

Figure 1-1. Representation of a vector.

Vector quantities can be distinguished from scalar quantities either by placing an arrow over the symbol for the vector quantity or by printing it in **bold** type. In this book, vector quantities are represented by bold letters. When a symbol represents only the magnitude of the vector, it is not bold.

Vector Addition

The sum of two or more vectors is not the sum of their magnitudes. Instead, the sum of two or more vectors is another vector called the **resultant**. The resultant vector is the result of the combined effect of the vectors. Like all vectors, the resultant has a magnitude and a direction. There are several methods for determining a resultant.

The Head-to-Tail Method

In the *head-to-tail method* of vector addition, a diagram is drawn in which the vectors are positioned so that the tail of one vector is placed at the head of the other vector. Figure 1-2 shows a head-to-tail diagram of two vectors **A** and **B**. The resultant of these two vectors, labeled **C**, is obtained by drawing a new vector from the tail of the first vector to the head of the last. The magnitude of **C** is found by measuring the length of **C** and using the given conversion scale. The direction of **C** can be found with a protractor.

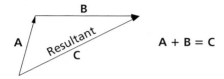

Figure 1-2. The head-to-tail method of vector addition.

The Parallelogram Method

To use the *parallelogram method* of vector addition, begin by drawing a diagram in which the tails of the two vectors to be added (**A** and **B**) are placed at a common point. Then a parallelogram is constructed with the two vectors as adjacent sides (Figure 1-3). The resultant vector **C** is the diagonal of the parallelogram drawn from the point where the tails meet. As with the head-to-tail method, the most direct way of finding the magnitude and direction of the resultant is with a ruler and protractor. Using the specified conversion scale—such as 1 cm = 1 newton (in the case of force vectors), or 1 cm = 1 m/s (in the case of velocity vectors)—the magnitude of the resultant is found by measuring its length, and its direction is found by measuring the angle between the resultant and any chosen direction.

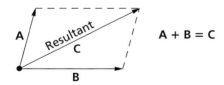

Figure 1-3. The parallelogram method of vector addition.

Mathematical methods of vector addition are particularly convenient for cases involving 0°, 180°, and 90° angles between the vectors to be added. Consider two vectors, one with a magnitude of 5 m and the other with a magnitude of 2 m, and find their resultant at these angles.

At 0° The simplest case of vector addition occurs when the two vectors to be added point in the same direction (Figure 1-4). In this case, the head-to-tail method must be used. The magnitude of the resultant is simply the sum of the magnitudes of the two vectors (2 m + 5 m, or 7 m), and the direction of the resultant is the same as the direction of the two vectors.

Figure 1-4. Vectors added at 0°.

At 180° In this case the vectors are pointing in opposite directions, and again, the head-to-tail method must be used. Suppose the vector with a magnitude of 5 m points to the right, and the vector with a magnitude of 2 m points to the left. Using the head-to-tail method, we find that the magnitude of the resultant is equal to the difference of the magnitudes of the two vectors (5 m − 2 m, or 3 m), and the direction of the resultant is the direction of the larger vector—in this case to the right (Figure 1-5).

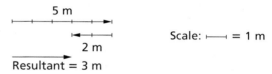

Figure 1-5. Vectors added at 180°.

At 90° If the vectors are at a 90° angle from one to another, the resultant can be found by using either the head-to-tail method or the parallelogram method. Once the resultant has been drawn using one of these methods, its magnitude can be calculated by using the Pythagorean theorem, since the two vectors to be added and their resultant form a right triangle (Figure 1-6). The resultant is the hypotenuse of the triangle. The Pythagorean theorem states that the square of the hypotenuse of a right triangle is equal to the sum of the squares of the other two sides ($a^2 + b^2 = c^2$). In our case we can write

$$R^2 = (5 \text{ m})^2 + (2 \text{ m})^2 = 29 \text{ m}^2$$

$$R = \sqrt{29 \text{ m}^2} = 5.4 \text{ m}$$

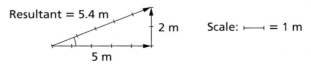

Figure 1-6. Vectors added at 90°.

The direction of the resultant can be found with a protractor or by using certain trigonometric operations. If the vector with a magnitude of 5 m points due east and the vector with a magnitude of 2 m points due north, the direction of the resultant is approximately 22° north of east.

The magnitude of the resultant of any two vectors is a maximum when the vectors point in the same direction—that is, with a 0° angle between them. It is a minimum when the vectors point in opposite directions—with a 180° angle between them. For vectors at angles between 0° and 180°, the magnitude of the resultant lies between the maximum and minimum values. As the angle between two vectors increases from 0° to 180°, their resultant decreases from its maximum to its minimum value.

Resolution of Vectors

In the previous section we learned how two or more vectors can be added to obtain the resultant. The *resolution of vectors* is the reverse procedure: a single vector is regarded as the resultant of two or more vectors, called the *component vectors*, which must be determined. A vector can be resolved into an infinite number of components. Figure 1-7 shows a vector **R** resolved into four components, **A**, **B**, **C**, and **D**.

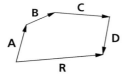

Figure 1-7. Resolution for vector **R** into four components: **A** + **B** + **C** + **D** = **R**.

It is often useful to resolve a vector into two perpendicular components. Consider a force vector with a magnitude of 60. N, directed 30° from the horizontal. Let us resolve this vector into two components—one horizontal and the other vertical. The first step is to draw the 60.-N vector, making sure, with a ruler and protractor and using the appropriate conversion scale, that the length of the vector represents 60. N and its direction is 30° from the horizontal (Figure 1-8).

The next step is to draw a vertical line through the head of the vector and a horizontal line through the tail of the vector, forming a right triangle. The 60.-N vector to be resolved is now the hypotenuse, and its vertical and horizontal components are the legs of the triangle.

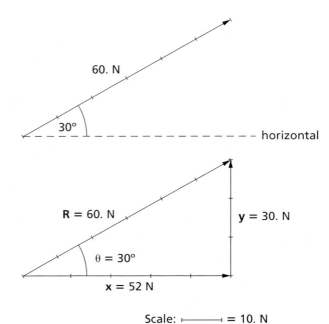

Figure 1-8. Resolution of vector **R** into two perpendicular components.

Finally, we measure the legs of the triangle and, using the specified scale, determine the magnitudes of the vertical and horizontal components. In this case, the horizontal component has a magnitude of 52 N and the vertical component has a magnitude of 30. N.

Another way to determine the magnitude of the vertical and horizontal components is by using trigonometry. We know from trigonometry that $\cos \theta$ is equal to the magnitude of the adjacent leg divided by the magnitude of the hypotenuse. (See the *Physics Reference Tables*.) Therefore, the magnitude of the horizontal component can be found by using the formula

$$A_x = A \cos \theta$$

where A_x is the magnitude of the horizontal component, A is the magnitude of the resultant, and θ is the angle between them. In the example, the horizontal component becomes

$$A_x = 60. \text{ N} \cos 30°$$

$$A_x = 60. \text{ N} \times 0.866 = 52 \text{ N}$$

Similarly, $\sin \theta$ is equal to the magnitude of the opposite leg divided by the hypotenuse. Thus, the magnitude of the vertical component can be found by using the formula

$$A_y = A \sin \theta$$

where A_y is the magnitude of the vertical component. In the example above, the vertical component becomes

$$A_y = 60. \text{ N} \sin 30°$$

$$A_y = 60. \text{ N} \times 0.500 = 30. \text{ N}$$

Displacement

Displacement is a vector quantity, symbolized by **d**, indicating a change of position in a particular direction. An object's displacement vector is always a straight line drawn from the object's initial position to the object's final position no matter what the object's actual path was between these two points (Figure 1-9). The magnitude of the displacement vector represents the shortest distance between the two points. The actual distance the object has traveled is a scalar quantity that represents the length of its path. For example, a boomerang may be thrown a distance of several meters. If it returns to its original starting point, however, the magnitude of its displacement is zero. If the path of an object consists of several displacement vectors, its total displacement vector is the resultant of all the various displacement vectors.

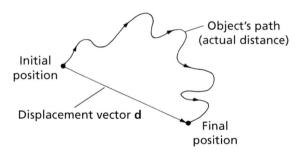

Figure 1-9. Displacement.

Sample Problem

1. A bird flies 6 km east, then 4 km west. Find the total distance traveled and displacement after this flight.

    Solution: The scalar distance traveled is 6 km + 4 km, or 10 km. To determine the displacement, the head-to-tail method of vector addition must be used (see Figure 1-10). The total displacement vector is 2 km east. It is the resultant of the two displacement vectors.

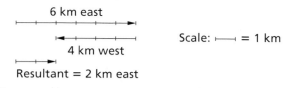

Figure 1-10.

QUESTIONS

1. A girl leaves a history classroom and walks 10. meters north to a drinking fountain. Then she turns and walks 30. meters south to an art classroom. What is the girl's total displacement from the history classroom to the art classroom? (1) 20. m south (2) 20. m north (3) 40. m south (4) 40. m north

2. A projectile is fired with an initial velocity of 120. meters per second at an angle θ above the horizontal. If the projectile's initial horizontal speed is 55 meters per second, then angle θ measures approximately (1) 13° (2) 27° (3) 63° (4) 75°

3. A student on her way to school walks four blocks south, five blocks west, and another four blocks south, as shown in the diagram.

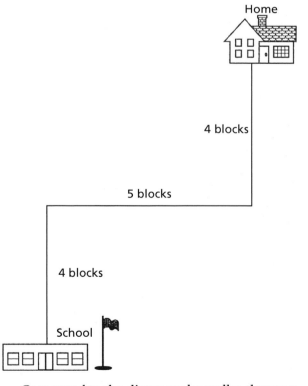

Compared to the distance she walks, the magnitude of her displacement from home to school is (1) less (2) greater (3) the same

4. The diagram below shows a resultant vector, **R**.

Which diagram best represents a pair of component vectors, **A** and **B**, that would combine to form resultant vector **R**?

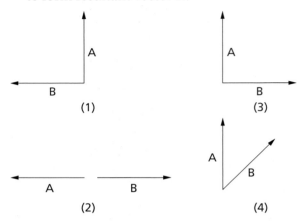

5. A vector makes an angle θ with the horizontal. The horizontal and vertical components of the vector will be equal in magnitude if the angle is (1) 30° (2) 45° (3) 60° (4) 90°

PART B-1
---

6. Which pair of forces acting concurrently on an object will produce the resultant of greatest magnitude?

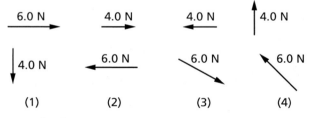

7. Which vector diagram represents the greatest magnitude of displacement for an object?

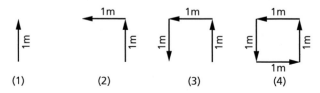

KINEMATICS

Displacement vectors are the basis of two other important vector quantities, velocity and acceler-

ation. Together, these three vectors are used to describe the motion of an object in space.

Velocity and Speed

Velocity is a vector quantity defined as the change in displacement per unit time. This is stated mathematically as

$$v = \frac{\Delta d}{\Delta t}$$

where Δd is the change in displacement that occurs during the time interval Δt. **Speed** is a scalar quantity equal to the magnitude of the velocity vector. **Average speed** is defined as the distance traveled per unit time and is symbolized by $\bar{v}$:

$$\bar{v} = \frac{\text{total distance}}{\text{total time}}$$

$$\bar{v} = \frac{d}{t}$$

The unit for velocity, speed, and average speed is meters per second (m/s).

Sample Problems
---

2. A car travels 2.1 km south in 7.0 minutes. Find its average velocity in meters/second.

Solution:

Convert kilometers to meters.

$$d = 2.1 \text{ km} \times 1000 \text{ m/km} = 2100 \text{ m south}$$

Convert minutes to seconds.

$$t = 7.0 \text{ min} \times 60. \text{ s/min} = 420 \text{ s}$$

$$\bar{v} = \frac{d}{t} = \frac{2100 \text{ m}}{420 \text{ s}} = 5.0 \text{ m/s south}$$

3. A car travels 600. m one way with a speed of 40. m/s and then travels 600. m in another way at 20. m/s. What is the car's average speed? (*Caution*: The answer is not 30. m/s.)

Solution:

$$\bar{v} = \frac{\text{total distance}}{\text{total time}} = \frac{1200. \text{ m}}{\Delta t}$$

The time for trip one is

$$\Delta t_1 = \frac{600. \text{ m}}{40. \text{ m/s}} = 15 \text{ s}$$

The time for trip two is

$$\Delta t_2 = \frac{600.\ \text{m}}{20.\ \text{m/s}} = 30\ \text{s}$$

Thus,

$$\bar{v} = \frac{1200.\ \text{m}}{45\ \text{s}} = 27\ \text{m/s}$$

Uniform Velocity

Uniform velocity occurs when the speed and the direction of a moving object are constant. Table 1-2 lists the displacements from the starting point at various times for an object moving with uniform velocity. Note that for every second that passes, there is an equal change in the displacement.

Table 1-2								
Time (s)	0	1	2	3	4	5	6	7
Displacement (m)	0	3	6	9	12	15	18	21

Using the data in Table 1-2, the velocity of the object can be found from any of its d and t values. For instance, at $t = 0$ s and $t = 4$ s, $d = 0$ m and $d = 12$ m, respectively. Plugging these values into the equation for velocity, we obtain

$$v = \frac{\Delta d}{\Delta t} = \frac{d_f - d_i}{t_f - t_i} = \frac{12\ \text{m} - 0\ \text{m}}{4\ \text{s} - 0\ \text{s}} = 3\ \text{m/s}$$

The same velocity will be found using any of the other pairs of data. For instance, at $t = 1$ and $t = 6$, $d = 3$ m and $d = 18$ m, respectively. Thus,

$$v = \frac{\Delta d}{\Delta t} = \frac{d_f - d_i}{t_f - t_i} = \frac{18\ \text{m} - 3\ \text{m}}{6\ \text{s} - 1\ \text{s}} = \frac{15\ \text{m}}{5\ \text{s}} = 3\ \text{m/s}$$

When all the data pairs are graphed, we obtain the typical displacement-time or d-t graph for an object moving at uniform velocity—a straight line with a constant slope (Figure 1-11). The slope of the line $\left(\dfrac{\Delta y}{\Delta x}\right)$ is equal to the magnitude of the object's velocity.

$$\frac{\Delta y}{\Delta x} = \frac{\Delta d}{\Delta t} = \frac{6\ \text{m}}{2\ \text{s}} = 3\ \text{m/s}$$

Interpreting *d* vs. *t* Graphs

The distance vs. time graph, or d-t graph, representing the motion of an object moving with uniform velocity is always a straight line.

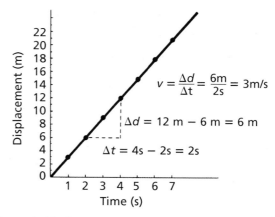

Figure 1-11. Slope equals velocity on a distance vs. time graph.

Figure 1-12 is the d-t graph for an object at rest. As time changes, there is no change in the object's position. An object at rest is represented by a line whose slope equals zero.

Figure 1-12.

Figure 1-13 represents two objects moving at different uniform velocities. Since the slope of the d-t graph is equal to the velocity, the faster object is represented by the line with the steeper slope. Remember slope, $m = \dfrac{\Delta y}{\Delta x}$.

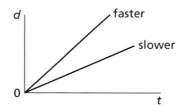

Figure 1-13.

Figure 1-14 represents uniform velocity, but the slope of the line is negative. This means that **d** is decreasing with time and the object is moving toward (rather than away from) the starting point.

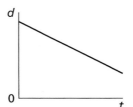

Figure 1-14.

The *d-t* graph in Figure 1-15 is not a straight line and represents nonuniform velocity. The slope increases with time, indicating that the object's speed is increasing as time goes by.

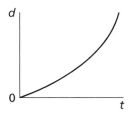

Figure 1-15.

These graphs may be combined to describe the entire journey of an object. For example, in Figure 1-16, an object is initially at rest and remains motionless until time t_1. The object then accelerates until time t_2, after which its velocity is uniform until t_3. Its velocity then jumps to a higher value, and remains constant until time t_4. The object then decelerates (the slope decreases) to a stop and remains at rest between times t_5 and t_6. Finally, the object reverses direction and returns to its starting point with uniform velocity.

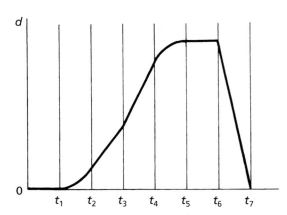

Figure 1-16.

Acceleration

Acceleration is a vector quantity defined as the time-rate of change of velocity. Acceleration is denoted by *a* and is expressed mathematically as

$$a = \frac{\Delta v}{\Delta t} = \frac{v_f - v_i}{t_f - t_i}$$

where Δv is the change in the object's velocity during the time interval Δt. It is equal to the difference between the final velocity v_f at time t_f and the initial velocity v_i at time t_i divided by Δt.

The unit for acceleration is m/s/s (meters per second per second). This indicates that acceleration is the change in velocity (in m/s) per unit time (in s). The standard practice is to combine the time units and report acceleration in m/s² (meters per second squared).

Uniform Acceleration

An object whose velocity is changing by a fixed amount per second is said to be *accelerating uniformly*. The average velocity $\bar{v}$ of an object undergoing **uniform acceleration** can be found from the formula

$$\bar{v} = \frac{v_i + v_f}{2}$$

$$\text{or} \ \ \bar{v} = \frac{d}{t}$$

This expression provides a formula for finding the total distance traveled by an object during uniformly accelerated motion:

$$d = \bar{v}t$$

This distance can also be expressed as a function of initial velocity, acceleration, and time.

$$d = v_i t + \tfrac{1}{2}at^2$$

When the object starts from rest, $v_i = 0$ and this formula becomes $d = \tfrac{1}{2}at^2$.

By combining $a = \frac{\Delta v}{t}$ with $\Delta v = v_f - v_i$, we can express the final velocity of a uniformly accelerated object as

$$v_f = v_i + at$$

We can also express the final velocity as a time-independent equation:

$$v_f^2 = v_i^2 + 2ad$$

When $v_f = 0$ these equations become

$$v_f = at \ \ \text{and} \ \ v_f^2 = 2ad$$

Sample Problems

4. A racing car increases its speed at a uniform rate from 30 m/s to 40 m/s in 5 s. Find its acceleration.

   Solution:

   $$\Delta v = v_f - v_i = 40 \text{ m/s} - 30 \text{ m/s} = 10 \text{ m/s}$$

   $$a = \frac{\Delta v}{t} = \frac{10 \text{ m/s}}{5 \text{ s}} = 2 \text{ m/s}^2$$

5. A cart moving at 9 m/s decelerates uniformly at the rate of 0.16 m/s² to 6 m/s. How much time has elapsed?

Solution:

$$\Delta v = v_f - v_i$$
$$= 6 \text{ m/s} - 9 \text{ m/s} = -3 \text{ m/s}$$

$$t = \frac{\Delta v}{a}$$

$$t = \frac{-3 \text{ m/s}}{-0.16 \text{ m/s}^2} = 19 \text{ s}$$

6. A jet starts from rest and accelerates uniformly at 3 m/s². What is its final velocity and the distance it has traveled from the starting point after 20 s?

Solution:

$$v_f = v_i + at$$

$$v_i = 0$$

$$v_f = at$$

$$v_f = (3 \text{ m/s})(20 \text{ s}) = 60 \text{ m/s}$$

$$d = v_i t + \tfrac{1}{2} at^2$$

$$v_i = 0, \, v_i t = 0$$

$$d = \tfrac{1}{2} at^2$$

$$d = \tfrac{1}{2}(3 \text{ m/s}^2)(20 \text{ s})^2 = 600 \text{ m}$$

7. A plane must achieve a speed of 95 m/s to take off. If the runway is 350 m long, what acceleration is required?

Solution:

$$v_f^2 = v_i^2 + 2ad$$

$$v_i = 0$$

$$v_f^2 = 2ad$$

$$a = \frac{v_f^2}{2d}$$

$$a = \frac{(95 \text{ m/s})^2}{2 \times 350 \text{ m}}$$

$$a = \frac{9025 \text{ m}^2/\text{s}^2}{700 \text{ m}} = 13 \text{ m/s}^2$$

8. An object accelerates uniformly from 12 m/s to 40 m/s in 7.0 s. How far does it travel during this interval?

Solution:
Method I
Find a, then find d.

$$a = \frac{\Delta v}{t}$$

$$a = \frac{40. \text{ m/s} - 12 \text{ m/s}}{7.0 \text{ s}} = 4.0 \text{ m/s}^2$$

$$d = v_i t + \tfrac{1}{2} at^2$$

$$d = (12 \text{ m/s})(7.0 \text{ s}) + \tfrac{1}{2}(4.0 \text{ m/s}^2)(7.0 \text{ s})^2$$

$$d = 84 \text{ m} + 98 \text{ m} = 182 \text{ m} = 180 \text{ m}$$

Method II
Find $\bar{v}$, then find d.

$$\bar{v} = \frac{v_i + v_f}{2}$$

$$\bar{v} = \frac{12 \text{ m/s} + 40. \text{ m/s}}{2} = 26 \text{ m/s}$$

$$\bar{v} = \frac{d}{t}$$

$$d = \bar{v} t$$

$$d = (26 \text{ m/s})(7.0 \text{ s}) = 182 \text{ m} = 180 \text{ m}$$

Accelerated Motion Graphs

The velocity-time data in Table 1-3 for an object accelerating uniformly from rest show that there are equal increments in velocity for equal increments of time.

Table 1-3

Time (s)	0	1.0	2.0	3.0	4.0	5.0	6.0	7.0
Velocity (m/s)	0	1.5	3.0	4.5	6.0	7.5	9.0	10.5

The object's acceleration can be obtained by dividing the change in velocity Δv between any two points in time by the corresponding change in time t. Thus

$$a = \frac{\Delta v}{t} = \frac{9.0 \text{ m/s} - 3.0 \text{ m/s}}{6.0 \text{ s} - 2.0 \text{ s}} = 1.5 \text{ m/s}^2$$

When all the data pairs are graphed, we obtain the typical velocity vs. time, or v-t, graph for an object undergoing uniform acceleration: a straight line with a constant slope equal to the acceleration (Figure 1-17).

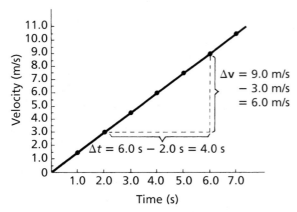

Figure 1-17. Slope equals acceleration on a velocity vs. time graph.

$$a = \frac{\Delta v}{t} = \frac{6.0 \text{ m/s}}{4.0 \text{ s}} = 1.5 \text{ m/s}^2$$

The *d-t* (displacement-time) data for this object can be calculated by using the formula $d = v_i t + \frac{1}{2}at^2$, with $v_i = 0$ and $a = 1.5$ m/s². When these data are graphed, we obtain the typical *d-t* graph for an object undergoing uniform acceleration: a parabola (Figure 1-18). The slope of a line tangent to the *d-t* graph at any point is equal to the **instantaneous velocity** at that point.

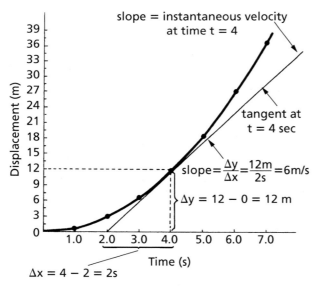

Figure 1-18.

Figure 1-19 represents an object traveling with uniform velocity. The slope of the *v-t* graph is equal to zero, indicating no acceleration.

Figure 1-19.

In Figure 1-20, the straight line has a negative slope, indicating that the object is decelerating, slowing uniformly.

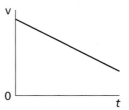

Figure 1-20.

In Figure 1-21, an object decelerates uniformly, until it comes to a stop on reaching the *x*-axis, when $v = 0$. Then it reverses direction (v is negative) and travels at ever increasing speed.

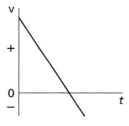

Figure 1-21.

The area under a *v-t* graph represents the distance an object has traveled. Figure 1-22 illustrates the simplest case in which the velocity is constant throughout the given time interval.

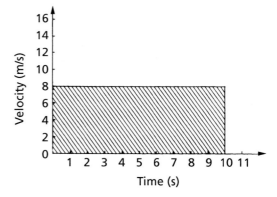

Figure 1-22.

Area = (base)(height)
      = (time)(velocity)
      = distance
      = (8 s)(10 m/s) = 80 m

When the area under the *v vs. t* graph is a right triangle, the area is found using the equation

$$\text{Area} = \tfrac{1}{2}\text{base} \times \text{height}$$

QUESTIONS

PART A

8. A car travels 90. meters due north in 15 seconds. Then the car turns around and travels 40. meters due south in 5.0 seconds. What is the magnitude of the average velocity of the car during this 20-second interval? (1) 2.5 m/s (2) 5.0 m/s (3) 6.5 m/s (4) 7.0 m/s

9. A skater increases her speed uniformly from 2.0 meters per second to 7.0 meters per second over a distance of 12 meters. The magnitude of her acceleration as she travels this 12 meters is (1) 1.9 m/s^2 (2) 2.2 m/s^2 (3) 2.4 m/s^2 (4) 3.8 m/s^2

10. In a 4.0-kilometer race, a runner completes the first kilometer in 5.9 minutes, the second kilometer in 6.2 minutes, the third kilometer in 6.3 minutes, and the final kilometer in 6.0 minutes. The average speed of the runner for the race is approximately (1) 0.16 km/min (2) 0.33 km/min (3) 12 km/min (4) 24 km/min

11. A golf ball is hit with an initial velocity of 15 meters per second at an angle of 35 degrees above the horizontal. What is the vertical component of the golf ball's initial velocity? (1) 8.6 m/s (2) 9.8 m/s (3) 12 m/s (4) 15 m/s

12. The speed of a wagon increases from 2.5 meters per second to 9.0 meters per second in 3.0 seconds as it accelerates uniformly down a hill. What is the magnitude of the acceleration of the wagon during this 3.0-second interval? (1) 0.83 m/s^2 (2) 2.2 m/s^2 (3) 3.0 m/s^2 (4) 3.8 m/s^2

13. An object with an initial speed of 4.0 meters per second accelerates uniformly at 2.0 meters per second^2 in the direction of its motion for a distance of 5.0 meters. What is the final speed of the object? (1) 6.0 m/s (2) 10. m/s (3) 14 m/s (4) 36 m/s

14. An astronaut drops a hammer from 2.0 meters above the surface of the moon. If the acceleration due to gravity on the moon is 1.62

meters per second^2, how long will it take for the hammer to fall to the moon's surface? (1) 0.62 s (2) 1.2 s (3) 1.6 s (4) 2.5 s

15. The average speed of a runner in a 400.-meter race is 8.0 meters per second. How long did it take the runner to complete the race? (1) 80. s (2) 50. s (3) 40. s (4) 32. s

16. Which statement about the movement of an object with zero acceleration is true? (1) The object must be at rest. (2) The object must be slowing down. (3) The object may be speeding up. (4) The object may be in motion.

17. An object travels for 8.00 seconds with an average speed of 160. meters per second. The distance traveled by the object is (1) 20.0 m (2) 200. m (3) 1280 m (4) 2460 m

18. An object is displaced 12 meters to the right and then 16 meters upward. The magnitude of the resultant displacement is (1) 1.3 m (2) 4.0 m (3) 20 m (4) 28 m

19. An object moves a distance of 10 meters in 5 seconds. The average speed of the object is (1) 0.5 m/s (2) 2.0 m/s (3) 40 m/s (4) 50 m/s

20. The graph following represents the relationship between velocity and time for an object moving in a straight line.

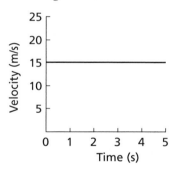

What is the acceleration of the object? (1) 0 m/s^2 (2) 5 m/s^2 (3) 3 m/s^2 (4) 15 m/s^2

21. Acceleration is a vector quantity that represents the time-rate of change in (1) momentum (2) velocity (3) distance (4) energy

22. A moving body must undergo a change of (1) velocity (2) acceleration (3) position (4) direction

23. What is the magnitude of the vertical component of the velocity vector shown below?

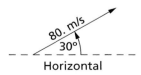

(1) 10. m/s (2) 69 m/s (3) 30. m/s (4) 40. m/s

24. The maximum number of components that a single force may be resolved into is (1) one (2) two (3) three (4) unlimited

25. Which quantity has both magnitude and direction? (1) distance (2) speed (3) mass (4) velocity

26. If a man walks 17 meters east then 17 meters south, the magnitude of the man's displacement is (1) 17 m (2) 24 m (3) 30. m (4) 34 m

PART B-1

27. Which graph best represents the motion of a block accelerating uniformly down an inclined plane?

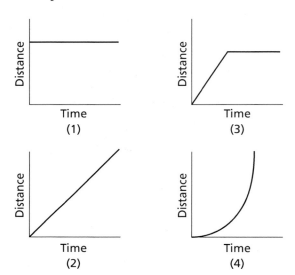

28. Which pair of graphs represents the same motion of an object?

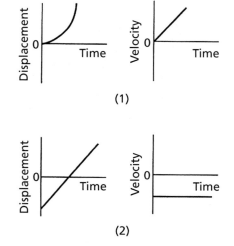

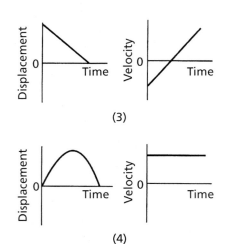

(3)

(4)

29. The graph below represents the relationship between speed and time for an object moving along a straight line.

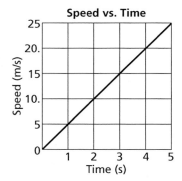

What is the total distance traveled by the object during the first 4 seconds? (1) 5 m (2) 20 m (3) 40 m (4) 80 m

30. The graph below shows the velocity of a race car moving along a straight line as a function of time.

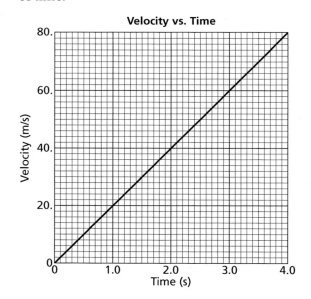

What is the magnitude of the displacement of the car from $t = 2.0$ seconds to $t = 4$ seconds? (1) 40. m (2) 60. m (3) 120 m (4) 160 m

31. The displacement-time graph below represents the motion of a cart initially moving forward along a straight line.

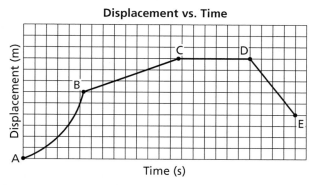

During which interval is the cart moving forward at constant speed? (1) *AB* (2) *BC* (3) *CD* (4) *DE*

32. Which graph best represents the relationship between velocity and time for an object that accelerates uniformly for 2 seconds, then moves at a constant velocity for 1 second, and finally decelerates for 3 seconds?

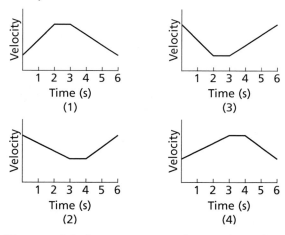

33. The graph below represents the motion of an object traveling in a straight line as a function of time. What is the average speed of the object during the first 4 seconds? (1) 1 m/s (2) 2 m/s (3) 0.5 m/s (4) 0 m/s

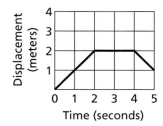

34. Which graph represents an object moving at a constant speed for the entire time interval?

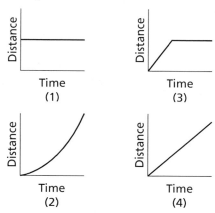

35. The graph shows the relationship between speed and time for two objects, *A* and *B*. Compared with the acceleration of object *B*, the acceleration of object *A* is (1) one-third as great (2) twice as great (3) three times as great (4) the same

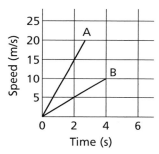

Base your answers to questions 36 through 41 on the graph below, which represents the relationship between speed and time for an object in motion along a straight line.

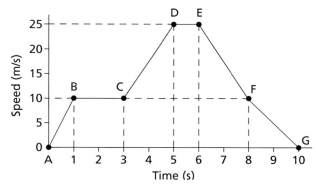

36. What is the acceleration of the object during the time interval $t = 3$ s to $t = 5$ s? (1) 5.0 m/s^2 (2) 7.5 m/s^2 (3) 12.5 m/s^2 (4) 17.5 m/s^2

37. What is the average speed of the object during the time interval $t = 6$ s to $t = 8$ s? (1) 7.5 m/s (2) 10 m/s (3) 15 m/s (4) 17.5 m/s

38. What is the total distance traveled by the object during the first 3 seconds? (1) 15 m (2) 20 m (3) 25 m (4) 30 m

39. During which interval is the object's acceleration the greatest? (1) *AB* (2) *CD* (3) *DE* (4) *EF*

40. During the interval $t = 8$ s to $t = 10$ s, the speed of the object is (1) zero (2) increasing (3) decreasing (4) constant, but not zero

41. What is the maximum speed reached by the object during the 10 seconds of travel? (1) 10 m/s (2) 25 m/s (3) 150 m/s (4) 250 m/s

Base your answers to questions 42 through 46 on the accompanying graph, which represents the motions of four cars on a straight road.

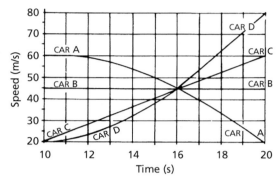

42. The speed of car *C* at time $t = 20$ s is closest to (1) 60 m/s (2) 45 m/s (3) 3.0 m/s (4) 0.6 m/s

43. Which car has zero acceleration? (1) *A* (2) *B* (3) *C* (4) *D*

44. Which car is decelerating? (1) *A* (2) *B* (3) *C* (4) *D*

45. Which car moves the greatest distance in the time interval $t = 10$ s to $t = 16$ s? (1) *A* (2) *B* (3) *C* (4) *D*

46. Which graph best represents the relationship between distance and time for car *C*?

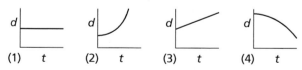

Statics

Statics is the study of the effect of forces on stationary objects.

Force

Force is defined as a push or a pull. It is a vector quantity with magnitude and direction. We speak of the downward force of gravity, or the upwardly directed force of buoyancy. A force may act on an object at a distance, without physical contact. The unit of force is the **newton** (N).

Addition of Concurrent Forces

When more than one force acts on an object at the same time and at the same point, the forces are said to be **concurrent forces**. The effect on the object is the same as if it were acted upon by the resultant of the concurrent forces. Vector addition must be used to find the resultant, or **net force**.

If two tractors pull on a tree, each with 5000. N of force and with an angle of 60° between their ropes (Figure 1-23), the tree experiences an applied force equal to the resultant of these two forces. Since concurrent forces share a common vertex, it is easiest to find the resultant by using the parallelogram method of vector addition. The resultant of the forces applied by the tractors on the tree has a magnitude of 8660. N (not 5000. N + 5000. N), and its direction is 30° from either of the force vectors.

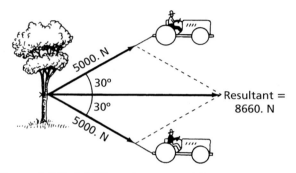

Figure 1-23. Addition of concurrent forces.

From the previous discussion of vector addition, we know that if the angle between the forces is 0°, the resultant is a maximum. If the tractors were pulling in the same direction, the tree would experience an applied force of 5000 + 5000 N, or 10,000 N. Conversely, if the angle between the forces is 180°, the force is a minimum. In this case, if the tractors were pulling in opposite directions, the resultant force on the tree would be zero.

Equilibrium

An object is in **equilibrium** when the vector sum of the concurrent forces acting on the object is zero. When there is no net force on an object at rest, it will remain at rest, or in **static equilibrium**. When there is no net force on an object already in motion, the object maintains a constant velocity and is said to be in **dynamic equilibrium**.

If the resultant of the concurrent forces on an object is not equal to zero, the object is not in a state of equilibrium. In order for equilibrium to be maintained, a force equal in magnitude and opposite in direction to the resultant must be applied to the object. The balancing force that creates equilibrium is called the **equilibrant**.

9. A sign is suspended from a horizontal pole by two wires (Figure 1-24). Each wire is at a 45° angle from the horizontal and pulls on the sign with a force of 28.3 N. (a) Find the resultant force on the sign due to the forces exerted by the two wires. (b) Find the weight of the sign.

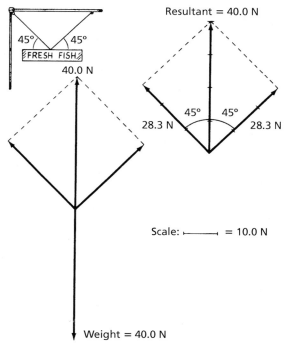

Figure 1-24.

Solution: Using the parallelogram method of vector addition and the given scale, we find that the resultant force has a magnitude of 40.0 N and pulls the sign directly upward.

Since the sign is in static equilibrium, the weight of the sign (acting vertically downward) is the equilibrant of the resultant of the two forces exerted by the wires. What must this weight be? The object's weight must therefore be equal in magnitude to the upward force of the resultant. Thus, the object must weigh 40.0 N. If we now add the three forces acting on the sign, the resultant is zero.

Resolution of Forces

Like all vectors, forces can be resolved into an unlimited number of components. In many situations, the effect of a force on an object can best be ascertained by resolving the force into two perpendicular components.

For example, in Figure 1-25 a woman is pulling a sled by a rope with a force of 35.0 N in a direction 30° from the horizontal. The horizontal component of the force acts to pull the sled in the horizontal direction while the vertical component of the force acts to lift the sled vertically upward. To determine the magnitude of the force in each of these directions, we must resolve the 35.0-N force into its horizontal and vertical components.

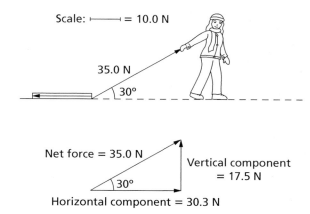

Figure 1-25.

By drawing a vector diagram and using either a ruler and protractor or trigonometry, we find that the vertical component has a magnitude of 17.5 N and the horizontal component has a magnitude of 30.3 N.

In some cases it is useful to resolve a vector into perpendicular components in directions other than horizontal and vertical. For example, consider a 90.-N box at rest on an inclined plane at a 30° angle from the horizontal (Figure 1-26). The weight of the box acts vertically downward.

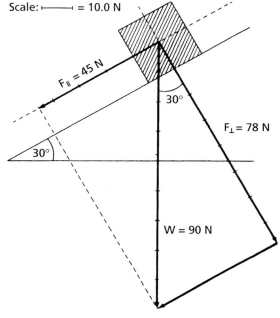

Figure 1-26.

We would like to determine the magnitude of the force that acts to push the box down the incline, and the magnitude of the force that acts to press the box against the surface of the plane. We must resolve the weight vector **w** into a component parallel to the surface and a component perpendicular to the surface. A vector diagram enables us to calculate the magnitude of these components. The magnitude of the perpendicular component $F_\perp$ can be expressed mathematically as

$$F_\perp = w \cos \theta$$

where θ is the angle between the inclined plane and the horizontal. In the example, $F_\perp$ becomes

$$F_\perp = w \cos \theta$$

$$F_\perp = (90. \text{ N})(\cos 30°)$$

$$F_\perp = (90. \text{ N})(0.866) = 78 \text{ N}$$

The component of weight parallel to the plane, $F_\parallel$, can be expressed mathematically as

$$F_\parallel = w \sin \theta$$

In the example $F_\parallel$ becomes

$$F_\parallel = w \sin \theta$$

$$F_\parallel = (90. \text{ N})(\sin 30°)$$

$$F_\parallel = (90. \text{ N})(0.500) = 45 \text{ N}$$

Sample Problem

10. In Figure 1-27 a person is pushing the handle of a lawnmower with a force of 200. N in a direction 45° from the horizontal. How much force acts to push the lawn mower in the horizontal direction, and how much force acts to push the lawnmower downward into the ground?

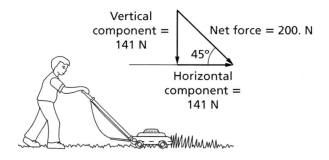

Scale: ⊢——⊣ = 100. N

Figure 1-27.

Solution: Drawing a vector diagram and using trigonometry, we find the vertical component:

$$A_y = A \sin \theta$$

$$A_y = (200. \text{ N})(\sin 45°)$$

$$A_y = (200. \text{ N})(0.707) = 141 \text{ N}$$

The horizontal component:

$$A_x = A \cos \theta$$

$$A_x = (200. \text{ N})(\cos 45°)$$

$$A_x = (200. \text{ N})(0.707) = 141 \text{ N}$$

QUESTIONS

PART A

47. The diagram below shows a worker using a rope to pull a cart.

The worker's pull on the handle of the cart can best be described as a force having (1) magnitude, only (2) direction, only (3) both magnitude and direction (4) neither magnitude nor direction

48. Which is a vector quantity? (1) distance (2) speed (3) mass (4) force

49. A 5.0-newton force and a 7.0-newton force act concurrently on a point. As the angle between the forces is increased from 0° to 180°, the magnitude of the resultant of the two forces changes from (1) 0.0 N to 12.0 N (2) 2.0 N to 12.0 N (3) 12.0 N to 2.0 N (4) 12.0 N to 0.0 N

50. A 5.0-newton force could have perpendicular components of (1) 1.0 N and 4.0 N (2) 2.0 N and 3.0 N (3) 3.0 N and 4.0 N (4) 5.0 N and 5.0 N

51. Forces **A** and **B** have a resultant **R**. Force **A** and resultant **R** are represented in the following diagram.

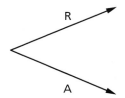

Which vector best represents force *B*?

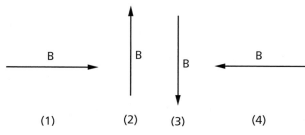

52. Two 10.0-newton forces act concurrently on a point at an angle of 180° to each other. The magnitude of the resultant of the two forces is (1) 0.00 N (2) 10.0 N (3) 18.0 N (4) 20.0N

53. A force of 3 newtons and a force of 5 newtons act concurrently to produce a resultant of 8 newtons. The angle between the forces may be (1) 0° (2) 60° (3) 90° (4) 180°

54. A table exerts a 2.0-newton force on a book lying on the table. The force exerted by the book on the table is (1) 20. N (2) 2.0 N (3) 0.20 N (4) 0 N

55. The diagram represents two concurrent forces acting on a point. Which vector best represents their resultant?

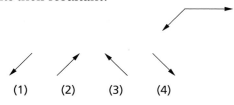

56. The resultant of two forces acting on the same point at the same time will be greatest when the angle between the forces is (1) 0° (2) 45° (3) 90° (4) 180°

57. What is the magnitude of the vector sum of the two concurrent forces represented in the diagram? (1) 2.0 N (2) 2.5 N (3) 3.5 N (4) 4.0 N

2.0 N

1.5 N

58. The resultant of two concurrent forces is minimum when the angle between them is (1) 0° (2) 45° (3) 90° (4) 180°

59. As the angle between two concurrent forces of 10 newtons and 12 newtons changes from 180° to 0°, the magnitude of their resultant changes from (1) 0 N to 12 N (2) 2.0 N to 22 N (3) 22 N to 2.0 N (4) 22 N to 0 N

60. Two concurrent forces act at right angles to each other. If one of the forces is 40 newtons and the resultant of the two forces is 50 newtons, the magnitude of the other force must be (1) 10 N (2) 20 N (3) 30 N (4) 40 N

61. If two 10.-newton concurrent forces have a resultant of zero, the angle between the forces must be (1) 0° (2) 45° (3) 90° (4) 180°

PART B-1

62. The diagram below shows a force of magnitude *F* applied to a mass at angle θ relative to a horizontal frictionless surface.

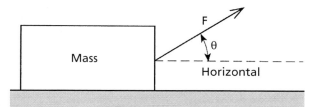

Frictionless surface

As the angle is increased, the horizontal acceleration of the mass (1) decreases (2) increases (3) remains the same

63. The diagram below represents a 5.0-newton force and a 12-newton force acting on point *P*.

The resultant of the two forces has a magnitude of (1) 5.0 N (2) 7.0 N (3) 12 N (4) 13 N

64. The vector diagram below represents two forces, **F₁** and **F₂**, simultaneously acting on an object.

Which vector best represents the resultant of the two forces?

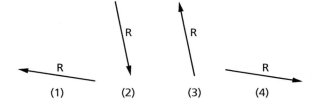

65. Two 30.-newton forces act concurrently on an object. In which diagram would the forces produce a resultant with a magnitude of 30. newtons?

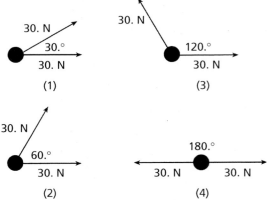

Dynamics

Our understanding of the laws of nature regarding motion have evolved over time. For a long time the dominant view was the one attributed to Aristotle. It claimed that in the absence of a force (a push, pull or any influence) acting on a body, the body remains at rest. Rest was presumed to be the "natural state" of a body. If an unbalanced force does act to influence a body, the body responds by moving. The stronger the force, the faster the motion of the body it acts on.

While this view sounds quite appealing and seems logical, it was eventually recognized that it conflicted with various observations. Objects continue to move, slide and roll even after the push that got them going no longer exists. This happened even in a vacuum, so air pressure (air rushing in to fill the vacuum left behind by the moving object) could not be the cause. Also, objects subject to a constant unbalanced force, such as a freely falling body, continue to accelerate as long as the force is active. So greater speed cannot be associated with the influence of a stronger force.

The First Law of Motion

The Aristotelian view was therefore replaced by the Newtonian system. In this view of things, the absence of a force leads to no change in motion. If the object is at rest, it remains at rest; if it is in motion, it maintains its speed and direction, forever, so long as no net force influences its motion. It is difficult to demonstrate this principle on Earth because there are always forces present (such as friction, air resistance, and gravity) that act to slow, and ultimately stop, a moving object. However, space vehicles continue to move for many years, even after their engines have been shut off.

Newton's first law of motion is also known as the *law of inertia*. **Inertia** is the property of matter that resists change in motion. The inertia of an object is proportional to its mass. A large mass tends to resist changes in its motion to a greater extent than a smaller mass.

The Second Law of Motion

Newton's second law of motion states: An unbalanced force acting on an object causes the object to accelerate in the direction of the force. In other words, if the resultant of all the forces, or the net force, acting on an object is not zero, the object's velocity will change in magnitude or direction or both. The acceleration of the object is directly proportional to the net force acting on it and inversely proportional to its mass. This relationship may be stated mathematically as

$$a = \frac{F_{net}}{m}$$

where a is the acceleration in m/s², F_{net} is the net force acting on the object in newtons (N), and m is the mass of the object in kilograms.

One newton (N) is defined as the amount of force that imparts an acceleration of one meter per second squared to a mass of one kilogram. Newton's second law also can be stated mathematically as

$$F_{net} = ma$$

where m is mass, in kg; a is acceleration in m/s²; and F_{net} is the force, in N. The newton, therefore, is a derived unit: $1 \text{ N} = 1 \text{ kg} \cdot \text{m/s}^2$.

The graph in Figure 1-28 illustrates the direct proportionality between force and acceleration. Each line is the force vs. acceleration graph for an object with a particular mass. The slope of each line is equal to the object's mass (since $F/a = m$).

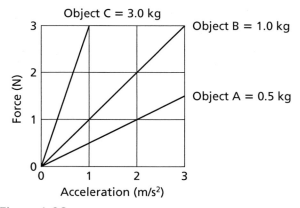

Figure 1-28.

Object *B* has twice as much mass as object *A*, so twice as much force is required to accelerate object *B* at the rate of 1 m/s² than to accelerate object *A* at the same rate. Object *C* has three times as much mass as object *B* and so it takes three times the amount of force to accelerate *C* at the rate of 1 m/s² than it does to accelerate object *B* at the same rate.

Sample Problems

11. A 120-N net force acts upon a 68-kg cart at rest. What acceleration results?

    Solution:

    $$a = \frac{F_{net}}{m}$$

    $$a = \frac{120 \text{ N}}{68 \text{ kg}} = \frac{124 \text{ kg} \cdot \text{m/s}}{68 \text{ kg}}$$

    $$a = 1.8 \text{ m/s}^2$$

12. A force of *F* newtons causes a mass of *m* kilograms to accelerate at 24 m/s². What acceleration will occur under the following conditions?

    (*a*) The force is doubled to 2*F* newtons, and the mass remains the same.

    Solution: Since acceleration is directly proportional to force, if the force is doubled the acceleration will double. Hence,

    $$a = 2(24 \text{ m/s}^2) = 48 \text{ m/s}^2$$

    (*b*) The force is *F* newtons and the mass is tripled.

    Solution: Since acceleration is inversely proportional to mass, if the mass is tripled the acceleration is divided by 3. Hence,

    $$a = \frac{24 \text{ m/s}^2}{3} = 8 \text{ m/s}^2$$

The Third Law of Motion

When a cannon is fired, it recoils as the cannonball flies out. Air rushing out of a balloon in one direction forces the balloon to move in the opposite direction. Such effects are explained by *Newton's third law of motion*, which states: For every action there is an equal and opposite reaction. The words "action" and "reaction" refer to forces. The law means that when one object exerts a force on another, the second object exerts an equally strong but oppositely directed force on the first. Forces always occur in pairs that are equal in magnitude and opposite in direction. The members of the pair do not act on the same object, are not concurrent, and cannot be added together. For example, when you exert a downward force of 500 N on a chair, the chair exerts an upward force of 500 N on you.

The Law of Universal Gravitation

The ancient natural philosophers believed that objects fall because Earth pulls on them. Newton insightfully proposed that Earth cannot be unique in having the ability to attract objects. All matter must attract other matter. This attraction is universal and is referred to as the **gravitational force**.

The gravitational forces between ordinary objects are extremely weak and therefore often undetectable. However, if one or both of the objects contain an enormous amount of mass, the gravitational force between them is significant. For example, Earth, sun, and moon pull on each other with tremendous gravitational force, creating Earth's orbit and the tides. The gravitational force between Earth and an object on its surface is also strong enough to be noticed. It is referred to as the object's **weight**.

Newton found that the gravitational force between any two objects is directly proportional to the product of their masses and inversely proportional to the square of the distance between them. This relationship is stated mathematically as

$$F \propto \frac{m_1 m_2}{r^2}$$

where m_1 and m_2 are the objects' masses and *r* is the distance between them. This proportionality can be changed to an equality by inserting a constant of proportionality into the equation

$$F = \frac{G m_1 m_2}{r^2}$$

The constant *G* is called the *universal gravitational constant*. Careful experiments have determined that when the masses are expressed in kilograms, the distance in meters, and the force in newtons, the value of *G* is 6.67×10^{-11} N·m²/kg².

This gravitational law applies only to (1) masses whose sizes are small compared to the distance between them (point masses); and (2) spherical masses of uniform density—when the distance between them is measured from the center of one to the center of the other.

The relationship between gravitational force and distance is an example of an *inverse square relationship*. As the distance between the masses increases, the gravitational force between them decreases rapidly. For example, if the distance between two masses is doubled, the gravitational force between them is decreased to one-fourth its original value. If the distance between the masses is halved, the gravitational force between them is quadrupled. This inverse square relationship is illustrated by the graph in Figure 1-29.

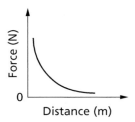

Figure 1-29. Inverse square relationship.

Sample Problems

13. Find the gravitational pull exerted by the moon on Earth (which is identical to the pull of Earth on the moon).

    Solution:

$$\text{Mass of Earth} = 5.98 \times 10^{24} \text{ kg}$$

$$\text{Mass of the moon} = 7.35 \times 10^{22} \text{ kg}$$

$$\text{Earth-moon distance} = 3.84 \times 10^8 \text{ m}$$

$$F_g = \frac{Gm_1m_2}{r^2}$$

$$F_g = \frac{(6.67 \times 10^{-11} \text{ N} \cdot \text{m}^2/\text{kg}^2)}{(3.84 \times 10^8 \text{ m})^2}$$

$$F_g = 1.99 \times 10^{20} \text{ N}$$

14. Find the gravitational force exerted by Earth on a 1-kg mass on its surface.

    Solution: Since Earth is large and spherical the appropriate distance to use is the distance between Earth's center and the center of 1-kg mass. This distance is approximately equal to Earth's radius.

$$F_g = \frac{Gm_1m_2}{r^2}$$

$$F_g = \frac{(6.67 \times 10^{-11} \text{ N} \cdot \text{m}^2/\text{kg}^2)(5.98 \times 10^{24} \text{ kg})(1 \text{ kg})}{(6.37 \times 10^6 \text{ m})^2}$$

$$F_g = 9.83 \text{ N}$$

This is the force exerted by Earth on the 1-kg object that causes the object to fall. According to Newton's third law, the 1-kg object also pulls Earth toward it with the same force, but due to Earth's enormous mass, no noticeable acceleration of Earth results.

Gravitational Field

The fact that a mass exerts a force on another mass some distance away means that the space between the masses is unlike that where no masses are present. We say that every mass sets up a **gravitational field** in the space around itself. This field then acts on other masses so that an attraction results. The gravitational field strength g at any point in the field is defined as the force experienced by a 1-kg mass at that point in the field. This is expressed mathematically as

$$g = \frac{F_g}{m}$$

where F_g is the gravitational force in newtons experienced by a mass m. If the force exerted on a 5-kg object by a gravitational field at some point is 20 N, the intensity of the field at that point is 20 N/5 kg = 4 N/kg.

A gravitational field is a vector quantity. The direction of the gravitational field at any given point is the direction of the gravitational force exerted by the field on a mass placed in the field at that point.

Freely Falling Objects

Like all forces, a gravitational force will cause an object to accelerate. An object is said to be in **free fall** if only Earth's gravitational pull acts on it and no other forces (such as friction or air resistance) are present. The acceleration of any freely falling object near Earth's surface is 9.81 m/s², regardless of the object's mass, size, or shape. The symbol g is used to represent this important constant.

$$g = 9.81 \text{ m/s}^2$$

The motion of freely falling objects can be described by using the equations for uniform acceleration where $a = g$.

The value $g = 9.81$ m/s² applies only to objects near sea level. At higher or lower elevations, the gravitational force will vary according to the inverse square relation and, therefore, the acceleration due to gravity will also vary.

Sample Problems

15. Using the equations for uniform acceleration, determine the velocity and distance traveled by a freely falling object after 4.0 seconds.

Solution:

$$v_f = v_i + at$$

$$v_i = 0, a = g$$

$$v_f = gt$$

$$v_f = 9.81 \text{ m/s}^2 \times 4.0 \text{ s} = 39 \text{ m/s}$$

$$d = v_i t + \tfrac{1}{2} at^2$$

$$v_i = 0, a = g$$

$$d = \tfrac{1}{2} gt^2 = \tfrac{1}{2}(9.81 \text{ m/s}^2)(4.0 \text{ s})^2$$

$$d = 78 \text{ m}$$

16. An object hits the ground after dropping 5000. m in free fall. What is its velocity as it hits the ground, and how long did it fall?

Solution:

Final velocity:

$$v_f^2 = v_i^2 + 2ad$$

$$v_i = 0, a = g$$

$$v_f^2 = 2(9.81 \text{ m/s}^2)(5000. \text{ m})$$

$$v = \sqrt{98,000 \text{ m}^2/\text{s}^2} = 313 \text{ m/s}$$

Time of drop:

$$d = v_i t + \tfrac{1}{2} at^2$$

$$v_i = 0, a = g$$

$$d = \tfrac{1}{2} gt^2$$

$$t^2 = \frac{2d}{g}$$

$$t^2 = \frac{2(5000. \text{ m})}{9.81 \text{ m}/\text{s}^2} = 1020 \text{ s}^2$$

$$t = \sqrt{1020 \text{ s}^2} = 32 \text{ s}$$

Weight

The *weight* F_g of an object is defined as the gravitational force it experiences. Since all freely falling objects at sea level experience the same acceleration g, we can use Newton's second law.

$$g = \frac{F_g}{m}$$

Where g is the acceleration due to gravity, F_g is the weight or force due to gravity, and m is the mass of the object.

Rearranging symbols, we get

$$F_g = mg$$

Like all forces, weight is a vector quantity whose magnitude is expressed in newtons. The direction of the weight vector is always vertically downward toward the center of Earth. Note that since the magnitude of the gravitational force experienced by an object varies with distance from Earth, the weight of the object and its acceleration vary in the same way. A graph of weight vs. mass (Figure 1-30) for objects at sea level is a straight line whose slope is g.

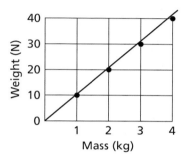

Figure 1-30. $g = 9.81 \text{ N/kg} = 9.81 \text{ m/s}^2$.

Sample Problems

17. What is the weight of a 100.-kg mass?

Solution:

$$g = \frac{F_g}{m}$$

$$F_g = mg$$

$$F_g = (100. \text{ kg})(9.81 \text{ m/s}^2) = 981 \text{ N}$$

18. What is the mass of a 100.-N weight?

Solution:

$$g = \frac{F_g}{m}$$

$$m = \frac{F_g}{g} = \frac{100. \text{ N}}{9.81 \text{ m/s}^2}$$

$$m = \frac{100. \text{ kg} \cdot \text{m/s}^2}{9.81 \text{ m/s}^2} = 10.2 \text{ kg}$$

QUESTIONS

66. How far will a brick starting from rest fall freely in 3 seconds? (1) 15 m (2) 29 m (3) 44 m (4) 88 m

67. If the sum of all the forces acting on a moving object is zero, the object will (1) slow and stop (2) change the direction of its motion (3) accelerate uniformly (4) continue moving with constant velocity

68. A net force of 10. newtons accelerates an object at 5.0 meters per second². What net force would be required to accelerate the same object at 1 meter per second²? (1) 1.0 N (2) 2.0 N (3) 5.0 N (4) 50.0 N

69. The graph below represents the relationship between gravitational force and mass for objects near the surface of Earth.

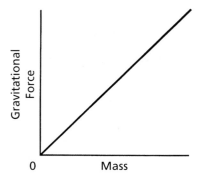

The slope of the graph represents the (1) acceleration due to gravity (2) universal gravitational constant (3) momentum of objects (4) weight of objects

70. A 1200-kilogram car traveling at 10. meters per second hits a tree and is brought to rest in 0.10 second. What is the magnitude of the average force acting on the car to bring it to rest? (1) 1.2×10^2 N (2) 1.2×10^3 N (3) 1.2×10^4 N (4) 1.2×10^5 N

71. A spring scale reads 20. newtons as it pulls a 5.0-kilogram mass across a table. What is the magnitude of the force exerted by the mass on the spring scale? (1) 49 N (2) 20. N (3) 5.0 N (4) 4.0 N

72. An object weighs 100. newtons on Earth's surface. When it is moved to a point one Earth radius above Earth's surface, it will weigh (1) 25.0 N (2) 50.0 N (3) 100. N (4) 400. N

73. A ball thrown vertically upward reaches a maximum height of 30. meters above the surface of Earth. At its maximum height, the speed of the ball is (1) 0.0 m/s (2) 3.1 m/s (3) 9.8 m/s (4) 24 m/s

74. Which object has the most inertia? (1) a 0.001-kilogram bumblebee traveling at 2 meters per second (2) a 0.1-kilogram baseball traveling at 20 meters per second (3) a 5-kilogram bowling ball traveling at 3 meters per second (4) a 10.-kilogram sled at rest

75. A 40.-kilogram mass is moving across a horizontal surface at 5.0 meters per second. What is the magnitude of the net force required to bring the mass to a stop in 8.0 seconds? (1) 1.0 N (2) 5.0 N (3) 25 N (4) 40. N

76. A ball dropped from rest falls freely until it hits the ground with a speed of 20 meters per second. The time during which the ball is in free fall is approximately (1) 1 s (2) 2 s (3) 0.5 s (4) 10 s

77. Two carts are pushed apart by an expanding spring, as shown in the diagram below.

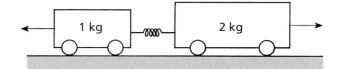

If the average force on the 1-kilogram cart is 1 newton, what is the average force on the 2-kilogram cart? (1) 1 N (2) 0.0 N (3) 0.5 N (4) 4 N

78. A lab cart is loaded with different masses and moved at various velocities. Which diagram shows the cart-mass system with the greatest inertia?

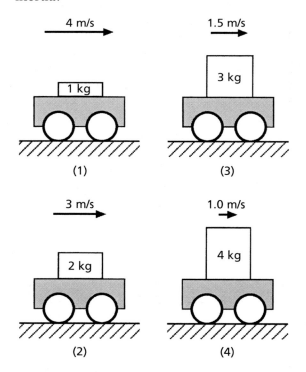

79. A 1.0-kilogram ball is dropped from the roof of a building 40. meters tall. What is the approximate time of fall? (Neglect air resistance.) (1) 2.9 s (2) 2.0 s (3) 4.1 s (4) 8.2 s

80. A 2.0-kilogram laboratory cart is sliding across a horizontal frictionless surface at a constant velocity of 4.0 meters per second east. What will be the cart's velocity after a 6.0-newton westward force acts on it for 2.0 seconds? (1) 2.0 m/s east (2) 2.0 m/s west (3) 10. m/s east (4) 10. m/s west

Base your answers to questions 81 and 82 on the diagram below, which shows a 1.0-newton metal disk resting on an index card that is balanced on top of a glass.

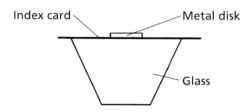

81. What is the net force acting on the disk? (1) 1.0 N (2) 2.0 N (3) 0 N (4) 9.8 N

82. When the index card is quickly pulled away from the glass in a horizontal direction, the disk falls straight down into the glass. This action is a result of the disk's (1) inertia (2) charge (3) shape (4) temperature

83. A 400-newton girl standing on a dock exerts a force of 100 newtons on a 10,000-newton sailboat as she pushes it away from the dock. How much force does the sailboat exert on the girl? (1) 25 N (2) 100 N (3) 400 N (4) 10,000 N

84. The diagram below represents two satellites of equal mass, *A* and *B*, in circular orbits around a planet.

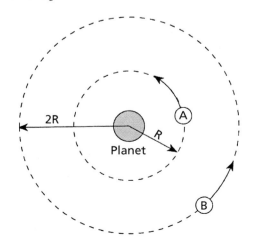

Compared to the magnitude of the gravitational force of attraction between satellite *A* and the planet, the magnitude of the gravitational force of attraction between satellite *B* and the planet is (1) half as great (2) twice as great (3) one-fourth as great (4) four times as great

85. After a model rocket reached its maximum height, it then took 5.0 seconds to return to the launch site. What is the approximate maximum height reached by the rocket? (Neglect air resistance.) (1) 49 m (2) 98 m (3) 120 m (4) 250 m

86. A 70.-kilogram astronaut has a weight of 560 newtons on the surface of planet Alpha. What is the acceleration due to gravity on planet Alpha? (1) 0.0 m/s² (2) 8.0 m/s² (3) 9.8 m/s² (4) 80. m/s²

87. The diagram below shows a horizontal 8.0-newton force applied to a 4.0-kilogram block on a frictionless table.

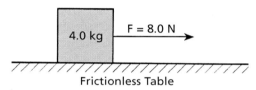

What is the magnitude of the block's acceleration? (1) 0.50 m/s² (2) 2.0 m/s² (3) 9.8 m/s² (4) 32 m/s²

88. An object is dropped from rest and falls freely 20. meters to Earth. When is the speed of the object 9.8 meters per second?
(1) during the entire first second of its fall
(2) at the end of its first second of fall
(3) during its entire time of fall
(4) after it has fallen 9.8 meters

89. Which cart has the greatest inertia?
(1) a 1-kilogram cart traveling at a speed of 4 m/s
(2) a 2-kilogram cart traveling at a speed of 3 m/s
(3) a 3-kilogram cart traveling at a speed of 2 m/s
(4) a 4-kilogram cart traveling at a speed of 1 m/s

90. A container of rocks with a mass of 65.0 kilograms is brought back from the moon's surface, where the acceleration due to gravity is 1.62 meters per second². What is the weight of the container of rocks on Earth's surface? (1) 638 N (2) 394 N (3) 105 N (4) 65.0 N

91. A satellite weighs 200 newtons on the surface of Earth. What is its weight at a distance of one

Earth radius above the surface of Earth?
(1) 50 N (2) 100 N (3) 400 N (4) 800 N

92. The diagram below shows a 5.00-kilogram block at rest on a horizontal, frictionless table.

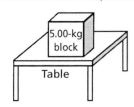

Which diagram best represents the force exerted on the block by the table?

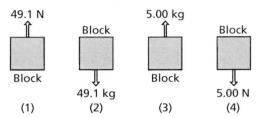

93. A person is standing on a bathroom scale in an elevator car. If the scale reads a value greater than the weight of the person at rest, the elevator car could be moving (1) downward at constant speed (2) upward at constant speed (3) downward at increasing speed (4) upward at increasing speed

94. The acceleration due to gravity on the surface of planet X is 19.6 meters per second². If an object on the surface of this planet weighs 980. newtons, the mass of the object is (1) 50.0 kg (2) 100. kg (3) 490. N (4) 908 N

95. A basketball player jumped straight up to grab a rebound. If she was in the air for 0.80 second, how high did she jump? (1) 78 m (2) 3.1 m (3) 1.2 m (4) 0.78 m

96. A man is pushing a baby stroller. Compared to the magnitude, or amount, of force exerted on the stroller by the man, the magnitude of the force exerted on the man by the stroller is (1) zero (2) smaller, but greater than zero (3) larger (4) the same

97. A man standing on a scale in an elevator notices that the scale reads 30 newtons greater than his normal weight. Which type of movement of the elevator could cause this greater-than-normal reading? (1) accelerating upward (2) accelerating downward (3) moving upward at constant speed (4) moving downward at constant speed

98. An object with a mass of 2 kilograms is accelerated at 5 meters per second². The net force acting on the mass is (1) 5 N (2) 2 N (3) 10 N (4) 20 N

99. The diagram below represents a constant force F acting on a box located on a frictionless horizontal surface. As the angle θ between the force and the horizontal increases, the acceleration of the box will (1) decrease (2) increase (3) remain the same

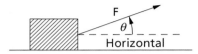

100. An object accelerates at 2.5 meters per second² when an unbalanced force of 10. newtons acts on it. What is the mass of the object? (1) 1.0 kg (2) 2.0 kg (3) 3.0 kg (4) 4.0 kg

101. An unbalanced force of 10.0 newtons causes an object to accelerate at 2.0 m/s². What is the mass of the object? (1) 0.2 kg (2) 5.0 kg (3) 8.0 kg (4) 20 kg

102. An unbalanced force of 10 newtons acts on a 20-kilogram mass for 5 seconds. The acceleration of the mass is (1) 0.5 m/s² (2) 2 m/s² (3) 40 m/s² (4) 200 m/s²

103. Two objects of equal mass are a fixed distance apart. If the mass of each object could be tripled, the gravitational force between the objects would (1) decrease by one-third (2) decrease by one-ninth (3) triple (4) increase 9 times

104. Which two quantities are measured in the same units? (1) velocity and acceleration (2) weight and force (3) mass and weight (4) force and acceleration

105. What is the weight of a 5.0-kilogram object at Earth's surface? (1) 5.0 kg (2) 25 N (3) 49 N (4) 49 kg

106. What is the gravitational acceleration on a planet where a 2-kilogram mass has a weight of 16 newtons on the planet's surface? (1) $\frac{1}{8}$ m/s² (2) 8 m/s² (3) 10 m/s² (4) 32 m/s²

107. Which graph represents the relationship between the mass of an object and its distance from Earth's surface?

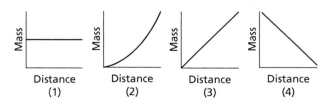

108. Which is constant for a freely falling object? (1) displacement (2) speed (3) velocity (4) acceleration

109. An object starting from rest falls freely near Earth's surface. Which graph best represents the motion of the object?

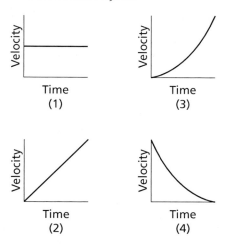

110. Starting from rest, an object rolls freely down an incline that is 10 meters long in 2 seconds. The acceleration of the object is approximately (1) 5 m/sec (2) 5 m/sec² (3) 10 m/sec (4) 10 m/sec²

111. An object, initially at rest, falls freely near Earth's surface. How long does it take the object to attain a speed of 98 meters per second? (1) 0.1 s (2) 10 s (3) 98 s (4) 960 s

112. Starting from rest, object *A* falls freely for 2.0 seconds, and object *B* falls freely for 4.0 seconds. Compared with object *A*, object *B* falls (1) one-half as far (2) twice as far (3) three times as far (4) four times as far

113. An object is thrown vertically upward from Earth's surface. Which graph best shows the relationship between velocity and time as the object rises?

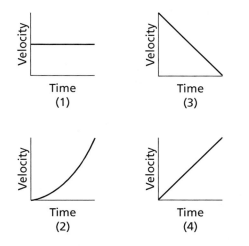

114. An astronaut drops a stone near the surface of the moon. Which graph best represents the motion of the stone as it falls toward the moon's surface?

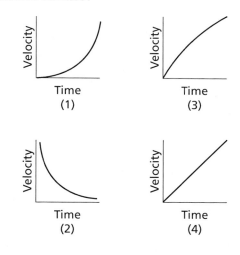

115. A constant unbalanced force is applied to an object for a period of time. Which graph best represents the acceleration of the object as a function of elapsed time?

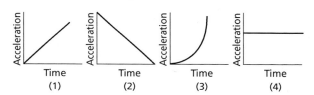

Base your answers to questions 116 and 117 on the information and table below.

The weight of an object was determined at five different distances from the center of Earth. The results are shown in the table below. Position *A* represents results for the object at the surface of Earth.

Position	Distance from Earth's Center (m)	Weight (N)
A	6.37×10^6	1.0×10^3
B	1.27×10^7	2.5×10^2
C	1.91×10^7	1.1×10^2
D	2.55×10^7	6.3×10^1
E	3.19×10^7	4.0×10^1

116. The approximate mass of the object is (1) 0.01 kg (2) 10 kg (3) 100 kg (4) 1000 kg

117. At what distance from the center of Earth is the weight of the object approximately 28 newtons? (1) 3.5×10^7 m (2) 3.8×10^7 m (3) 4.1×10^7 m (4) 4.5×10^7 m

Base your answers to questions 118 through 122 on the following information.

A 10.-kilogram object, starting from rest, slides down a frictionless incline with a constant acceleration of 2.0 m/s^2 for 4.0 seconds.

118. What is the velocity of the object at the end of the 4.0 seconds? (1) 16 m/s (2) 8.0 m/s (3) 4.0 m/s (4) 2.0 m/s

119. During the 4.0 seconds, the object moves a total distance of (1) 32 m (2) 16 m (3) 8.0 m (4) 4.0 m

120. To produce this acceleration, what is the force on the object? (1) 10. N (2) 2.0 × 10^1 N (3) 5.0 N (4) 2.0 × 10^2 N

121. What is the approximate weight of the object? (1) 1 N (2) 10 N (3) 100 N (4) 1,000 N

122. Which graph best represents the relationship between acceleration (*a*) and time (*t*) for the object?

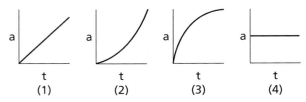

123. Which graph could represent the motion of an object with no unbalanced forced acting on it?

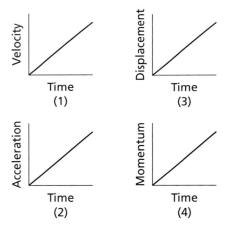

124. Two frictionless blocks having masses of 8.0 kilograms and 2.0 kilograms rest on a horizontal surface. If a force applied to the 8.0-kilogram block gives it an acceleration of 5.0 m/s^2, then the same force will give the 2.0-kilogram block an acceleration of (1) 1.2 m/s^2 (2) 2.5 m/s^2 (3) 10. m/s^2 (4) 20. m/s^2

125. The graph below shows the relationship between the acceleration of an object and the unbalanced force producing the acceleration. The ratio ($\Delta F/\Delta a$) of the graph represents the object's

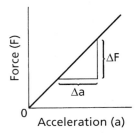

(1) mass (2) momentum (3) kinetic energy (4) displacement

Friction

Kinetic friction is the force that opposes the motion of one surface over another. This resistance to motion is the result of the contact between surfaces. Frictional force is always directed opposite to the motion and acts to slow and ultimately stop the motion. The magnitude of the frictional force depends on the nature of the surfaces in contact (their roughness and composition) and on the amount of force pressing the two surfaces together. It is independent of the amount of surface area in contact and of the relative speed of the objects. The nature of the surfaces is represented by a quantity referred to as the **coefficient of kinetic friction** (μ). Table 1-4 lists the approximate coefficients of friction for some selected materials. The force that presses the two surfaces together is called the **normal force**, because it acts in a direction perpendicular to the surfaces. When an object moves over a horizontal surface, the magnitude of the normal force is equal to the weight of the object. When an object slides on an inclined plane, the magnitude of the normal force is equal to the component of the object's weight perpendicular to the surface.

Table 1-4
Approximate Coefficients of Kinetic and Static Friction

Materials	Kinetic	Static
Rubber on concrete (dry)	0.68	0.90
Rubber on concrete (wet)	0.58	
Rubber on asphalt (dry)	0.67	0.85
Rubber on asphalt (wet)	0.53	
Rubber on ice	0.15	
Waxed ski on snow	0.05	0.14
Wood on wood	0.30	0.42
Steel on steel	0.57	0.74
Copper on steel	0.36	0.53
Teflon on Teflon	0.04	

The force of kinetic friction is calculated using the equation

$$F_f = \mu F_N$$

where μ is the coefficient of kinetic friction, F_f is the force of kinetic friction, and F_N is the normal force. (See Figure 1-31.)

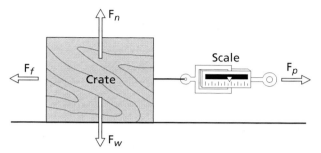

Figure 1-31. When the crate moves at constant velocity, the pulling force F_p must equal the force of kinetic friction F_f between the crate and the floor.

Sample Problem

19. A 60-N wooden box is pushed along a horizontal concrete sidewalk (Figure 1-32). The coefficient of friction between wood and concrete is 0.15. What frictional force must be overcome for the box to move at constant velocity?

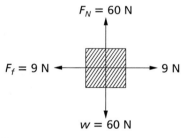

Figure 1-32.

Solution:

$$F_f = \mu F_N$$

$$F_f = (0.15)(60 \text{ N}) = 9 \text{ N}$$

Static Friction

The coefficient of friction between two surfaces is greatest when there is no relative motion between them and we wish to start one surface moving over the other. The frictional force that must be overcome to start one surface moving over another is called **static friction.** Static friction is equal in magnitude and opposite in direction to the force applied parallel to the surfaces that acts

to get this motion started (Figure 1-33), but it cannot exceed the quantity μF_N, where μ is the coefficient of static friction. Static friction is sometimes called *starting friction.*

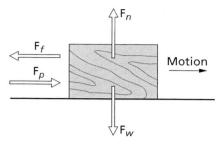

Figure 1-33. The force of static friction, F_f is equal in magnitude and opposite to the direction of F_p, the force acting to start the object moving.

Rolling Friction

When one object rolls over another, it experiences a force of **rolling friction**. Rolling friction is generally weaker than sliding friction. This is why ball bearings are often placed between two surfaces that must move past each other.

Fluid Friction

Fluid friction is the force that resists the motion of an object through a fluid, such as water or air. Swimmers, boats, planes, and parachutes all experience fluid friction as they move. Fluid friction can be large enough to equal the weight of a falling object. This is how a parachute works. At first, a parachutist accelerates downward due to his or her weight. At a certain velocity, the fluid friction of the air exerted on the parachute becomes equal and opposite to the weight and the net force is zero. When this occurs, the parachutist ceases to accelerate and falls at a constant velocity, called **terminal velocity**.

QUESTIONS

PART A

126. The diagram below shows a block sliding down a plane inclined at angle with the horizontal.

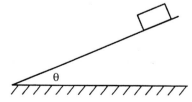

As the angle θ is increased, the coefficient of kinetic friction between the bottom surface of the block and the surface of the incline will (1) decrease (2) increase (3) remain the same

127. A box is pushed toward the right across a class room floor. The force of friction on the box is directed toward the (1) left (2) right (3) ceiling (4) floor

128. The diagram below shows a sled and rider sliding down a snow-covered hill that makes an angle of 30.° with the horizontal.

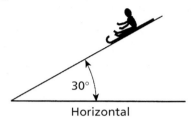

Which vector best represents the direction of the normal force, F_N, exerted by the hill on the sled?

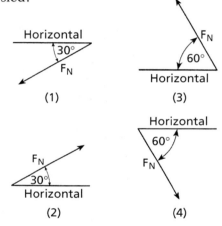

129. Compared to the force needed to start sliding a crate across a rough level floor, the force needed to keep it sliding once it is moving is (1) less (2) greater (3) the same

130. The diagram below shows a granite block being slid at constant speed across a horizontal concrete floor by a force parallel to the floor.

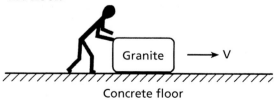

Which pair of quantities could be used to determine the coefficient of friction for the granite on the concrete? (1) mass and speed of the block (2) mass and normal force on the block (3) frictional force and speed of

the block (4) frictional force and normal force on the block

131. The force required to start an object sliding across a uniform horizontal surface is larger than the force required to keep the object sliding at a constant velocity. The magnitudes of the required forces are different in these situations because the force of kinetic friction
(1) is greater than the force of static friction
(2) is less than the force of static friction
(3) increases as the speed of the object relative to the surface increases
(4) decreases as the speed of the object relative to the surface increases

132. When a 12-newton horizontal force is applied to a box on a horizontal tabletop, the box remains at rest. The force of static friction acting on the box is (1) 0 N (2) between 0 N and 12 N (3) 12 N (4) greater than 12 N

133. In the diagram below, a box is at rest on an inclined plane.

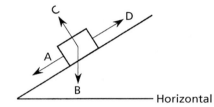

Which vector best represents the direction of the normal force acting on the box? (1) **A** (2) **B** (3) **C** (4) **D**

134. The strongest frictional force between two surfaces is (1) static friction (2) kinetic friction (3) sliding friction (4) rolling friction

135. A 40.-newton object requires 5.0 newtons to start moving over a horizontal surface. The coefficient of static friction is (1) 0.13 (2) 4.0 (3) 5.0 (4) 35

136. A 12-newton cart is moving on a horizontal surface with a coefficient of kinetic friction of 0.10. What force of friction must be overcome to keep the object moving at constant speed? (1) 0.10 N (2) 1.2 N (3) 12 N (4) 120 N

137. The coefficient of friction between two dry sliding surfaces is 0.05. What value is possible if these surfaces are lubricated? (1) 5.00 (2) 0.02 (3) 0.05 (4) 0.08

138. If the normal force between two surfaces is doubled, the static friction force will (1) be halved (2) remain the same (3) be doubled (4) be quadrupled

139. If the normal force between two surfaces is doubled, the coefficient of static friction will

(1) be halved (2) remain the same (3) be doubled (4) be quadrupled

140. In which situation is fluid friction not involved? (1) a boat moving in water (2) a parachute dropping (3) a plane flying (4) climbing up a pole

PART B-1

141. Which vector diagram best represents a cart slowing down as it travels to the right on a horizontal surface?

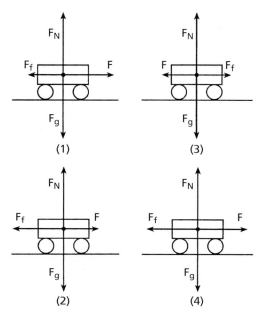

(1)   (3)

(2)   (4)

142. The diagram at the right represents a block at rest on an incline.

Block

Which diagram best represents the forces acting on the block? (F_f = frictional force, F_N = normal force, and F_w = weight.)

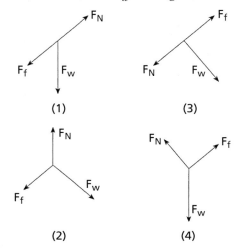

(1)   (3)

(2)   (4)

Questions 143 through 149 relate to this situation: A 6.0-N block is moving to the right on a horizontal surface. The friction force during the motion is found to be 1.0 N.

143. In which direction does the friction force act on the block? (1) to the right (2) to the left (3) upward (4) downward

144. What is the magnitude of the normal force on the block? (1) less than 6.0 N (2) 6.0 N (3) more than 6.0 N

145. What is the coefficient of friction in this example? (1) 1.0 (2) 0.17 (3) 5.0 (4) 6.0

146. If the block is at rest, how much force will be needed to get it moving? (1) 1.0 N (2) less than 1.0 N (3) more than 1.0 N

147. If a weight is placed on top of the block, the force of friction will (1) decrease (2) increase (3) remain the same

148. If the same block slides on the surface at greater speed, the force of friction will be (1) the same (2) weaker (3) stronger

149. If the block is turned so that it slides on a side whose surface is smaller, the force of friction will (1) increase (2) remain the same (3) decrease

150. A constant unbalanced force of friction acts on a 15.0-kilogram mass moving along a horizontal surface at 10.0 meters per second. If the mass is brought to rest in 1.50 seconds, what is the magnitude of the force of friction? (1) 10.0 N (2) 100. N (3) 147. N (4) 150. N

Momentum

An object's **momentum** is defined as the product of its mass and its velocity—that is, its mass multiplied by its velocity. A fast-moving car has more momentum than a slow-moving car of the same mass, and a heavy bus has more momentum than a small car moving with the same velocity. The symbol for momentum is p.

$$p = mv$$

where m is the mass, in kilograms; and v is the velocity, in meters per second. The unit of momentum is kg·m/s.

Momentum is a vector quantity whose direction is the same as the object's velocity. It is customary to assign a positive value to the velocity and momentum of an object that moves to the right and a negative value to the velocity and momentum of an object moving to the left.

20. What is the momentum of a 15-kg bicycle moving at 12 m/s?

    Solution:

    $$p = mv$$

    $$p = (15 \text{ kg})(12 \text{ m/s}) = 180 \text{ kg} \cdot \text{m/s}$$

21. A bullet with a mass of 0.0025 kg has a momentum of 1.8 kg · m/s. What is the magnitude of its velocity?

    Solution:

    $$p = mv$$

    $$v = \frac{p}{m} = \frac{1.8 \text{ kg} \cdot \text{m/s}}{0.0025 \text{ kg}} = 720 \text{ m/s}$$

Impulse

Impulse is defined as the product of the net force acting on an object and the time during which the force acts. The symbol for impulse is J.

$$J = F_{net}t$$

where F_{net} is the net force, in newtons; and t is the time interval in seconds. The unit for impulse is the newton second (N · s). Impulse is a vector quantity whose direction is the same as the net force acting on the object.

When a net force acts on an object, the object's velocity and momentum change. We can use Newton's second law to show that the change in momentum (Δp) is equal to the impulse imparted to the object.

Newton's second law:

$$a = \frac{F_{net}}{m} \tag{1}$$

Definition of acceleration:

$$a = \frac{\Delta v}{t} \tag{2}$$

Substituting (2) into (1):

$$F_{net} = m\frac{\Delta v}{t}$$

$$F_{net}t = m\Delta v$$

$$J = \Delta p$$

Therefore,

$$J = F_{net}t = \Delta p$$

Note that the unit for impulse and the unit for momentum are equivalent:

$$\text{N} \cdot \text{s} = (\text{kg} \cdot \text{m/s}^2) \times \text{s} = \text{kg} \cdot \text{m/s}$$

22. A gust of wind exerts a 300.-N net force on a 1000.-kg sailboat for 15 s. What is the change in momentum and velocity of the boat?

    Solution:

    $$J = F_{net}t$$

    $$J = (300. \text{ N})(15 \text{ s}) = 4500 \text{ N} \cdot \text{s}$$

    $$J = \Delta p = 4500 \text{ kg} \cdot \text{m/s}$$

    $$p = mv$$

    $$v = \frac{p}{m}$$

    $$v = \frac{4500 \text{ kg} \cdot \text{m/s}}{1000. \text{ kg}} = 4.5 \text{ m/s}$$

23. How long should a 10.-N force be applied to a 5.0-kg object to cause its velocity to slow from 7.0 m/s to 4.0 m/s?

    Solution:

    $$\Delta v = v_f - v_i$$

    $$\Delta v = 4.0 \text{ m/s} - 7.0 \text{ m/s} = -3.0 \text{ m/s}$$

    $$\Delta p = m\Delta v$$

    $$\Delta p = (5.0 \text{ kg})(-3.0 \text{ m/s}) = -15 \text{ kg} \cdot \text{m/s}$$

    $$Ft = \Delta p$$

    (*Note*: F is negative since it opposes the motion.)

    $$(-10. \text{ N})(t) = -15 \text{ kg} \cdot \text{m/s}$$

    $$t = 1.5 \text{ s}$$

Conservation of Momentum

The *law of conservation of momentum* states that if no external force is acting on a system, the total momentum of the system remains unchanged. This means that when a set of objects interact, the total momentum before the event equals the total momentum after the event. For example, consider the total momentum of a set of billiard balls before and after a collision. Before the event, only the cue ball is moving toward the other assembled balls.

After the cue ball collides with the set of balls, the balls scatter across the pool table. The resultant of the momentum vectors of all the moving balls after the collision is identical to the momentum vector of the cue ball before the collision.

In the simple case of a head-on interaction (collision or explosion) between two objects with masses m_1 and m_2, the law of conservation of momentum can be expressed mathematically as

Total momentum before = total momentum after

$$p_{before} = p_{after}$$

$$m_1 v_1 + m_2 v_2 = m_1 v_1' + m_2 v_2'$$

where v_1 and v_2 are the velocities of the objects before the event, and v_1' and v_2' are their velocities after the event.

Sample Problems

24. A 2-kg cart moving to the right at 5 m/s collides with an 8-kg cart at rest (Figure 1-34). As a result of the collision, the carts lock together.

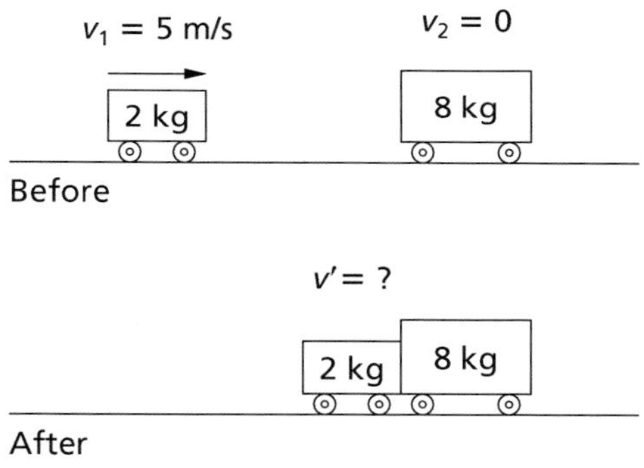

$v_1 = 5$ m/s $\qquad v_2 = 0$

2 kg $\qquad$ 8 kg

Before

$v' = ?$

2 kg 8 kg

After

Figure 1-34.

What is the velocity of the carts after the event?

Solution:

$$p_{before} = p_{after}$$

$$m_1 v_1 + m_2 v_2 = m_1 v_1' + m_2 v_2'$$

Since $v_1' = v_2'$ we can call them both v' and factor v' from the equation to obtain:

$$m_1 v_1 + m_2 v_2 = (m_1 + m_2) v'$$

$$v' = \frac{m_1 v_1 + m_2 v_2}{m_1 + m_2}$$

$$v' = \frac{(2 \text{ kg})(5 \text{ m/s}) + (8 \text{ kg})(0 \text{ m/s})}{2 \text{ kg} + 8 \text{ kg}}$$

$$v' = \frac{10 \text{ kg} \cdot \text{m/s}}{10 \text{ kg}} = 1 \text{ m/s (to the right)}$$

25. A standard physics problem is the "explosion" of two carts initially at rest, which sends them moving in opposite directions. For example, Figure 1-35 shows a 6-kg cart and a 4-kg cart at rest.

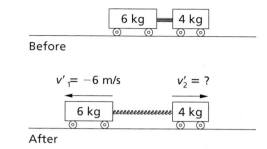

6 kg 4 kg

Before

$v'_1 = -6$ m/s $\qquad v'_2 = ?$

6 kg 4 kg

After

Figure 1-35.

Between the carts is a compressed spring. When the spring expands, it sends both carts in opposite directions. If the 6-kg cart moves to the left with a velocity of -6 m/s, what is the velocity of the other cart?

Solution:

$$p_{before} = p_{after}$$

$$m_1 v_1 + m_2 v_2 = m_1 v_2' + m_2 v_2'$$

$$v_2' = \frac{(m_1 v_1 + m_2 v_2) - m_1 v_1'}{m_2}$$

$$v_2' = \frac{(6 \text{ kg})(0 \text{ m/s}) + (4 \text{ kg})(0 \text{ m/s}) - (6 \text{ kg})(6 \text{ m/s})}{4 \text{ kg}}$$

$$v_2' = \frac{36 \text{ kg} \cdot \text{m/s}}{4 \text{ kg}}$$

$$v_2' = 9 \text{ m/s}$$

Even though both cars are moving after the event, the resultant of their momentum vectors is zero, just as it was before the event. The 4-kg cart has a momentum of $(4 \text{ kg})(9 \text{ m/s})$, or $36 \text{ kg} \cdot \text{m/s}$, to the right, while the 6-kg cart has a momentum of $(6 \text{ kg})(-6 \text{ m/s})$, or $-36 \text{ kg} \cdot \text{m/s}$ to the left.

Momentum and Newton's Third Law of Motion

Recall that Newton's third law of motion states that every action produces an equal and opposite reaction. Thus, when two objects explode or collide, the forces they experience must be equal in magnitude and opposite in direction. Since the time during which these forces act (t) must be the same for both, the impulses (Ft) imparted to each object are equal in magnitude and opposite in direction. Because impulse equals change in momentum, we conclude that *the changes in momentum experienced by the objects as a result of the event are equal in magnitude and opposite in direction.*

QUESTIONS

PART A

151. A 1.2-kilogram block and a 1.8-kilogram block are initially at rest on a frictionless, horizontal surface. When a compressed spring between the blocks is released, the 1.8-kilogram block moves to the right at 2.0 meters per second, as shown.

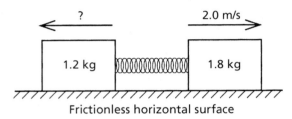

Frictionless horizontal surface

What is the speed of the 1.2-kilogram block after the spring is released? (1) 1.4 m/s (2) 2.0 m/s (3) 3.0 m/s (4) 3.6 m/s

152. One car travels 40. meters due east in 5.0 seconds, and a second car travels 64 meters due west in 8.0 seconds. During their periods of travel, the cars definitely had the same (1) average velocity (2) total displacement (3) change in momentum (4) average speed

153. What is the speed of a 1.0×10^3 kilogram car that has a momentum of 2.0×10^4 kilogram • meters per second east? (1) 5.0×10^{-2} m/s (2) 2.0×10^1 m/s (3) 1.0×10^4 m/s (4) 2.0×10^7 m/s

154. In the diagram below, a 60.-kilogram roller skater exerts a 10.-newton force on a 30.-kilogram roller skater for 0.20 second.

60. kg   30. kg

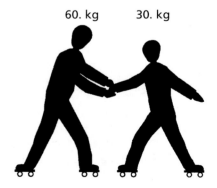

What is the magnitude of the impulse applied to the 30.-kilogram roller skater? (1) 50. N•s (2) 2.0 N•s (3) 6.0 N•s (4) 12 N•s

155. A 2.0-kilogram body is initially traveling at a velocity of 40. meters per second east. If a constant force of 10. newtons due east is applied to the body for 5.0 seconds, the final speed of the body is (1) 15 m/s (2) 25 m/s (3) 65 m/s (4) 130 m/s

156. Which is a scalar quantity? (1) acceleration (2) momentum (3) speed (4) displacement

157. A 0.10-kilogram model rocket's engine is designed to deliver an impulse of 6.0 newton-seconds. If the rocket engine burns for 0.75 second, what average force does it produce? (1) 4.5 N (2) 8.0 N (3) 45 N (4) 80. N

Base your answers to questions 158 and 159 on the information and diagram below.

The diagram shows a compressed spring between two carts initially at rest on a horizontal frictionless surface. Cart A has a mass of 2 kilograms and cart B has a mass of 1 kilogram. A string holds the carts together.

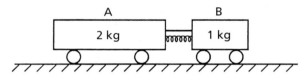

158. What occurs when the string is cut and the carts move apart?
(1) The magnitude of the acceleration of cart A is one-half the magnitude of the acceleration of cart B.
(2) The length of time that the force acts on cart A is twice the length of time the force acts on cart B.
(3) The magnitude of the force exerted on cart A is one-half the magnitude of the force exerted on cart B.
(4) The magnitude of the impulse applied to cart A is twice the magnitude of the impulse applied to cart B.

159. After the string is cut and the two carts move apart, the magnitude of which quantity is the same for both carts? (1) momentum (2) velocity (3) inertia (4) kinetic energy

160. At the circus, a 100.-kilogram clown is fired at 15 meters per second from a 500.-kilogram cannon. What is the recoil speed of the cannon? (1) 75 m/s (2) 15 m/s (3) 3.0 m/s (4) 5.0 m/s

161. Velocity is to speed as displacement is to (1) acceleration (2) time (3) momentum (4) distance

162. A 50.-kilogram student threw a 0.40-kilogram ball with a speed of 20. meters per second. What was the magnitude of the impulse that the student exerted on the ball? (1) 8.0 N · s (2) 78 N · s (3) 4.0 × 10² N · s (4) 1.0 × 10³ N · s

163. Ball *A* of mass 5.0 kilograms moving at 20. meters per second collides with ball *B* of unknown mass moving at 10. meters per second in the same direction. After the collision, ball *A* moves at 10. meters per second and ball *B* at 15 meters per second, both still in the same direction. What is the mass of ball *B*? (1) 6.0 kg (2) 2.0 kg (3) 10. kg (4) 12 kg

164. Which is an acceptable unit for impulse? (1) N · m (2) J/s (3) J · s (4) kg · m/s

165. A 20-kilogram mass moving at a speed of 3.0 meters per second is stopped by a constant force of 15 newtons. How many seconds must the force act on the mass to stop it? (1) 0.20 s (2) 1.3 s (3) 4.0 s (4) 5.0 s

166. An object traveling at 4.0 meters per second has a momentum of 16 kilogram-meters per second. What is the mass of the object? (1) 64 kg (2) 20 kg (3) 12 kg (4) 4.0 kg

167. Two carts resting on a frictionless surface are forced apart by a spring. One cart has a mass of 2 kilograms and moves to the left at a speed of 3 meters per second. If the second cart has a mass of 3 kilograms, it will move to the right at a speed of (1) 1 m/s (2) 2 m/s (3) 3 m/s (4) 6 m/s

168. A 15-newton force acts on an object in a direction due east for 3.0 seconds. What will be the change in momentum of the object? (1) 45 kg-m/s due east (2) 45 kg-m/s due west (3) 5.0 kg-m/s due east (4) 0.20 kg-m/s due west

169. A 5.0-kilogram cart moving with a velocity of 4.0 meters per second is brought to a stop in 2.0 seconds. The magnitude of the average force used to stop the cart is (1) 2.0 N (2) 4.0 N (3) 10. N (4) 20. N

170. A 5.0-newton force imparts an impulse of 15 newton-seconds to an object. The force acted on the object for a period of (1) 75 s (2) 20. s (3) 3.0 s (4) 0.33 s

171. A net force of 12 newtons acting north on an object for 4.0 seconds will produce an impulse of (1) 48 kg-m/s north (2) 48 kg-m/s south (3) 3.0 kg-m/s north (4) 3.0 kg-m/s south

172. Two disk magnets are arranged at rest on a frictionless horizontal surface as shown in the diagram. When the string holding them together is cut, they move apart under a magnetic force of repulsion. When the 1.0-kilogram disk reaches a speed of 3.0 meters per second, what is the speed of the 0.5-kilogram disk?

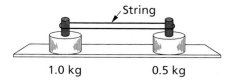

(1) 0.50 m/s (2) 1.0 m/s (3) 3.0 m/s (4) 6.0 m/s

173. The diagram represents two identical carts, attached by a cord moving to the right at speed *V*. If the cord is cut, what would be the speed of cart *A*?

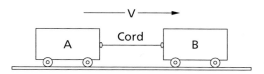

(1) 0 (2) *V*/2 (3) *V* (4) 2*V*

174. If a 3.0-kilogram object moves 10. meters in 2.0 seconds, its average momentum is (1) 60. kg · m/s (2) 30. kg · m/s (3) 15 kg · m/s (4) 10. kg · m/s

175. An impulse of 30.0 newton-seconds is applied to a 5.00-kilogram mass. If the mass had a speed of 100. meters per second before the impulse, its speed after the impulse could be (1) 250. m/s (2) 106 m/s (3) 6.00 m/s (4) 0 m/s

176. Two carts of masses of 5.0 kilograms and 1.0 kilogram are pushed apart by a compressed spring. If the 5.0-kilogram cart moves westward at 2.0 meters per second, the magnitude of the velocity of the 1.0-kilogram cart will be (1) 2.0 kg-m/s (2) 2.0 m/s (3) 10. kg-m/s (4) 10. m/s

177. The direction of an object's momentum is always the same as the direction of the object's (1) inertia (2) potential energy (3) velocity (4) weight

178. An unbalanced force of 20 newtons is applied to an object for 10 seconds. The change in the momentum of the object will be (1) 200 kg · m/s (2) 2000 kg · m/s (3) 10,000 kg · m/s (4) 20,000 kg · m/s

179. A 4.0-kilogram mass is moving at 3.0 meters per second toward the right and a 6.0-kilogram mass is moving at 2.0 meters per second toward the left on a horizontal frictionless table. If the two masses collide and remain together after the collision, their final momentum is (1) 24 kg·m/s (2) 12 kg·m/s (3) 1.0 kg·m/s (4) 0 kg·m/s

180. In the diagram below, a block of mass M initially at rest on a frictionless horizontal surface is struck by a bullet of mass m moving with horizontal velocity v.

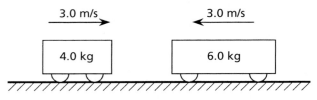

What is the velocity of the bullet-block system after the bullet embeds itself in the block?

$$\left(\frac{M+v}{M}\right)m \qquad \left(\frac{m+M}{m}\right)v \qquad \left(\frac{m+v}{M}\right)m \qquad \left(\frac{m}{m+M}\right)v$$
(1) \qquad\qquad (2) \qquad\qquad (3) \qquad\qquad (4)

181. The diagram below shows a 4.0-kilogram cart moving to the right and a 6.0-kilogram cart moving to the left on a horizontal frictionless surface.

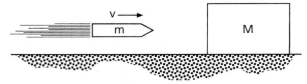

Frictionless surface

When the two carts collide they lock together. The magnitude of the total momentum of the two-cart system after the collision is (1) 0.0 kg·m/s (2) 6.0 kg·m/s (3) 15 kg·m/s (4) 30. kg·m/s

Two-Dimensional Motion and Trajectories

Up to now we have discussed the motion of objects traveling along a straight line, or in one dimension. Many objects, however, move within a plane, or in two dimensions. Any object that is launched by some force and continues to move by its own inertia is called a **projectile**. The path of a projectile is referred to as its trajectory.

The motion of a projectile is best described by separating its motion into horizontal (x) and vertical (y) components of the vector quantities displacement, velocity, and acceleration. Using the equations of motion you learned earlier, you can calculate how far a projectile will travel in the horizontal direction and how long it will take to hit the ground. The motion equations are applied independently to the horizontal motion and the vertical motion. Those equations are listed in Table 1-5.

Table 1-5

Horizontal Components	
$a_x = 0$	Ignoring air resistance, there is no force in the horizontal direction to produce an acceleration.
$v_x = $ constant	The horizontal component of velocity remains constant.
$d_x = v_x t$	The horizontal displacement or range.

Vertical Components	
$a_y = g = 9.81$ m/s^2	Acceleration due to gravity near Earth is constant.
$v_{fy} = v_{iy} + a_y t$	When $v_{iy} = 0$, this equation reduces to $v_{fy} = a_y t$.
$d_y = v_{iy}t + \frac{1}{2}a_y t^2$	When $v_{iy} = 0$, this equation reduces to $d_y = \frac{1}{2}a_y t^2$.

Projectiles Launched Horizontally

An object launched horizontally with a velocity v_x starts its motion with an initial vertical velocity of zero ($v_{iy} = 0$). Since no forces act in the horizontal direction (ignoring air resistance), the horizontal component of the object's velocity remains constant. The vertical component, however, is affected by Earth's gravitational pull and accelerates at the rate of 9.81 m/s^2. The resulting path of the object is a parabola.

The time it takes the object to reach the ground depends only on the height above the ground from which it is launched and the acceleration due to gravity. The horizontal component of its motion has no effect on this time. Balls A, B, and C in Figure 1-36 start with different hori-

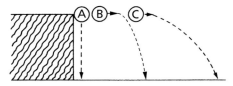

Figure 1-36. Ball A is released and falls straight down. Ball B is launched horizontally at 5 m/s. Ball C is launched horizontally at 10 m/s.

zontal velocities, but all three balls reach the ground at the same time. On the other hand, the horizontal distance traveled, or **range**, is dependent on the horizontal velocity and the time of flight.

Sample Problems

26A. A ball is tossed horizontally with a velocity of 15 m/s from a height of 100. m above the ground (Figure 1-37). How long will it take the ball to hit the ground?

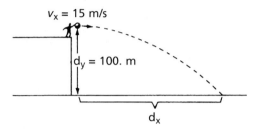

$v_x = 15$ m/s

$d_y = 100.$ m

d_x

Figure 1-37.

Solution:

$$d_y = v_{iy}t + \tfrac{1}{2}a_y t^2$$

$$v_{iy} = 0, \, a_y = g$$

$$t^2 = \frac{d_y}{\tfrac{1}{2}g}$$

$$t = \sqrt{\frac{100. \text{ m}}{4.9 \text{ m/s}^2}} = 4.5 \text{ s}$$

26B. What are the horizontal and vertical components of the ball's velocity as it strikes the ground?

Solution:

$$v_x = 15 \text{ m/s}$$

$$v_{fy} = v_{iy} + a_y t$$

$$v_{fy} = 0 \text{ m/s} + \left[(9.81 \text{ m/s}^2)(4.5 \text{ m/s})\right] = 44 \text{ m/s}$$

26C. How far did the ball travel horizontally in its flight?

Solution:

$$d_x = v_x t$$

$$d_x = (15 \text{ m/s})(4.5 \text{ s}) = 68. \text{ m}$$

26D. What is the magnitude of the resultant velocity vector as the ball strikes the ground?

Solution: The resultant velocity vector is the vector sum of the ball's vertical and horizontal components. Thus, using the Pythagorean theorem, $c^2 = a^2 + b^2$, we find

$$v_f^2 = v_{fx}^2 + v_{fy}^2$$

$$v_f^2 = (15 \text{ m/s})^2 + (44 \text{ m/s})^2$$

$$v_f = \sqrt{2161 \text{ m}^2/\text{s}^2} = 46 \text{ m/s}$$

27. In Figure 1-38, an airplane flying horizontally at 150 m/s drops a bomb from a height of 2500 m. How far in front of the target must the bomb be released?

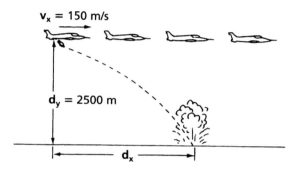

$v_x = 150$ m/s

$d_y = 2500$ m

d_x

Figure 1-38.

Solution: First, find the time it takes for the bomb to hit the ground.

$$d_y = v_{iy} + \tfrac{1}{2}a_y t^2$$

$$v_{iy} = 0$$

$$t^2 = \frac{d_y}{\tfrac{1}{2}a} = \frac{2500 \text{ m}}{\tfrac{1}{2}(9.81 \text{ m/s}^2)}$$

$$t = \sqrt{\frac{2500 \text{ m}}{4.9 \text{ m/s}^2}} = 23 \text{ s}$$

Now find the range.

$$\bar{v} = \frac{d_x}{t}$$

$$d_x = \bar{v}t$$

$$d_x = (150 \text{ m/s})(23 \text{ s}) = 3.5 \times 10^3 \text{ m}$$

Thus the bomb must be released 3.5×10^3 m in front of the target.

Projectiles Launched at an Angle

Golf balls, baseballs, and missiles are examples of projectiles launched at an angle to the surface of Earth. The maximum height achieved by such projectiles and their total horizontal displacement, or range, depend on the magnitude and direction of the velocity vector at the time of launch and on the acceleration due to gravity. Once the initial vertical and horizontal components of velocity have been determined, the motion of the projectile can be treated as two independent linear motion problems.

The horizontal and vertical components of the initial velocity vector can be determined by vector resolution. Figure 1-39 is a vector diagram of the initial velocity vector v_i of a projectile launched at some angle, θ. You can also use the following equations given in *The Reference Tables for Physics*.

$$A_y = A \sin \theta$$

$$A_x = A \cos \theta$$

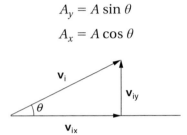

Figure 1-39.

The *maximum range* of a projectile launched at a given speed is achieved when the angle with the horizontal is 45°.

Sample Problem

28A. A cannonball is fired from ground level at an angle of 60° with the ground at a speed of 72 m/s (Figure 1-40). What are the vertical and horizontal components of the velocity at the time of launch?

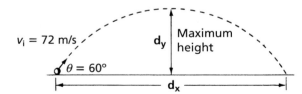

Figure 1-40.

Solution:

$$A_y = A \sin \theta$$

$$A_y = v_{iy}$$

$$v_{iy} = v_i \sin \theta = (72 \text{ m/s})(\sin 60°)$$

$$v_{iy} = (72 \text{ m/s})(0.866) = 62 \text{ m/s}$$

$$A_x = A \cos \theta$$

$$A_x = v_{ix}$$

$$v_{ix} = v_i \cos \theta = (72 \text{ m/s})(\cos 60°)$$

$$v_{ix} = (72 \text{ m/s})(0.500) = 36 \text{ m/s}$$

28B. How long does it take the cannonball to reach its maximum height?

Solution: When the ball is at its highest point the vertical component of the velocity is zero. At that point the cannonball has stopped moving upward and is just about to begin moving downward. Thus,

$$v_{fy} = v_{iy} + a_y t$$

$$v_{fy} = 0$$

$$t = \frac{-v_{iy}}{a}$$

$$t = \frac{-62 \text{ m/s}}{-9.81 \text{ m/s}^2} = 6.3 \text{ s}$$

28C. How high does the cannonball rise?

Solution:

$$d_y = v_{iy}t + \tfrac{1}{2}a_y t^2$$

$$d_y = (62 \text{ m/s})(6.3 \text{ s}) + \tfrac{1}{2}(-9.81 \text{ m/s}^2)(6.3 \text{ s})^2$$

$$d_y = 196 \text{ m}$$

28D. What is the cannonball's total time of flight?

Solution: We have already found that the time needed to reach the maximum height is 6.3 s. It will take the same amount of time to return to the ground. Therefore, the total time of flight is

$$2 \times 6.3 \text{ s} = 12.6 \text{ s}$$

The answer can also be obtained by calculating the time to return to the ground and adding it to the time it took to reach its maximum height:

$$d_y = v_{iy}t + \tfrac{1}{2}at^2$$

$$v_{iy} = 0$$

$$t^2 = \frac{d_{iy}}{\frac{1}{2}a} = \frac{196 \text{ m}}{\frac{1}{2}(9.81 \text{ m/s}^2)}$$

$$t = \sqrt{\frac{196 \text{ m}}{4.9 \text{ m/s}}} = 6.3 \text{ s}$$

Total time of flight 6.3 s (up) + 6.3 s (down) = 12.6 s

28E. What is the cannonball's range?

Solution:

$$v_x = \frac{d_x}{t}$$

$$d_x = v_x t$$

$$d_x = (36 \text{ m/s})(12.6 \text{ s}) = 453 \text{ m}$$

Uniform Circular Motion

An object undergoes **uniform circular motion** if it moves along a circular path at a constant speed. Although the magnitude of the object's velocity is constant, the velocity *direction* is constantly changing. A change in the velocity vector of an object means that the object is accelerating. The acceleration experienced by an object in uniform circular motion is called **centripetal acceleration**. Centripetal acceleration a_c is a vector quantity, directed toward the center of the circle. The magnitude of this acceleration vector is directly proportional to the square of the object's speed and inversely proportional to the radius of its path. Thus,

$$a_c = \frac{v^2}{r}$$

where v is the speed of the object along the circular path and r is the radius of the circle.

Since a circling object is accelerating, there must be a net force acting on it—it is not in equilibrium. The force that causes the centripetal acceleration is called the **centripetal force F_c**. This force acts in the same direction as the centripetal acceleration—toward the center of the circle (Figure 1-41). The magnitude of the centripetal force is obtained from Newton's second law.

$$F_c = ma_c$$

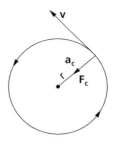

Figure 1-41.

The centripetal force is the net force on the circling object that acts to change the object's direction. The object's speed is constant because there are no tangential forces acting on the object.

Sample Problem

29. A 1.0-kg ball attached to the end of a rope 0.50 m long is swung in a circle. Its speed along the circular path is 6.0 m/s. Find the centripetal acceleration and force.

Solution:

$$a_c = \frac{v^2}{r} = \frac{(6.0 \text{ m/s})^2}{0.50 \text{ m}} = 72 \text{ m/s}^2$$

$$F_c = ma_c = (1.0 \text{ kg})(72 \text{ m/s}^2) = 72 \text{ N}$$

Questions

PART A

Base your answers to questions 182 and 183 on the information and diagram below.

A child kicks a ball with an initial velocity of 8.5 meters per second at an angle of 35° with the horizontal, as shown. The ball has an initial vertical velocity of 4.9 meters per second and a total time of flight of 1.0 second. (Neglect air resistance.)

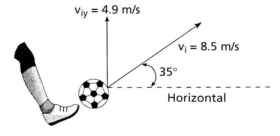

182. The horizontal component of the ball's initial velocity is approximately (1) 3.6 m/s (2) 4.9 m/s (3) 7.0 m/s (4) 13 m/s

183. The maximum height reached by the ball is approximately (1) 1.2 m (2) 2.5 m (3) 4.9 m (4) 8.5 m

184. The diagram below shows a student throwing a baseball horizontally at 25 meters per second from a cliff 45 meters above the level ground.

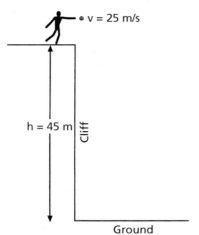

Approximately how far from the base of the cliff does the ball hit the ground? [Neglect air resistance.] (1) 45 m (2) 75 m (3) 140 m (4) 230 m

185. A projectile is fired from a gun near the surface of Earth. The initial velocity of the projectile has a vertical component of 98 meters per second and a horizontal component of 49 meters per second. How long will it take the projectile to reach the highest point in its path? (1) 5.0 s (2) 10. s (3) 20. s (4) 100. s

186. A golf ball is hit at an angle of 45° above the horizontal. What is the acceleration of the golf ball at the highest point in its trajectory? (Neglect friction.) (1) 9.8 m/s² upward (2) 9.8 m/s² downward (3) 6.9 m/s² horizontal (4) 0.0 m/s²

187. A ball is thrown horizontally at a speed of 24 meters per second from the top of a cliff. If the ball hits the ground 4.0 seconds later, approximately how high is the cliff? (1) 6.0 m (2) 39 m (3) 78 m (4) 96 m

188. A 0.2-kilogram red ball is thrown horizontally at a speed of 4 meters per second from a height of 3 meters. A 0.4-kilogram green ball is thrown horizontally from the same height at a speed of 8 meters per second. Compared to the time it takes the red ball to reach the ground, the time it takes the green ball to reach the ground is (1) one-half as great (2) twice as great (3) the same (4) four times as great

189. A ball is thrown at an angle of 38° to the horizontal. What happens to the magnitude of the ball's vertical acceleration during the total time interval that the ball is in the air? (1) It decreases, then increases. (2) It decreases, then remains the same. (3) It increases, then decreases. (4) It remains the same.

Base your answers to questions 190 and 191 on the information below.

Projectile *A* is launched horizontally at a speed of 20. meters per second from the top of a cliff and strikes a level surface below, 3.0 seconds later. Projectile *B* is launched horizontally from the same location at a speed of 30. meters per second.

190. The time it takes projectile *B* to reach the level surface is (1) 4.5 s (2) 2.0 s (3) 3.0 s (4) 10. s

191. Approximately how high is the cliff? (1) 29 m (2) 44 m (3) 60. m (4) 104 m

Base your answers to questions 192 and 193 on the information below.

A 2.0×10^3 kilogram car travels at a constant speed of 12 meters per second around a circular curve of radius 30. meters.

192. What is the magnitude of the centripetal acceleration of the car as it goes around the curve? (1) 0.40 m/s² (2) 4.8 m/s² (3) 800 m/s² (4) 9600 m/s²

193. As the car goes around the curve, the centripetal force is directed
(1) toward the center of the circular curve
(2) away from the center of the circular curve
(3) tangent to the curve in the direction of motion
(4) tangent to the curve opposite the direction of motion

194. What is the magnitude of the centripetal force on a 2.0×10^3 kilogram car as it goes around a 25-meter curve with an acceleration of 5 m/s²? (1) 125 N (2) 1.0×10^3 N (3) 1.0×10^4 N (4) 5.0×10^4 N

195. A ball of mass *M* at the end of a string is swung in a horizontal circular path of radius *R* at constant speed *V*. Which combination of changes would require the greatest increase in the centripetal force acting on the ball? (1) doubling *V* and doubling *R* (2) doubling *V* and halving *R* (3) halving *V* and doubling *R* (4) halving *V* and halving *R*

196. The following diagram shows a 5.0-kilogram bucket of water being swung in a horizontal circle of 0.70-meter radius at a constant speed of 2.0 meters per second.

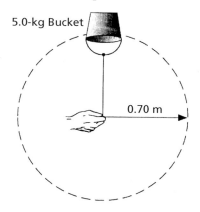

5.0-kg Bucket

0.70 m

The magnitude of the centripetal force on the bucket of water is approximately (1) 5.7 N (2) 14 N (3) 29 N (4) 200 N

Base your answers to questions 197 and 198 on the diagram and information below.

The diagram shows a student seated on a rotating circular platform, holding a 2.0-kilogram block with a spring scale. The block is 1.2 meters from the center of the platform. The block has a constant speed of 8.0 meters per second. (Frictional forces on the block are negligible.)

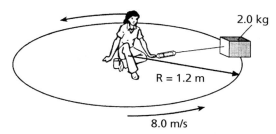

2.0 kg

R = 1.2 m

8.0 m/s

197. Which statement best describes the block's movement as the platform rotates?
(1) Its velocity is directed tangent to the circular path, with an inward acceleration.
(2) Its velocity is directed tangent to the circular path, with an outward acceleration.
(3) Its velocity is directed perpendicular to the circular path, with an inward acceleration.
(4) Its velocity is directed perpendicular to the circular path, with an outward acceleration.

198. The reading on the spring scale is approximately (1) 20. N (2) 53 N (3) 110 N (4) 130 N

Base your answers to questions 199 through 202 on the following diagram, which represents a 2.0-kilogram mass moving in a circular path on the end of a string 0.50 meter long. The mass moves in a horizontal plane at a constant speed of 4.0 meters per second.

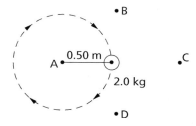

• B

A • 0.50 m ○ • C

2.0 kg

• D

199. The force exerted on the mass by the string is (1) 8 N (2) 16 N (3) 32 N (4) 64 N

200. In the position shown in the diagram, the momentum of the mass is directed toward point (1) A (2) B (3) C (4) D

201. The centripetal force acting on the mass is directed toward point (1) A (2) B (3) C (4) D

202. The speed of the mass is changed to 2.0 meters per second. Compared to the centripetal acceleration of the mass when moving at 4.0 meters per second, its centripetal acceleration when moving at 2.0 meters per second would be (1) half as great (2) twice as great (3) one-fourth as great (4) four times as great

Base your answers to questions 203 through 207 on the diagram below, which represents a 5.0-kilogram object revolving around a circular track in a horizontal plane at a constant speed. The radius of the track is 20. meters and the centripetal force on the object is 4.0×10^2 newtons.

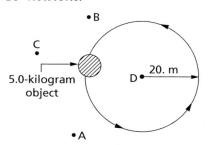

• B

C •

5.0-kilogram object

D • 20. m

• A

203. In the position shown, the object's centripetal acceleration is directed toward point (1) A (2) B (3) C (4) D

204. In the position shown, the object's velocity is directed toward point (1) A (2) B (3) C (4) D

205. The object's centripetal acceleration is (1) 0.012 m/s² (2) 20. m/s² (3) 80. m/s² (4) 1.0×10^2 m/s²

206. The object's speed is (1) 20. m/s (2) 40. m/s (3) 60. m/s (4) 90. m/s

207. If the radius of the track is increased, the centripetal force necessary to keep the object revolving at the same speed would (1) decrease (2) increase (3) remain the same

208. At what angle from the horizontal must a projectile be launched in order to achieve the greatest range? (1) 20° (2) 30° (3) 45° (4) 57.3°

209. At the same moment that a baseball is thrown horizontally by a pitcher, a ring drops vertically off his hand. Which statement about the baseball and the ring is correct, neglecting air resistance? (1) The baseball hits the ground first. (2) The ring hits the ground first. (3) They both hit the ground at the same time.

210. In the diagram below, a cart travels clockwise at constant speed in a horizontal circle.

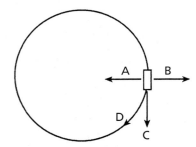

At the position shown in the diagram, which arrow indicates the direction of the centripetal acceleration of the cart? (1) *A* (2) *B* (3) *C* (4) *D*

211. In the diagram below, *S* is a point on a car tire rotating at a constant rate.

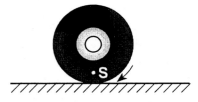

Which graph best represents the magnitude of the centripetal acceleration of point *S* as a function of time?

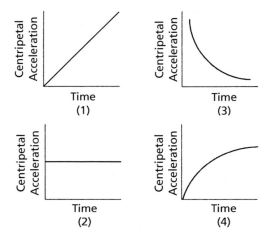

Base your answers to questions 212 through 216 on the diagram below, which represents a flat racetrack as viewed from above, with the radii of its two curves indicated. A car with a mass of 1000 kilograms moves counterclockwise around the track at a constant speed of 20 meters per second.

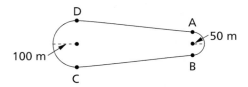

212. The net force acting on the car while it is moving from *A* to *D* is (1) 0 N (2) 400 N (3) 8000 N (4) 20,000 N

213. The net force acting on the car while it is moving from *D* to *C* is (1) 0 N (2) 200 N (3) 4000 N (4) 20,000 N

214. If the car moved from *C* to *B* in 20 seconds, the distance *CB* is (1) 100 m (2) 200 m (3) 300 m (4) 400 m

215. Compared with the centripetal acceleration of the car while moving from *B* to *A*, the centripetal acceleration of the car while moving from *D* to *C* is (1) the same (2) twice as great (3) one-half as great (4) 4 times greater

Note that question 216 has only three choices

216. Compared with the speed of the car while moving from *A* to *D*, the speed of the car while moving from *D* to *C* is (1) less (2) greater (3) the same

Free Fall

"What goes up, must come down," goes the old saying. This is true because of gravity. Gravity is the force that acts to pull objects toward Earth's center. When the only force acting on a falling object is gravity, the object is said to be in free fall. An object in free fall accelerates as it falls because gravity acts as an unbalanced force in the downward direction. The acceleration due to gravity is 9.81 m/s².

Does this mean that a skydiver accelerates at a rate of 9.81 m/s² after jumping from an aircraft? The answer is no. In many cases, such as this, there is another force acting on falling objects—air resistance. Air resistance is a friction force that acts in the direction opposite to the direction of motion of the skydiver. In these cases, the objects are not falling freely.

Air resistance, or air friction, is not a constant value. Instead, it depends on the surface area and velocity of the falling object. Generally, the greater the surface area of the object perpendicular to the direction of motion, the greater is the air resistance on the object. And, generally, the greater the velocity of the object, the greater is the air resistance on the object. So as a falling object speeds up due to the force of gravity, the air resistance on the object increases. At some point, air resistance becomes equal to the force of gravity. When this happens, the downward force on the object is balanced by the upward force on the object. When forces are balanced, there is no acceleration. Does the object stop falling? Of course, not. The object continues to fall. However, its velocity no longer increases. The velocity at which this occurs is called *terminal velocity*.

Questions

1. Under what conditions is an object in free fall?
2. What is air resistance?
3. Describe the force(s) acting on a skydiver before and after terminal velocity is reached. Include diagrams in your description.
4. How is a skydiver's parachute related to the motion of the skydiver? How does a parachute affect the motion of a skydiver?

Chapter Review Questions

PART A

1. The diagram below represents the path of an object after it was thrown.

What happens to the object's acceleration as it travels from *A* to *B*? (Neglect friction.) (1) It decreases. (2) It increases. (3) It remains the same.

2. The speed of a car is increased uniformly from 20. meters per second to 30. meters per second in 4.0 seconds. The magnitude of the car's average acceleration in this 4.0-second interval is (1) 0.40 m/s² (2) 2.5 m/s² (3) 10 m/s² (4) 13 m/s²

3. A roller coaster car, traveling with an initial speed of 15 meters per second, decelerates uniformly at −7.0 meters per second² to a full stop. Approximately how far does the roller coaster car travel during its deceleration? (1) 1.0 m (2) 2.0 m (3) 16 m (4) 32 m

4. If the magnitude of the gravitational force of Earth on the moon is *F*, the magnitude of the

gravitational force of the moon on Earth is (1) smaller than F (2) larger than F (3) equal to F

5. Which term represents a scalar quantity? (1) distance (2) displacement (3) force (4) weight

6. The centers of two 15.0-kilogram spheres are separated by 3.00 meters. The magnitude of the gravitational force between the two spheres is approximately (1) 1.11×10^{-10} N (2) 3.34×10^{-10} N (3) 1.67×10^{-9} N (4) 5.00×10^{-9} N

7. During a collision, an 84-kilogram driver of a car moving at 24 meters per second is brought to rest by an inflating air bag in 1.2 seconds. The magnitude of the force exerted on the driver by the air bag is approximately (1) 7.0×10^1 N (2) 8.2×10^2 N (3) 1.7×10^3 N (4) 2.0×10^3 N

8. An apple weighing 1 newton on the surface of Earth has a mass of approximately (1) 1×10^{-1} kg (2) 1×10^0 kg (3) 1×10^1 kg (4) 1×10^2 kg

9. A car initially traveling at a speed of 16 meters per second accelerates uniformly to a speed of 20. meters per second over a distance of 36 meters. What is the magnitude of the car's acceleration? (1) 0.11 m/s² (2) 2.0 m/s² (3) 0.22 m/s² (4) 9.0 m/s²

10. A net force of 25 newtons is applied horizontally to a 10.-kilogram block resting on a table. What is the magnitude of the acceleration of the block? (1) 0.0 m/s² (2) 0.26 m/s² (3) 0.40 m/s² (4) 2.5 m/s²

11. A child is riding on a merry-go-round. As the speed of the merry-go-round is doubled, the magnitude of the centripetal force acting on the child (1) remains the same (2) is doubled (3) is halved (4) is quadrupled

12. The diagram below represents a 0.40-kilogram stone attached to a string. The stone is moving at a constant speed of 4.0 meters per second in a horizontal circle having a radius of 0.80 meter.

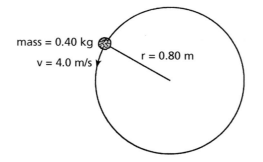

The magnitude of the centripetal acceleration of the stone is (1) 0.0 m/s² (2) 2.0 m/s² (3) 5.0 m/s² (4) 20. ms/²

13. In which situation is the net force on the object equal to zero?
    (1) a satellite moving at constant speed around Earth in a circular orbit
    (2) an automobile braking to a stop
    (3) a bicycle moving at constant speed on a straight, level road
    (4) a pitched baseball being hit by a bat

14. A 60-kilogram skydiver is falling at a constant speed near the surface of Earth. The magnitude of the force of air friction acting on the skydiver is approximately (1) 0 N (2) 6 N (3) 60 N (4) 600 N

15. An astronaut weighs 8.00×10^2 newtons on the surface of Earth. What is the weight of the astronaut 6.37×10^6 meters above the surface of Earth? (1) 0.00 N (2) 2.00×10^2 N (3) 1.60×10^3 N (4) 3.2×10^3 N

16. A 1.5-kilogram lab cart is accelerated uniformly from rest to a speed of 2.0 meters per second in 0.50 second. What is the magnitude of the force producing this acceleration? (1) 0.70 N (2) 1.5 N (3) 3.0 N (4) 6.0 N

17. Which person has the greatest inertia? (1) a 110-kg wrestler resting on a mat (2) a 90-kg man walking at 2 m/s (3) a 70-kg long-distance runner traveling at 5 m/s (4) a 50-kg girl sprinting at 10 m/s

PART B-1

18. Which graph best represents the motion of an object that is *not* in equilibrium as it travels along a straight line?

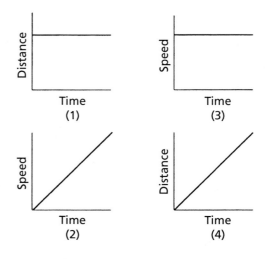

19. Three forces act on a box on an inclined plane as shown in the diagram below. (Vectors are not drawn to scale.)

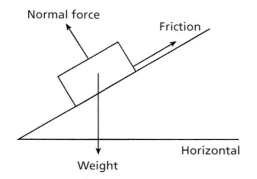

If the box is at rest, the net force acting on it is equal to (1) the weight (2) the normal force (3) friction (4) zero

20. A 1.0×10^3 kilogram car travels at a constant speed of 20. meters per second around a horizontal circular track. Which diagram correctly represents the direction of the car's velocity (v) and the direction of the centripetal force (F_c) acting on the car at one particular moment?

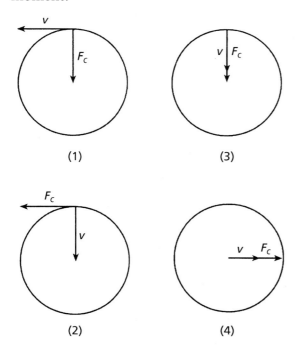

21. An archer uses a bow to fire two similar arrows with the same string force. One arrow is fired at an angle of 60.° with the horizontal, and the other is fired at an angle of 45° with the horizontal. Compared to the arrow fired at 60.°, the arrow fired at 45° has a
(1) longer flight time and longer horizontal range

(2) longer flight time and shorter horizontal range
(3) shorter flight time and longer horizontal range
(4) shorter flight time and shorter horizontal range

22. The graph below shows the velocity of a race car moving along a straight line as a function of time.

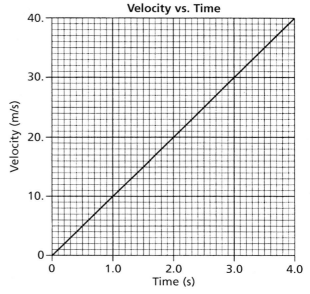

What is the magnitude of the displacement of the car from $t = 2.0$ seconds to $t = 4.0$ seconds? (1) 20. m (2) 40. m (3) 60. m (4) 80. m

23. A force vector was resolved into two perpendicular components, F_1 and F_2, as shown in the diagram at the right.

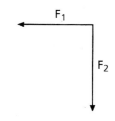

Which vector best represents the original force?

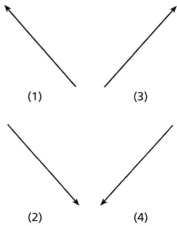

24. The diagram below shows a 10.0-kilogram mass held at rest on a frictionless 30.0° incline by force *F*.

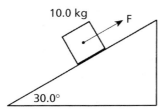

What is the approximate magnitude of force *F*? (1) 9.81 N (2) 49.1 N (3) 85.0 N (4) 98.1 N

Base your answers to questions 25 through 29 on the four graphs below, which represent the relationship between speed and time of four different objects, A, B, C, and D.

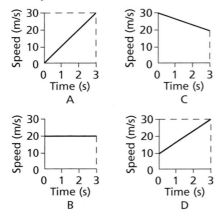

25. Which object was slowing? (1) *A* (2) *B* (3) *C* (4) *D*

26. Which object was neither accelerating nor decelerating? (1) *A* (2) *B* (3) *C* (4) *D*

27 Which object traveled the greatest distance in the 3.0-second interval? (1) *A* (2) *B* (3) *C* (4) *D*

28. Which object had the greatest acceleration? (1) *A* (2) *B* (3) *C* (4) *D*

29. Compared with the average velocity of object *A*, the average velocity of object *D* is (1) less (2) greater (3) the same

Base your answers to questions 30 through 33 on the following diagram, which represents a car of mass 1000 kilograms traveling around a horizontal circular track of radius 200 meters at a constant speed of 20 meters per second.

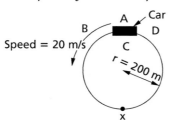

30. When the car is in the position shown, the direction of its centripetal acceleration is toward (1) *A* (2) *B* (3) *C* (4) *D*

31. The magnitude of the centripetal force acting on the car is closest to (1) 100 N (2) 1000 N (3) 2000 N (4) 4000 N

32. If the speed of the car were doubled, the centripetal acceleration of the car would be (1) the same (2) doubled (3) one-half as great (4) 4 times as great

33. If additional passengers were riding in the car, at the original speed, the car's centripetal acceleration would be (1) less (2) greater (3) the same

Base your answers to questions 34 and 35 on the information below.

An outfielder throws a baseball to the first baseman at a speed of 19.6 meters per second and an angle of 30.° above the horizontal.

34. Which pair represents the initial horizontal velocity (v_x) and initial vertical velocity (v_y) of the baseball?
(1) $v_x = 17.0$ m/s, $v_y = 9.80$ m/s
(2) $v_x = 9.80$ m/s, $v_y = 17.0$ m/s
(3) $v_x = 19.4$ m/s, $v_y = 5.90$ m/s
(4) $v_x = 19.6$ m/s, $v_y = 19.6$ m/s

35. If the ball is caught at the same height from which it was thrown, calculate the amount of time the ball was in the air. Show all work, including the equation and substitution with units.

Base your answers to questions 36 and 37 on the information below.

A soccer player accelerates a 0.50-kilogram soccer ball by kicking it with a net force of 5.0 newtons.

36. Calculate the magnitude of the acceleration of the ball. Show all work, including the equation and substitution with units.

37. What is the magnitude of the force of the soccer ball on the player's foot?

38. State the *two* general characteristics that are used to define a vector quantity.

39. An airplane is moving with a constant velocity in level flight. Compare the magnitude of the forward force provided by the engines to the magnitude of the backward frictional drag force.

Base your answers to questions 40 through 42 on the information and diagram below.

An object was projected horizontally from a tall cliff. The diagram below represents the path of the object, neglecting friction.

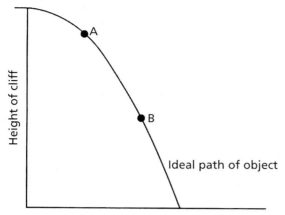

40. How does the magnitude of the horizontal component of the object's velocity at point *A* compare with the magnitude of the horizontal component of the object's velocity at point *B*?

41. How does the magnitude of the vertical component of the object's velocity at point *A* compare with the magnitude of the vertical component of the object's velocity at point *B*?

42. Sketch a likely path of the horizontally projected object, assuming that it was subject to air resistance.

Base your answers to questions 43 through 45 on the information and diagram below.

In the scaled diagram, two forces, F_1 and F_2, act on a 4.0-kilogram block at point *P*. Force F_1 has a magnitude of 12.0 newtons, and is directed toward the right.

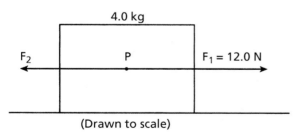

(Drawn to scale)

43. Using a ruler and the scaled diagram, determine the magnitude of F_2 in newtons.

44. Determine the magnitude of the net force acting on the block.

45. Calculate the magnitude of the acceleration of the block. Show all work, including the equation and substitution with units.

46. The coefficient of kinetic friction between a 780.-newton crate and a level warehouse floor is 0.200. Calculate the magnitude of the horizontal force required to move the crate across the floor at constant speed. Show all work, including the equation and substitution with units.

47. Objects in free fall near the surface of Earth accelerate downward at 9.81 meters per second². Explain why a feather does *not* accelerate at this rate when dropped near the surface of Earth.

48. A skier on waxed skis is pulled at constant speed across level snow by a horizontal force of 39 newtons. Calculate the normal force exerted on the skier. Show all work, including the equation and substitution with units.

49. A 1000-kilogram car traveling due east at 15 meters per second is hit from behind and receives a forward impulse of 6000 newton-seconds. Determine the magnitude of the car's change in momentum due to this impulse.

50. Using dimensional analysis, show that the expression v^2/d has the same units as acceleration. Show all the steps used to arrive at your answer.

Base your answers to questions 51 through 53 on the information and diagram below.

A 1.50-kilogram cart travels in a horizontal circle of radius 2.40 meters at a constant speed of 4.00 meters per second.

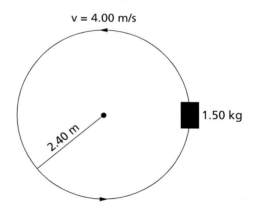

51. Calculate the time required for the cart to make one complete revolution. Show all work, including the equation and substitution with units.

52. Describe a change that would quadruple the magnitude of the centripetal force.

53. Copy the diagram, then draw an arrow to represent the direction of the acceleration of the cart in the position shown. Label the arrow *a*.

Base your answers to questions 54 and 55 on the information below.

A car traveling at a speed of 13 meters per second accelerates uniformly to a speed of 25 meters per second in 5.0 seconds.

54. Calculate the magnitude of the acceleration of the car during this 5.0-second time interval. Show all work, including the equation and substitution with units.

55. A truck traveling at a constant speed covers the same total distance as the car in the same 5.0-second time interval. Determine the speed of the truck.

56. The gravitational force of attraction between Earth and the sun is 3.52×10^{22} newtons. Calculate the mass of the sun. Show all work, including the equation and substitution with units.

Base your answers to questions 57 through 59 on the information below.

The combined mass of a race car and its driver is 600. kilograms. Traveling at constant speed, the car completes one lap around a circular track of radius 160 meters in 36 seconds.

57. Calculate the speed of the car. Show all work, including the equation and substitution with units.

58. Copy the figure below, then draw an arrow to represent the direction of the net force acting on the car when it is in position *A*.

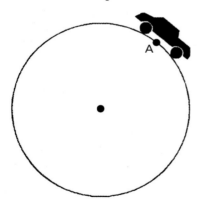

59. Calculate the magnitude of the centripetal acceleration of the car. Show all work, including the equation and substitution with units.

Base your answers to questions 60 and 61 on the information below.

An 8.00-kilogram ball is fired horizontally from a 1.00×10^3 kilogram cannon initially at rest. After having been fired, the momentum of the ball is 2.40×10^3 kilogram·meters per second east. Neglect friction.

60. Calculate the magnitude of the cannon's velocity after the ball is fired. Show all work, including the equation and substitution with units.

61. Identify the direction of the cannon's velocity after the ball is fired.

62. During a 5.0-second interval, an object's velocity changes from 25 meters per second east to 15 meters per second east. Determine the magnitude and direction of the object's acceleration.

63. A projectile has an initial horizontal velocity of 15 meters per second and an initial vertical velocity of 25 meters per second. Determine the projectile's horizontal displacement if the total time of flight is 5.0 seconds. Neglect friction.

Base your answers to questions 64 and 65 on the information below.

A hiker walks 5.00 kilometers due north and then 7.00 kilometers due east.

64. What is the magnitude of her resultant displacement?

65. What total distance has she traveled?

66. When a child squeezes the nozzle of a garden hose, water shoots out of the hose toward the east. What is the compass direction of the force being exerted on the child by the nozzle?

Base your answers to questions 67 through 70 on the information below.

A force of 6.0×10^{-15} newton due south and a force of 8.0×10^{-15} newton due east act concurrently on an electron, e^-.

67. Draw a force diagram to represent the *two* forces acting on the electron with the electron represented by a dot. Use a metric ruler and the scale of 1.0 centimeter = 1.0×10^{-15} newton. Begin each vector at the dot representing the electron and label its magnitude in newtons.

68. Determine the resultant force on the electron, *graphically*. Label the resultant vector **R**.

69. Determine the magnitude of the resultant vector **R**.

70. Determine the angle between the resultant and the 6.0×10^{-15} newton vector.

Base your answers to questions 71 through 75 on the information below.

A force of 10. newtons toward the right is exerted on a wooden crate initially moving to the right on a horizontal wooden floor. The crate weighs 25 newtons.

71. Calculate the magnitude of the force of friction between the crate and the floor. Show all work, including the equation and substitution with units.

72. Draw and label all vertical forces acting on the crate.

73. Draw and label all horizontal forces acting on the crate.

74. What is the magnitude of the net force acting on the crate?

75. Is the crate accelerating? Explain your answer.

Base your answers to questions 76 through 80 on the information below.

A manufacturer's advertisement claims that their 1,250-kilogram (12,300-newton) sports car can accelerate on a level road from 0 to 60.0 miles per hour (0 to 26.8 meters per second) in 3.75 seconds.

76. Determine the acceleration, in meters per second^2, of the car according to the advertisement.

77. Calculate the net force required to give the car the acceleration claimed in the advertisement. Show all work, including the equation and substitution with units.

78. What is the normal force exerted by the road on the car?

79. The coefficient of friction between the car's tires and the road is 0.80. Calculate the maximum force of friction between the car's tires and the road. Show all work, including the equation and substitution with units.

80. Using the values for the forces you have calculated, explain whether the manufacturer's claim for the car's acceleration is possible.

81. Two physics students were selected by NASA to accompany astronauts on a future mission to the moon. The students are to design and carry out a simple experiment to measure the acceleration due to gravity on the surface of the moon.
Describe an experiment that the students could conduct to measure the acceleration due to gravity on the moon. Your description must include:
   • the equipment needed
   • what quantities would be measured using the equipment

   • what procedure the students should follow in conducting their experiment
   • what equations and/or calculations the students would need to do to arrive at a value for the acceleration due to gravity on the moon

Base your answers to questions 82 through 85 on the information and diagram below.

In an experiment, a rubber stopper is attached to one end of a string that is passed through a plastic tube before weights are attached to the other end. The stopper is whirled in a horizontal circular path at constant speed.

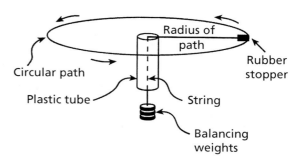

82. Copy the figure below of the top view of the circular path, then draw the path of the rubber stopper if the string breaks at the position shown.

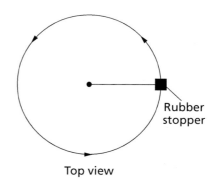

Top view

83. Describe what would happen to the radius of the circle if the student whirls the stopper at a greater speed without changing the balancing weights.

84. List *three* measurements that must be taken to show that the magnitude of the centripetal force is equal to the balancing weights. (Neglect friction.)

85. The rubber stopper is now whirled in a vertical circle at the same speed. Copy the side-view figure below then draw and label vectors to indicate the direction of the weight (F_g) and

the direction of the centripetal force (F_c) at the position shown in the figure.

Vertical Circle (side view)

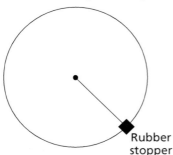

Rubber stopper

Base your answers to questions 86 through 88 on the information below.

A projectile is fired from the ground with an initial velocity of 250. meters per second at an angle of 60.° above the horizontal.

86. Using a protractor and a ruler, draw a vector to represent the initial velocity of the projectile. Use a scale of 1.0 centimeter = 50. meters per second.

87. Determine the horizontal component of the initial velocity.

88. Explain why the projectile has *no* acceleration in the horizontal direction. (Neglect air friction.)

89. Explain how to find the coefficient of kinetic friction between a wooden block of unknown mass and a tabletop in the laboratory. Include the following in your explanation:

- Measurements required
- Equipment needed
- Procedure
- Equation(s) needed to calculate the coefficient of friction

Base your answers to questions 90 through 92 on the information and diagram below.

A 10.-kilogram box, sliding to the right across a rough horizontal floor, accelerates at −2.0 meters per second² due to the force of friction.

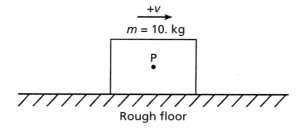

Rough floor

90. Calculate the magnitude of the net force acting on the box. Show all work, including the equation and substitution with units.

91. Copy the diagram, then draw a vector representing the net force acting on the box. Begin the vector at point *P* and use a scale of 1.0 centimeter = 5.0 newtons.

92. Calculate the coefficient of kinetic friction between the box and the floor. Show all work, including the equation and substitution with units.

Base your answers to questions 93 through 95 on the information and diagram below.

A projectile is launched horizontally at a speed of 30. meters per second from a platform located a vertical distance *h* above the ground. The projectile strikes the ground after time *t* at horizontal distance *d* from the base of the platform. (Neglect friction.)

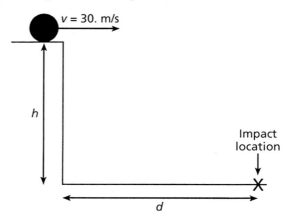

93. Copy the diagram, then sketch the theoretical path of the projectile.

94. Calculate the horizontal distance, *d*, if the projectile's total time of flight is 2.5 seconds. Show all work, including the equation and substitution with units.

95. Express the projectile's total time of flight, *t*, in terms of the vertical distance, *h*, and the acceleration due to gravity, *g*. Write an appropriate equation and solve it for *t*.

Base your answers to questions 96 and 97 on the information below.

A physics class is to design an experiment to determine the acceleration of a student on in-line skates coasting straight down a gentle incline. The incline has a constant slope. The students have tape measures, traffic cones, and stopwatches.

96. Describe a procedure to obtain the measurements necessary for this experiment.

97. Indicate which equation(s) they should use to determine the student's acceleration.

Base your answers to questions 98 through 100 on the information and diagram below.

A child is flying a kite, *K*. A student at point *B*, located 100. meters away from point *A* (directly underneath the kite), measures the angle of elevation of the kite from the ground as 30.°.

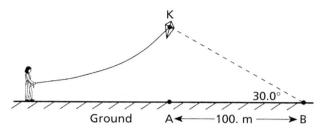

98. Using a metric ruler and protractor, draw a triangle representing the positions of the kite, *K*, and point *A* relative to point *B* that is given. Label points *A* and *K*. Use a scale of 1.0 centimeter = 10 meters.

99. Use a metric ruler and your scale diagram to determine the height, *AK*, of the kite.

100. A small lead sphere is dropped from the kite. Calculate the amount of time required for the sphere to fall to the ground. (Show all calculations, including the equation and substitution with units. Neglect air resistance.)

Base your answers to questions 101 and 102 on the information given below.

Friction provides the centripetal force that allows a car to round a circular curve.

101. Find the minimum coefficient of friction needed between the tires and the road to allow a 1600-kilogram car to round a curve of radius 80. meters at a speed of 20. meters per second. Show all work, including formulas and substitutions with units.

102. If the mass of the car were increased, how would that affect the maximum speed at which it could round the curve?

Enrichment
Mechanics

KEPLER'S LAWS

Johannes Kepler deduced three laws describing planetary motion. These laws inspired Newton and led to his equations of motion and gravitation.

Kepler's First Law

Kepler's first law states that the path of each planet is an ellipse with the sun at one focus.

An **ellipse** is defined as a closed curve such that the sum of the distances from any point p on the curve to two fixed points called the foci is constant (Figure 1-E1). A circle is an ellipse in which the two foci coincide at the center.

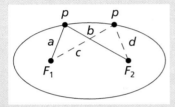

Figure 1-E1. F_1 and F_2 are the foci and $a + b = c + d$.

The orbits of the planets are nearly circular while the orbits of the other objects in the solar system, such as Pluto, are distinctly elliptical. Pluto's elliptical orbit causes it to be closer to the sun than Neptune for about 20 years during its 248-year path. This last occurred from 1979 to 1999. Earth's elliptical orbit brings it closest to the sun in January and farthest away in July. Many comets have elongated elliptical paths with the sun at one focus and the other far beyond the solar system (Figure 1-E2).

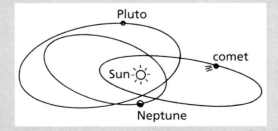

Figure 1-E2.

Kepler's Second Law

Kepler's second law states that each planet moves in such a way that an imaginary line drawn from the sun to the planet sweeps out equal areas in equal periods of time. For example, the wedge-shaped sectors A_1, A_2, and A_3 in the ellipse illustrated in Figure 1-E3 are all equal in area. If an imaginary line connecting the planet to the sun passes through area A_1 in one week, it passes through areas A_2 and A_3 in one week each as well. This means that the planet moves faster when it is closer to the sun. As a planet gets closer to the sun, its gravitational potential energy decreases and its kinetic energy increases.

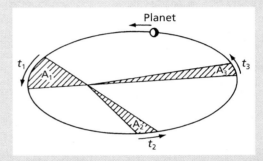

Figure 1-E3. The planet moves faster when it is closer to the sun; $t_1 = t_2 = t_3$ and $A_1 = A_2 = A_3$.

Kepler's Third Law

Kepler's third law states that the ratio of the mean radius of orbit cubed (R^3) to the orbital period squared (T^2) is a constant for all the planets. This relationship is expressed mathematically as

$$\frac{R^3}{T^2} = k$$

where k is a constant. Kepler's third law is applicable to any group of satellites that orbit around a given body. The value of k depends on the mass of the particular body being orbited. In the case of the planets orbiting the sun, k is equal to 3.35×10^{18} m³/s² As the mass of the object being orbited decreases, k decreases. For example, the value of k for objects orbiting Earth is 1.02×10^{13} m³/s².

SATELLITE MOTION

A **satellite** is defined as any body revolving around a larger body. The planets are satellites of the sun, and the moons of Jupiter are satellites of Jupiter. Earth's satellites include the moon and the artificial objects placed in orbit about Earth.

Newton was the first to compare the motion of a satellite around Earth to the motion of a projectile. He concluded that a satellite is simply a projectile that "falls freely" toward Earth. We know that the greater the horizontal component of a projectile's velocity, the greater the horizontal distance it will travel before hitting the ground. It follows that if a projectile has a great enough horizontal velocity, the curved path of its motion will match Earth's curvature and it will rotate around, rather than fall to, Earth. If a satellite is too close to Earth, the drag of the atmosphere will slow it, and it will spiral inward toward Earth. If, on the other hand, its speed is too great, the satellite will spiral outward and escape from Earth's gravitational pull. The minimum speed an object must have to escape the influence of a body's gravitational pull is called the **escape velocity**.

Apparent Weightlessness

Astronauts in orbiting spaceships experience a state of "weightlessness" even though Earth's gravity still pulls them toward Earth. Indeed, it is this pull that provides the centripetal force that maintains the orbit. Why then does an astronaut's weight not register on a scale, and why do all the objects in the spaceship float freely? These effects occur because the spaceship is falling toward Earth together with the astronaut, the scale, and all the objects aboard. If the astronaut released a glove, for example, it would not fall to the floor because the floor, the glove, and the astronaut are all falling at the same rate.

Geosynchronous Orbit

Communication and weather satellites are usually placed in **geosynchronous orbits**. A geosynchronous orbit is one in which the satellite's orbital period is equal to the period of Earth's rotation about its own axis (24 hours). A satellite in geosynchronous orbit remains over the same spot on Earth's equator.

Kepler's third law can be used to determine the radius of orbit needed for an artificial satellite to have a desired orbital period. Since the value of R^3/T^2 is the same for all Earth satellites (both natural and artificial), we can simply equate R^3/T^2 for the moon and R^3/T^2 for the satellite. For example,

$$\frac{R^3_{shuttle}}{T^2_{shuttle}} = \frac{R^3_{moon}}{T^2_{moon}} = k_{Earth} = 1.02 \times 10^{13} \text{ m}^3/\text{s}^2$$

We conclude that in order to be in geosynchronous orbit, a spaceship must orbit at a distance from Earth's center equal to approximately 6.6 times Earth's radius.

Questions

E1. The path of a planet around the sun is best described as (1) a circle (2) an ellipse (3) a parabola (4) an epicycle

E2. With respect to a planet's orbit, the sun is situated at (1) the center (2) one of the foci (3) a nodal point (4) a geocentric point

E3. Until 1999 Neptune was farther from the sun than Pluto because Pluto (1) has switched orbits with Neptune temporarily (2) orbits Neptune (3) is denser than Neptune (4) has a very elongated orbit

Questions E4 through E10 are based on the following diagram of a planet orbiting the sun. It takes the planet one month to travel from point A to point B, and one month to travel from point C to point D.

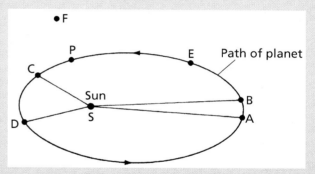

E4. Which of the following statements is correct? (1) arc AB = arc CD (2) length SC = length SA (3) area = SCD = area SAB (4) potential energy at C = potential energy at A

E5. At which point does the planet have the greatest kinetic energy and speed? (1) A (2) C (3) E (4) the speed is the same at all of the above

E6. If this planet is Earth, when is it located at point C? (1) January (2) June (3) July (4) September

E7. If the planet's mass were suddenly doubled, the period of its revolution in orbit would (1) decrease (2) increase (3) remain the same.

E8. When the planet is at point P, the direction of the planet's velocity is toward point (1) S (2) F (3) C (4) B

E9. The direction of the planet's acceleration at point P is toward point (1) S (2) E (3) F (4) B

E10. If the mass of the sun were suddenly to increase, then the value of R^3/T^2 would (1) decrease (2) increase (3) remain the same

E11. What is true of a satellite in geosynchronous orbit? (1) It remains in the same position over a point on the equator. (2) It remains in a fixed position between Earth and the moon. (3) It remains in a fixed position between Earth and the sun. (4) Its period of revolution does not equal Earth's period of rotation.

E12. The radius of orbit for an artificial satellite around Earth may be determined by equating R^3/T^2 for the satellite with R^3/T^2 for the (1) Earth (2) moon (3) sun (4) other planets

E13. What occurs if an orbiting satellite's speed exceeds escape velocity? (1) It spirals back toward earth. (2) It achieves a geosynchronous orbit. (3) It spirals outward away from Earth. (4) Any of the above are possible.

CHAPTER 2 Energy

Energy is needed to do work. The relationship between work and energy and the study of the various forms of energy are the subjects of this chapter.

Work

In physics, **work** is done whenever a force acts on an object in the direction of the object's motion. For example, a weight lifter does work as she lifts a weight from the ground to a position over her head because she pushes up as the weight goes up (Figure 2-1). When she holds the weight at rest above her head, even though she may be straining to keep it there, she is not doing any work because the weight is not in motion.

Figure 2-1.

Work is a scalar quantity; it has magnitude but no direction. The amount of work W is the product of the magnitude of the applied force F and the object's displacement d in the direction of the force. A short person, therefore, does less work in lifting a weight from the floor to a position over his head than a tall person. In situations where the force acts in the same direction as the object's motion we can write

$$W = Fd$$

If only a component of the force acts in the direction of the object's motion, the amount of work done is equal to the product of the magnitude of that component and the displacement of the object. Figure 2-2A illustrates the resolution of the force acting at an angle to move a box along a horizontal surface. To calculate the component of the force exerted in the direction of motion, use the equation

$$A_x = A \cos \theta$$

substituting F_x for A_x and F for A. To calculate the work done, multiply F_x by the distance the box was moved. To calculate the work done using only one step, you can use the equation

$$W = F \cos \theta \times d$$

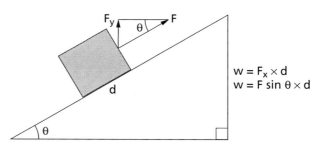

Figure 2-2. Only the component of the force that acts in the direction of motion is used to determine the work done.

The **joule** is the unit of work. One joule of work is done on an object when the force is one newton and the displacement is one meter. The joule (J) is a derived unit that can be expressed in terms of fundamental units.

$$1 \text{ joule} = 1 \text{ newton} \cdot \text{m} = 1 \text{ N} \cdot \text{m}$$

$$1 \text{ J} = \frac{1 \text{ kg} \cdot \text{m}}{\text{s}^2} \cdot \text{m} = \frac{1 \text{ kg} \cdot \text{m}^2}{\text{s}^2}$$

Sample Problems

1. A crate is pushed along the floor with a force of 9.0 N for a distance of 5.0 m. How much work is done?

   Solution:

   $$W = Fd$$

   $$W = (9.0 \text{ N})(5.0 \text{ m}) = 45 \text{ N} \cdot \text{m} = 45 \text{ J}$$

2. A 5.0-kg weight is lifted a vertical distance of 2.0 m. How much work is done?

   Solution: The force needed to lift an object at constant speed is equal to its weight, which in turn is equal to the product of its mass m and the acceleration due to gravity g. Thus,

   $$g = \frac{F_g}{m}$$

   $$F_g = mg$$

   $$F_g = (5.0 \text{ kg})(9.81 \text{ m/s}^2) = 49 \text{ kg} \cdot \text{m/s}^2 = 49 \text{ N}$$

   $$W = Fd = (49 \text{ N})(2.0 \text{ m}) = 98 \text{ J}$$

3. As shown in Figure 2-3, a 250-N force directed 60.° from the horizontal acted on a moving 120-N box as it went 11 m from point A to point B. How much work was done in moving the box?

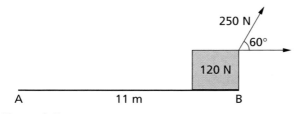

Figure 2-3.

   Solution:

   $$A_x = A \cos \theta$$

   $$F_x = F \cos 60.°$$

$$F_x = 250 \text{ N} \times 0.500 = 125 \text{ N}$$

$$W = F_x \times d$$

$$W = 125 \text{ N} \times 11 \text{ m} = 1375 \text{ N} \cdot \text{m} = 1380 \text{ J}$$

4. A 120.-newton box sits on a 8.0-meter long frictionless plane inclined at an angle of 30.° to the horizontal as shown in Figure 2-4. Force F applied to the box causes it to move with constant speed up the incline. Calculate the amount of work done in moving the box from the bottom to the top of the inclined plane.

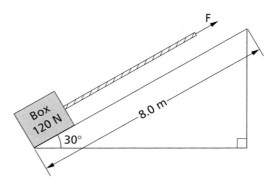

Figure 2-4.

   Solution:

   $$A_y = A \sin \theta$$

   $$F_y = F \sin 30.°$$

   $$F_y = 120 \text{ N} \times 0.500 = 60. \text{ N}$$

   $$W = F_y \times d$$

   $$W = 60. \text{ N} \times 8.0 \text{ m} = 480 \text{ N} \cdot \text{m} = 480 \text{ J}$$

Power

Power is the amount of work done per unit time. Like work, it is a scalar quantity. The symbol for power is P.

$$P = \frac{W}{t}$$

where W is work done in time t. Since $W = Fd$, we can write

$$P = \frac{W}{t} = \frac{Fd}{t}$$

Since $\bar{v} = \dfrac{d}{t}$ we may write

$$P = \frac{W}{t} = \frac{Fd}{t} = F\bar{v}$$

Thus, power can also be expressed as the product of the applied force and the velocity of the object.

The unit of power is the **watt** (W). One watt of power equals one joule of work done in one second. The watt is a derived unit and can be expressed in terms of fundamental units.

$$1 \text{ W} = 1\frac{\text{J}}{\text{s}} = 1\frac{\text{N} \cdot \text{m}}{\text{s}} = 1\frac{\text{kg} \cdot \text{m}^2}{\text{s}^3}$$

Sample Problems

5A. A hoist lifts an object weighing 1200 N a vertical distance of 15 m in 15 s. What work is done if the object rises at constant speed?

Solution:

$$W = Fd = (1200 \text{ N})(15 \text{ m})$$

$$W = 18,000 \text{ J}$$

5B. What power is developed?

Solution:

$$P = \frac{W}{t} = \frac{18,000 \text{ J}}{15 \text{ s}} = 1200 \text{ W}$$

6. How long will it take an 800.-W motor to push a boat through the water a distance of 1000. m with a force of 7.5 N?

Solution:

$$P = \frac{W}{t} = \frac{Fd}{t}$$

$$t = \frac{Fd}{P}$$

$$t = \frac{(7.5 \text{ N})(1000. \text{ m})}{800. \text{ W}}$$

$$t = \frac{7500 \text{ N} \cdot \text{m}}{800 \text{ N} \cdot \text{m/s}} = 9.4 \text{ s}$$

7. A 60-N force exerted on an object causes it to move with a constant velocity of 5 m/s. What power is used?

Solution:

$$P = F\overline{v}$$

$$P = (60 \text{ N})(5 \text{ m/s}) = 300 \text{ J/s} = 300 \text{ W}$$

Questions

1. A student does 60. joules of work pushing a 3.0-kilogram box up the full length of a ramp that is 5.0 meters long. What is the magnitude of the force applied to the box to do this work? (1) 20. N (2) 15 N (3) 12 N (4) 4.0 N

2. A boat weighing 9.0×10^2 newtons requires a horizontal force of 6.0×10^2 newtons to move it across the water at 1.5×10^1 meters per second. The boat's engine must provide energy at the rate of (1) 2.5×10^{-2} J (2) 4.0×10^1 W (3) 7.5×10^3 J (4) 9.0×10^3 W

3. A motor used 120. watts of power to raise a 15-newton object in 5.0 seconds. Through what vertical distance was the object raised? (1) 1.6 m (2) 8.0 m (3) 40. m (4) 360 m

4. Through what vertical distance is a 50.-newton object moved if 250 joules of work is done against Earth's gravitational field? (1) 2.5 m (2) 5.0 m (3) 9.8 m (4) 25 m

5. A constant force of 1900 newtons is required to keep an automobile having a mass of 1.0×10^3 kilograms moving at a constant speed of 20. meters per second. The work done in moving the automobile a distance of 2.0×10^3 meters is (1) 2.0×10^4 J (2) 3.8×10^4 J (3) 2.0×10^6 J (4) 3.8×10^6 J

6. What is the maximum height to which a 1200-watt motor could lift an object weighing 200. newtons in 4.0 seconds? (1) 0.67 m (2) 1.5 m (3) 6.0 m (4) 24 m

7. A 95-kilogram student climbs 4.0 meters up a rope in 3.0 seconds. What is the power output of the student? (1) 1.3×10^2 W (2) 3.8×10^2 W (3) 1.2×10^3 W (4) 3.7×10^3 W

8. Two weightlifters, one 1.5 meters tall and one 2.0 meters tall, raise identical 50.-kilogram masses above their heads. Compared to the work done by the weightlifter who is 1.5 meters tall, the work done by the weightlifter who is 2.0 meters tall is (1) less (2) greater (3) the same

9. A 40.-kilogram student runs up a staircase to a floor that is 5.0 meters higher than her starting point in 7.0 seconds. The student's power output is (1) 29 W (2) 280 W (3) 1.4×10^3 W (4) 1.4×10^4 W

10. What is the average power developed by a motor as it lifts a 400.-kilogram mass at constant speed through a vertical distance of 10.0 meters in 8.0 seconds? (1) 320 W (2) 500 W (3) 4900 W (4) 32,000 W

11. A student develops 250 watts of power running up a 5-meter-high staircase in 10 seconds. How much does the student weigh? (1) 125 N (2) 250 N (3) 400 N (4) 500 N

12. When a student raises an object vertically at a constant speed of 2.0 meters per second, 10. watts of power is developed. The weight of the object is (1) 5.0 N (2) 20. N (3) 40. N (4) 50. N

13. A force of 80. newtons pushes a 50.-kilogram object across a level floor for 8.0 meters. The work done is (1) 10 J (2) 400 J (3) 640 J (4) 3,920 J

14. Which of the following units is used to measure work? (1) newton (2) watt (3) joule (4) joule • second

15. If 700 watts of power is needed to keep a boat moving through the water at a constant speed of 10 meters per second, what is the magnitude of the force exerted by the water on the boat? (1) 0.01 N (2) 70 N (3) 700 N (4) 7,000 N

16. A crane raises a 200-newton weight to a height of 50 meters in 5 seconds. The crane does work at the rate of (1) 8×10^{-1} W (2) 2×10^1 W (3) 2×10^3 W (4) 2×10^4 W

17. A constant force of 20. newtons applied to a box causes it to move at a constant speed of 4.0 meters per second. How much work is done on the box in 6.0 seconds? (1) 480 J (2) 240 J (3) 120 J (4) 80. J

18. An object has a mass of 8.0 kilograms. A 2.0-newton force displaces the object a distance of 3.0 meters to the east, and then 4.0 meters to the north. What is the total work done on the object? (1) 10. J (2) 14 J (3) 28 J (4) 56 J

19. What is the minimum power required for a conveyor to raise an 8.0-newton box 4.0 meters vertically in 8.0 seconds? (1) 260 W (2) 64 W (3) 32 W (4) 4.0 W

20. As the power of a machine is increased, the time required to move an object a fixed distance (1) decreases (2) increases (3) remains the same

21. One elevator lifts a mass a given height in 10 seconds and a second elevator does the same work in 5 seconds. Compared with the power developed by the first elevator, the power developed by the second elevator is (1) one-half as great (2) twice as great (3) the same (4) four times as great

PART B-1

22. The following graph represents the relationship between the work done by a student running up a flight of stairs and the time of ascent.

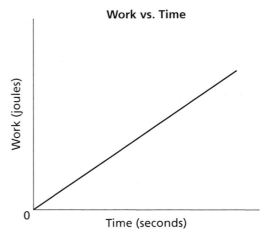

Work vs. Time

What does the slope of this graph represent? (1) impulse (2) momentum (3) speed (4) power

Base your answers to questions 23 and 24 on the following information.

You push against the handle of a lawn mower with a force of 200 newtons. The handle makes an angle of 60° with the ground. The mower moves at a constant rate of 0.5 meter per second.

23. How much work do you perform in one minute? (1) 100 J (2) 400 J (3) 1000 J (4) 3000 J

24. How much work is performed by the force of friction during this time? (1) 300 J (2) 750 J (3) 3000 J (4) 3500 J

Energy

Energy is the ability to do work. The amount of energy a system possesses is equal to the amount of work the system can do. Work can be done only by the transfer of energy from one object or system to another. For example, a hammer swinging downward has energy—it has the ability to do work on a nail. It can drive a nail a certain distance with force. As this work is done, the hammer's energy is transferred to the nail. Like work, energy is a scalar quantity and is measured in joules. This can be expressed by the equation

$$W = Fd = \Delta E_T$$

Energy is found in many forms: electromagnetic, chemical, mechanical, heat, nuclear, and sound. One form of energy can be converted into any other. For example, the energy in steam can be used to drive a turbine. In this process, heat energy is converted to mechanical energy. The energy from the turbine can be used to generate electricity, thus converting mechanical energy into electrical energy. Finally, the electrical energy can be used to run an assortment of appliances, such as a television, refrigerator, and computer.

Potential Energy

The energy that an object has owing to its position or condition is called **potential energy** (*PE*). For example, the position of an object above Earth's surface gives it gravitational potential energy because Earth's gravitational field can do work on it. The condition of a coiled spring, whether stretched or compressed, gives it elastic potential energy. Finally, energy stored in combustible fuel, such as gasoline, is an example of chemical potential energy.

The change in an object's potential energy is equal to the work required to bring the object to its new position or condition from its original position or condition (assuming no loss of energy due to friction or air resistance). For example, the elastic potential energy of a stretched spring is equal to the work done to stretch it.

$$\Delta PE = W$$

Gravitational Potential Energy

As an object is lifted from one position on Earth to another, work is done *against* Earth's gravitational attraction. After being lifted, there is an increase in the **gravitational potential energy** of the object. As an object is lowered from one position to another on Earth, work is done *by* Earth's gravitational field. This results in a decrease in the gravitational potential energy of the object. The change in gravitational potential energy is equal to the work done.

From the definition of work, we know that the amount of work done is equal to the product of the applied force and the displacement of the object in the direction of the force. Thus, we can write

$$\Delta PE = W = Fd$$

When an object is lowered or raised, the force exerted by or against Earth's gravitational field is equal to the object's weight *mg*. The displacement of the object is the vertical distance, Δh. Therefore, the change in the object's gravitational potential energy can be expressed as

$$\Delta PE = mg\,\Delta h$$

This equation applies only to displacements that are small compared with Earth's radius, when *g* is constant. Over larger distances, *g* varies and the equation is not applicable.

Frequently, we are interested only in the change in potential energy between two points and not in the true value of the potential energy at any point. It is convenient in such cases to choose a "base level"—a point that is arbitrarily assigned a potential energy value of zero. Base levels are usually the lowest point in an experiment, such as the ground, the floor in a room, a table top, or the lowest point in the swing of a pendulum. The change in the potential energy of an object between any point and the base level then becomes the potential energy value at that point, expressed by the formula

$$\Delta PE = mgh$$

where *PE* is equal to the potential energy of an object at any point *h* meters above the base level. At the base level, $h = 0$ and $PE = 0$

8. A 2.0-kg object is raised a vertical distance of 3.0 m. What is the resulting change in gravitational potential energy?

   Solution:

   $$\Delta PE = mg\Delta h$$

   $$\Delta PE = (2.0\text{ kg})(9.81\text{ m/s}^2)(3.0\text{ m})$$

   $$\Delta PE = 59\text{ kg} \cdot \text{m}^2\text{s}^2 = 59\text{ J}$$

9A. A 5.0-kg object sits at the top of a flight of stairs at a height of 6.0 m above the ground, as shown in Figure 2-5. What is the gravitational potential energy of the object with the ground as the base level?

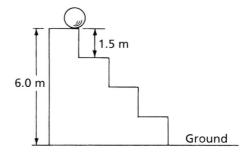

Figure 2-5.

   Solution:

   $$PE = mgh$$

   $$PE = (5.0\text{ kg})(9.81\text{ m/s}^2)(6.0\text{ m}) = 290\text{ J}$$

9B. How much potential energy will it lose if it falls down one step to a point 4.5 m above the ground?

Solution:

$$\Delta PE = mg \, \Delta h$$

$$\Delta PE = (5.0 \text{ kg})(9.81 \text{ m/s}^2)(6.0 \text{ m} - 4.5 \text{ m})$$

$$\Delta PE = 74 \text{ J}$$

Elastic Potential Energy

Elastic potential energy is the energy stored in a spring when it is compressed or stretched. The amount of potential energy stored in a spring is equal to the work done in stretching or compressing it from its original length. Springs are used in toys, scales, mattresses, and shock absorbers. For example, the jumping toy shown in Figure 2-6 is equipped with a spring. When the top is pushed against the base, the spring is compressed. The work done to compress the spring is stored in the spring as elastic potential energy. As the spring is released and returns to its original length, it does work on the top, sending the top flying upward.

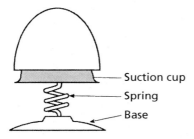

Figure 2-6. A jumping spring toy.

The force required to compress or stretch a spring is not constant. It is directly proportional to the distance that the spring has been compressed or stretched from its original length. The more a spring has been stretched or compressed, the harder it is to stretch or compress it further. The force F, in newtons, needed to keep a spring compressed or stretched at a distance x, in meters, is given by *Hooke's law*

$$F = kx$$

where k is a property of the spring known as the *spring constant*. The unit for the spring constant is the newton/meter (N/m).

The elastic potential energy stored in a spring is equal to one-half the product of the spring constant and the square of the distance that the spring has been stretched from its original length. This is expressed mathematically as

$$PE_s = \tfrac{1}{2}kx^2$$

10A. A spring with a spring constant of 40 N/m is stretched a distance of 0.5 m from its original length. What force is required to keep it stretched this distance?

Solution:

$$F_s = kx$$

$$F_s = (40\text{N/m})(0.5 \text{ m}) = 20 \text{ N}$$

10B. What is the elastic potential energy in the stretched spring?

Solution:

$$PE_s = \tfrac{1}{2}kx^2$$

$$PE_s = \tfrac{1}{2}(40 \text{ N/m})(0.5 \text{ m})^2 = 5 \text{ N} \cdot \text{m} = 5 \text{ J}$$

10C. What work was done to stretch the spring?

Solution:

$$W = \Delta PE_s = 5 \text{ J}$$

11A. The data in Table 2-1 was collected during a demonstration in which a spring was stretched by attaching various weights to it as in Figure 2-7. Using Hooke's law, determine the spring constant of the spring, using any of the given data pairs.

Table 2-1

Stretching a Spring

Force Applied (N)	Length Stretched (m)
0	0
40.	0.5
80.	1.0
120	1.5
160	2.0

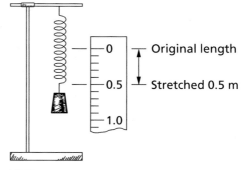

Figure 2-7.

Solution:

$$F_s = kx$$

$$k = \frac{F_s}{x} = \frac{40.0 \text{ N}}{0.5 \text{ m}} \text{ or } \frac{80. \text{ N}}{1.0 \text{ m}} \text{ or } \frac{120 \text{ N}}{1.5 \text{ m}} \text{ or } \frac{160 \text{ N}}{2.0 \text{ m}}$$

$$k = 80. \text{ N/m}$$

Refer to the force versus displacement graph of this data (Figure 2-8) and determine the slope of the line obtained.

$$\text{Slope} = \frac{\Delta y}{\Delta x}$$

$$\text{Slope} = \frac{160 \text{ N}}{2.0 \text{ m}} = 80. \text{ N/m}$$

Note that the slope of the force versus displacement curve is the spring constant k. The work done in stretching the spring is equal to the area under the curve.

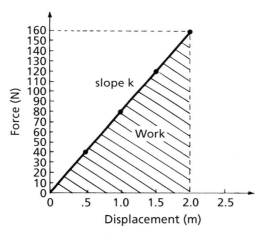

Figure 2-8. Graph of the data in Table 2-1.

11B. Determine the elastic potential energy in the spring for each length measured in the demonstration.

Solution:

$$PE_s = \frac{1}{2}kx^2$$

$$x = 0.5 \text{ m}, \quad PE_s = \frac{1}{2}(80. \text{ N/m})(0.5 \text{ m})^2 = 10. \text{ J}$$

$$x = 1.0 \text{ m}, \quad PE_s = \frac{1}{2}(80. \text{ N/m})(1.0 \text{ m})^2 = 40. \text{ J}$$

$$x = 1.5 \text{ m}, \quad PE_s = \frac{1}{2}(80. \text{ N/m})(1.5 \text{ m})^2 = 90. \text{ J}$$

$$x = 2.0 \text{ m}, \quad PE_s = \frac{1}{2}(80. \text{ N/m})(2.0 \text{ m})^2 = 160 \text{ J}$$

Kinetic Energy

Kinetic energy is the energy an object has because of its motion. Like all forms of energy, it is a scalar quantity. The symbol for kinetic energy is *KE* and its unit is the joule. The amount of kinetic energy an object of mass *m* has if it is moving with speed v is given by the formula

$$KE = \frac{1}{2}mv^2$$

The change in an object's kinetic energy is equal to the difference between its final kinetic energy and its initial kinetic energy. This is expressed as

$$\Delta KE = \frac{1}{2}mv_f^2 - \frac{1}{2}mv_i^2$$

The work done by a force as it acts on an object is equal to the sum of the change in the object's *KE* and *PE* and the work done against friction.

$$W = \Delta KE + \Delta PE + W_f$$

Sample Problems

12. Find the kinetic energy of a 4-kg object moving at 5 m/s and 10 m/s.

Solution:

$$KE = \frac{1}{2}mv^2 = \frac{1}{2}(4 \text{ kg})(5 \text{ m/s})^2 = 50 \text{ J}$$

$$KE = \frac{1}{2}mv^2 = \frac{1}{2}(4 \text{ kg})(10 \text{ m/s})^2 = 200 \text{ J}$$

Note that, since kinetic energy depends on the *square of the velocity*, it is *quadrupled* when the velocity is *doubled*.

13. How much work must be done to accelerate a 1200-kg car from rest to a speed of 2.0 m/s, assuming there is no friction?

Solution: ΔPE and W_f are each equal to zero.

$$W = \Delta KE = \frac{1}{2}mv_f^2 - \frac{1}{2}mv_i^2$$

$$W = \frac{1}{2}(1200 \text{ kg})(2.0 \text{ m/s})^2 - \frac{1}{2}(1200 \text{ kg})(0 \text{ m/s})^2$$

$$W = 2400 \text{ J}$$

14A. A new force of 5.0 N is applied to a 4.0-kg object over a distance of 6.0 m. What is the change in the object's kinetic energy, assuming there is no friction?

Solution: ΔPE and W_f are each equal to zero.

$$\Delta KE = W = Fd$$

$$\Delta KE = (5.0 \text{ N})(6.0 \text{ m})$$

$$\Delta KE = 30. \text{ J}$$

14B. If the object was moving at 2.0 m/s before the force was applied, what is its velocity after the force is applied?

Solution:

$$W = \Delta KE$$

$$KE = \tfrac{1}{2}mv_f^2 - \tfrac{1}{2}mv_i^2$$

$$30.\text{ J} = \tfrac{1}{2}(4.0\text{ kg})v_f^2 - \tfrac{1}{2}(4.0\text{ kg})(2.0\text{ m/s})^2$$

$$v_f^2 = 19\text{ m}^2\text{s}^2$$

$$v_f = 4.4\text{ m/s}$$

QUESTIONS

PART A

25. An object weighing 15 newtons is lifted from the ground to a height of 0.22 meter. The increase in the object's gravitational potential energy is approximately (1) 310 J (2) 32 J (3) 3.3 J (4) 0.34 J

26. As an object falls freely, the kinetic energy of the object (1) decreases (2) increases (3) remains the same

27. If the direction of a moving car changes and its speed remains constant, which quantity must remain the same? (1) velocity (2) momentum (3) displacement (4) kinetic energy

28. What is the gravitational potential energy with respect to the surface of the water of a 75.0-kilogram diver located 3.00 meters above the water? (1) 2.17×10^4 J (2) 2.21×10^3 J (3) 2.25×10^2 J (4) 2.29×10^1 J

29. A 60.0-kilogram runner has 1920 joules of kinetic energy. At what speed is she running? (1) 5.66 m/s (2) 8.00 m/s (3) 32.0 m/s (4) 64.0 m/s

30. A vertical spring 0.100 meter long is elongated to a length of 0.119 meter when a 1.00-kilogram mass is attached to the bottom of the spring. The spring constant of this spring is (1) 9.8 N/m (2) 82 N/m (3) 98 N/m (4) 520 N/m

31. A 6.8-kilogram block is sliding down a horizontal, frictionless surface at a constant speed of 6.0 meters per second. The kinetic energy of the block is approximately (1) 20. J (2) 41 J (3) 120 J (4) 240 J

32. When a mass is placed on a spring with a spring constant of 15 newtons per meter, the spring is compressed 0.25 meter. How much elastic potential energy is stored in the spring? (1) 0.47 J (2) 0.94 J (3) 1.9 J (4) 3.8 J

33. Two students of equal weight go from the first floor to the second floor. The first student uses an elevator and the second student walks up a flight of stairs. Compared to the gravitational potential energy gained by the first student, the gravitational potential energy gained by the second student is (1) less (2) greater (3) the same

34. An object moving at a constant speed of 25 meters per second possesses 450 joules of kinetic energy. What is the object's mass? (1) 0.72 kg (2) 1.4 kg (3) 18 kg (4) 36 kg

35. A spring of negligible mass has a spring constant of 50. newtons per meter. If the spring is stretched 0.40 meter from its equilibrium position, how much potential energy is stored in the spring? (1) 20. J (2) 10. J (3) 8.0 J (4) 4.0 J

36. The spring in a scale in the produce department of a supermarket stretches 0.025 meter when a watermelon weighing 1.0×10^2 newtons is placed on the scale. The spring constant for this spring is (1) 3.2×10^5 N/m (2) 4.0×10^3 N/m (3) 2.5×10^2 N/m (4) 3.1×10^{-2} N/m

37. As shown in the diagram below, a student exerts an average force of 600. newtons on a rope to lift a 50.0-kilogram crate a vertical distance of 3.00 meters.

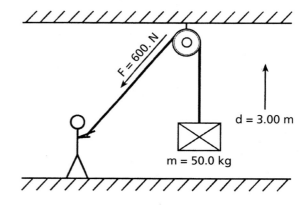

Compared to the work done by the student, the gravitational potential energy gained by the crate is (1) exactly the same (2) 330 J less (3) 330 J more (4) 150 J more

38. A 1.0-kilogram book resting on the ground is moved 1.0 meter at various angles relative to the horizontal. In which direction does the 1.0-meter displacement produce the greatest increase in the book's gravitational potential energy?

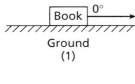

Book 0°

Ground
(1)

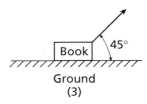

Book 20°

Ground
(2)

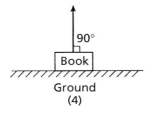

Book 45°

Ground
(3)

90°

Book

Ground
(4)

39. A 45.0-kilogram boy is riding a 15.0-kilogram bicycle with a speed of 8.00 meters per second. What is the combined kinetic energy of the boy and the bicycle? (1) 240. J (2) 480. J (3) 1440 J (4) 1920 J

40. A 5-newton force causes a spring to stretch 0.2 meter. What is the potential energy stored in the stretched spring? (1) 1 J (2) 0.5 J (3) 0.2 J (4) 0.1 J

41. A 10.-newton force is required to hold a stretched spring 0.20 meter from its rest position. What is the potential energy stored in the stretched spring? (1) 1.0 J (2) 2.0 J (3) 5.0 J (4) 50. J

42. As a spring is stretched, its elastic potential energy (1) decreases (2) increases (3) remains the same

PART B-1

43. The spring of a toy car is wound by pushing the car backward with an average force of 15 newtons through a distance of 0.50 meter. How much elastic potential energy is stored in the car's spring during this process? (1) 1.9 J (2) 7.5 J (3) 30. J (4) 56 J

44. The following graph shows elongation as a function of the applied force for two springs, A and B.

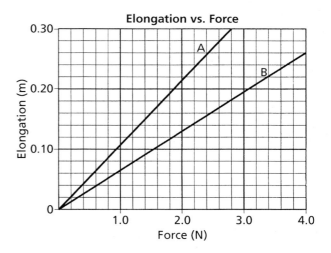

Compared to the spring constant for spring A, the spring constant for spring B is (1) smaller (2) larger (3) the same

45. As shown in the diagram below, a 0.50-meter-long spring is stretched from its equilibrium position to a length of 1.00 meter by a weight.

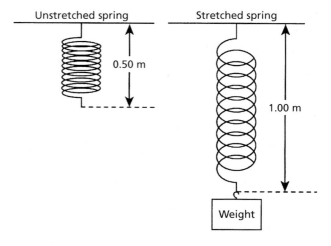

If 15 joules of energy are stored in the stretched spring, what is the value of the spring constant? (1) 30. N/m (2) 60. N/m (3) 120 N/m (4) 240 N/m

46. Which pair of quantities can be expressed using the same units? (1) work and kinetic energy (2) power and momentum (3) impulse and potential energy (4) acceleration and weight

47. An object falls freely near Earth's surface. Which graph best represents the relationship between the object's kinetic energy and its time of fall?

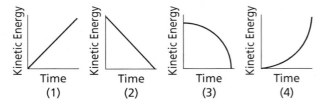

48. Which graph best represents the elastic potential energy stored in a spring (PE_s) as a function of its elongation, x?

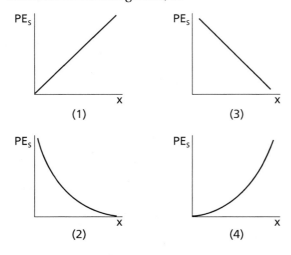

Conservation of Energy

The *law of conservation of energy* states that the total energy of a closed system remains constant. A closed system is one in which no external forces act and no external work is done on or by the system. Energy may be transferred between the objects in the system or be converted within the system from one form to another, but the total amount of energy within the system cannot change.

An example of conservation of energy is the conversion of potential to kinetic energy for a falling object on Earth. Table 2-2 lists the values of potential and kinetic energy at various heights as a 10-kg object falls to the ground from a height of

8 m (assuming no friction or air resistance). The sum of the kinetic energy and potential energy at any point is called the total mechanical energy. It is fixed throughout the downward trip (at 784 J).

At the point of release, the object has potential energy but no kinetic energy. As the object falls, its potential energy is converted to kinetic energy, but the total mechanical energy remains constant. Midway through the fall, half of its energy is potential and half is kinetic. Just before the object hits the ground, all of its potential energy has been converted to kinetic energy.

We can summarize by stating that *if there is no loss of energy due to friction or air resistance, the gain in kinetic energy equals the loss in potential energy, and the loss in kinetic energy equals the gain in potential energy.* In other words, for a freely falling object initially at rest, PE (top) $= KE$ (bottom) or $mg = \frac{1}{2}mv^2$.

Since g is a constant, we can use this formula to calculate the velocity of an object after any change in its height (Δh). Furthermore, if we know the velocity of an object as it hits the ground, we can determine the height, h, from which it fell.

Sample Problem

15. A ball is thrown up vertically with a velocity of 17.5 m/s. What height does it reach, assuming there is no friction?

Solution:

$$mg\ \Delta h = \tfrac{1}{2}mv^2$$

$$\Delta h = \tfrac{1}{2}\frac{v^2}{g} = \frac{v^2}{2g}$$

$$\Delta h = \frac{(17.5\ \text{m/s})^2}{2(9.81\ \text{m/s}^2)}$$

$$\Delta h = \frac{306.25\ \text{m}^2/\text{s}^2}{19.6\ \text{m/s}^2} = 15.6\ \text{m}$$

Table 2-2

Energy of a Falling Object

	Height h (m)	Velocity $v = \sqrt{2g\Delta h}$, (m/s)	Potential Energy $PE = mgh$, (J)	Kinetic Energy $KE = \frac{1}{2}mv^2$, (J)	Total mechanical energy $PE + KE$, (J)
(Release point)	8	0	784	0	784
	7	4.4	686	98	784
	4	8.9	392	392	784
	2	10.8	196	588	784
(Ground)	0	12.5	0	784	784

Conservation of energy can also be illustrated by the swing of a pendulum (Figure 2-9). If no energy is lost due to friction or air resistance, the motion and conversion of energy from potential to kinetic back to potential continues forever. At all points in the swing, the total energy of the pendulum remains constant. At the top of the swing the bob has only potential energy, and its velocity is zero. As the bob moves toward the bottom of the swing, it loses potential energy and gains kinetic energy as its velocity increases. The bottom of the swing is the base level; the potential energy there is zero and the kinetic energy is greatest. The kinetic energy at the bottom is equal to the potential energy at the top. The potential energy at any point in the swing is equal to mgh, where h is the vertical distance to the base level.

Sample Problems

16A. A 1.0-kg pendulum bob is released from a height of 0.50 m above the base level. Assuming there is no friction or air resistance, find the potential energy of the bob at the point of release.

Solution:

$$PE = mg\,\Delta h$$

$$PE = (1.0 \text{ kg})(9.81 \text{ m/s}^2)(0.50 \text{ m}) = 4.9 \text{ J}$$

16B. Find the velocity of the bob at the bottom of its swing.

Solution:

KE (on the bottom) $= PE$ (on the top)

$$KE = \tfrac{1}{2}mv^2$$

$$v^2 = \frac{2KE}{m}$$

$$v^2 = \frac{2(4.9 \text{ J})}{1.0 \text{ kg}} = \frac{9.8 \text{ N} \cdot \text{m}}{1 \text{ kg}}$$

$$v^2 = \frac{9.8\,(\text{kg} \cdot \text{m/s}^2)\,\text{m}}{1.0 \text{ kg}}$$

$$v = \sqrt{9.8 \text{ m}^2/\text{s}^2} = 3.1 \text{ m/s}$$

The Work-Energy Relationship

In our discussion of the conservation of energy in the previous section, we were careful to limit ourselves to cases in which the forces of friction and air resistance were absent. Of course, in most situations, the forces of friction and air resistance are present, opposing the motion of moving objects. In the case of a falling object, if air resistance is present, some of the potential energy the object loses on the way down is converted into heat, or internal energy Q. Some of the work done by gravity on the falling object will be spent fighting air resistance instead of accelerating the object and increasing its kinetic energy. Work done against air resistance or friction (W_f) is always converted into heat, or internal energy, which is discussed in the Enrichment section. The law of conservation of energy then means that the sum of kinetic energy, potential energy, and internal energy is constant.

Total energy = $PE + KE$ + Internal energy

or $E_T = PE + KE + Q$

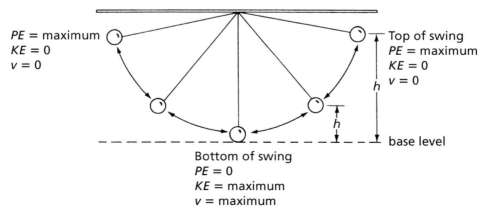

Figure 2-9. The swing of a pendulum.

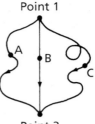

Point 1

A B

C

Point 2   **Figure 2-11.**

Sample Problem

17. In Figure 2-10 a 50.-kg jogger starts from rest at the bottom of a hill 6.0-m high (point *A*). When she reaches the top (point *B*) her velocity is 10. m/s. The work she does against air resistance and road friction between points *A* and *B* is 1500 J. Determine the total work required to jog straight up from point *A* to point *B*.

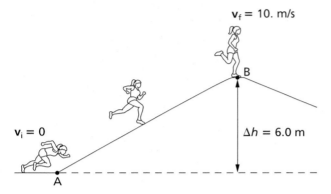

v_f = 10. m/s

B

v_i = 0

Δh = 6.0 m

A

Figure 2-10.

Solution:

$$W = \Delta PE + \Delta KE + W_f$$

$$\Delta PE = mg\ \Delta h = (50.\ \text{kg})(9.81\ \text{m/s}^2)(6.0\ \text{m})$$

$$\Delta PE = 2900\ \text{J}$$

$$\Delta KE = \tfrac{1}{2}mv_f{}^2 - \tfrac{1}{2}mv_i{}^2$$

$$\Delta KE = \tfrac{1}{2}(50.\ \text{kg})(10.\ \text{m/s})^2 - \tfrac{1}{2}(5.0\ \text{kg})(0\ \text{m/s})^2$$

$$\Delta KE = 2500\ \text{J}$$

$$W_f = 1500\ \text{J (given)}$$

$$W = 2900\ \text{J} + 2500\ \text{J} + 1500\ \text{J}$$

$$W = 6900\ \text{J}$$

Conservative Forces Forces, such as gravity, for which mechanical energy remains constant, are known as **conservative forces**. Forces, such as friction, that convert mechanical energy into internal energy and heat, are called **nonconservative forces**. In the example of the pendulum, we assumed that the kinetic energy at the bottom of the swing is equal to the potential energy at the top, even though the pendulum bob did not fall straight down but followed a curved path. This is true because the work done by or against gravity between two points is independent of the path taken between the points. Whether gravity pulls a

bead down wire *A*, *B*, or *C* in Figure 2-11, it does the same amount of work—as long as the bead starts at point 1 and ends at point 2. The work done by or against a *conservative force* is always independent of the path taken. On the other hand, the work done on an object by or against a *nonconservative force* depends on the path taken. For example, the work done in overcoming the force of friction as a block is dragged over a rough surface is greater when the path taken by the block is longer. Potential energy can be defined only for a conservative force. It is impossible to describe how much work a force can do if the amount of work will be different for different paths.

Questions

49. The diagram below shows points *A*, *B*, and *C* at or near Earth's surface. As a mass is moved from *A* to *B*, 100. joules of work are done against gravity.

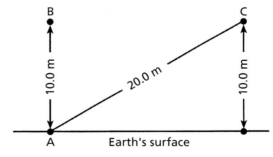

B

C

10.0 m

20.0 m

10.0 m

A   Earth's surface

What is the amount of work done against gravity as an identical mass is moved from *A* to *C*? (1) 100. J (2) 173 J (3) 200. J (4) 273 J

50. When a force moves an object over a rough, horizontal surface at a constant velocity, the work done against friction produces an increase in the object's (1) weight (2) momentum (3) potential energy (4) internal energy

51. A 55.0-kilogram diver falls freely from a diving platform that is 3.00 meters above the surface of the water in a pool. When she is 1.00 meter above the water, what are her gravitational potential energy and kinetic energy with respect to the water's surface? (1) *PE* = 1620 J and *KE* = 0 J (2) *PE* = 1080 J and *KE* = 540 J (3) *PE* = 810 J and *KE* = 810 J (4) *PE* = 540 J and *KE* = 1080 J

52. A 0.25-kilogram baseball is thrown upward with a speed of 30. meters per second. Neglecting friction, the maximum height reached by the baseball is approximately (1) 15 m (2) 46 m (3) 74 m (4) 92 m

53. A truck weighing 3.0×10^4 newtons was driven up a hill that is 1.6×10^3 meters long to a level area that is 8.0×10^2 meters above the starting point. If the trip took 480 seconds, what was the *minimum* power required? (1) 5.0×10^4 W (2) 1.0×10^5 W (3) 1.2×10^{10} W (4) 2.3×10^{10} W

54. The diagram below shows a moving, 5.00-kilogram cart at the foot of a hill 10.0 meters high. For the cart to reach the top of the hill, what is the minimum kinetic energy of the cart in the position shown? Neglect energy loss due to friction.

(1) 4.91 J (2) 50.0 J (3) 250. J (4) 491 J

55. In the diagram below, 400. joules of work is done raising a 72-newton weight a vertical distance of 5.0 meters.

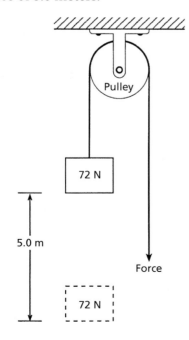

How much work is done to overcome friction as the weight is raised? (1) 40. J (2) 360 J (3) 400. J (4) 760 J

56. The diagram below shows a 50.-kilogram crate on a frictionless plane at angle θ to the horizontal. The crate is pushed at constant speed up the incline from point *A* to point *B* by force *F*.

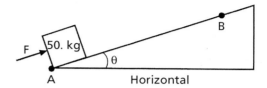

If angle θ were increased, what would be the effect on the magnitude of force *F* and the total work *W* done on the crate as it is moved from *A* to *B*?
(1) *W* would remain the same and the magnitude of *F* would decrease.
(2) *W* would remain the same and the magnitude of *F* would increase.
(3) *W* would increase and the magnitude of *F* would decrease.
(4) *W* would increase and the magnitude of *F* would increase.

57. As a ball falls freely (without friction) toward the ground, its total mechanical energy (1) decreases (2) increases (3) remains the same

58. A 0.50-kilogram ball is thrown vertically upward with an initial kinetic energy of 25 joules. Approximately how high will the ball rise? (Neglect air resistance.) (1) 2.6 m (2) 5.1 m (3) 13 m (4) 25 m

59. A 1.0-kilogram rubber ball traveling east at 4.0 meters per second hits a wall and bounces back toward the west at 2.0 meters per second. Compared to the kinetic energy of the ball before it hits the wall, the kinetic energy of the ball after it bounces off the wall is (1) one-fourth as great (2) one-half as great (3) the same (4) four times as great

60. A catapult with a spring constant of 1.0×10^4 newtons per meter is required to launch an airplane from the deck of an aircraft carrier. The plane is released when it has been displaced 0.50 meter from its equilibrium position by the catapult. The energy acquired by the airplane from the catapult during takeoff is approximately (1) 1.3×10^3 J (2) 2.0×10^4 J (3) 2.5×10^3 J (4) 1.0×10^4 J

Base your answers to questions 61 through 64 on the following diagram, which shows a

20-newton force pulling an object up a hill at a constant rate of 2 meters per second.

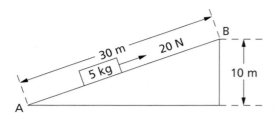

61. The work done by the force in pulling the object from *A* to *B* is (1) 50 J (2) 100 J (3) 500 J (4) 600 J

62. The kinetic energy of the moving object is (1) 5 J (2) 10 J (3) 15 J (4) 50 J

63. The work done against gravity in moving the object from point *A* to point *B* is approximately (1) 100 J (2) 200 J (3) 500 J (4) 600 J

64. Which graph best represents the relationship between velocity and time for the object?

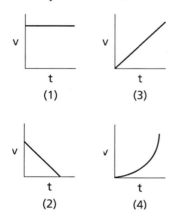

65. The work done on an object by a nonconservative force depends on: (1) the path taken (2) the object's mass (3) the object's velocity (4) the nature of the surfaces

66. An example of a nonconservative force is (1) gravitational force (2) elastic force (3) friction force (4) electrostatic force

PART B-1

67. A constant force is used to keep a block sliding at constant velocity along a rough horizontal track. As the block slides, there could be an increase in its
(1) gravitational potential energy, only
(2) internal energy, only
(3) gravitational potential energy and kinetic energy
(4) internal energy and kinetic energy

68. The graph below represents the kinetic energy, gravitational potential energy, and total mechanical energy of a moving block.

Energy vs. Distance Moved

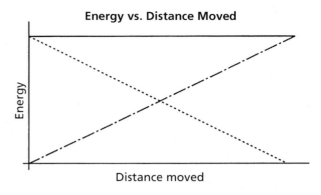

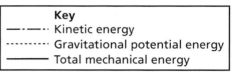

Which best describes the motion of the block? (1) accelerating on a flat horizontal surface (2) sliding up a frictionless incline (3) falling freely (4) being lifted at constant velocity

Base your answers to questions 69 through 73 on the diagram below, which represents a 2.0-kilogram mass placed on a frictionless track at point A *and released from rest. Assume the gravitational potential energy of the system to be zero at point* E.

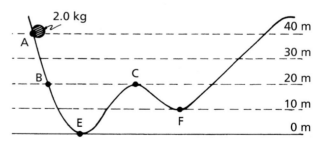

69. The gravitational potential energy of the system at point *A* is approximately (1) 80. J (2) 20. J (3) 8.0×10^2 J (4) 7.0×10^2 J

70. Compared with the kinetic energy of the mass at point *B*, the kinetic energy of the mass at point *E* is (1) one-half as great (2) twice as great (3) the same (4) 4 times greater

71. As the mass travels along the track, the maximum height it will reach above point *E* will be closest to (1) 10. m (2) 20. m (3) 30. m (4) 40. m

72. If the mass were released from rest at point *B*, its speed at point *C* would be (1) 0 m/s (2) 0.50 m/s (3) 10. m/s (4) 14 m/s

73. Compared with the total mechanical energy of the system at point *A*, the total mechanical energy of the system at point *F* is (1) less (2) more (3) the same

Base your answers to questions 74 through 77 on the diagram below, which represents a 10-kilogram object at rest at point A. *The object accelerates uniformly from point* A *to point* B *in 4 seconds, attaining a maximum speed of 10 meters per second at point* B. *The object then moves up the incline. (Neglect friction.)*

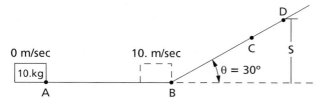

74. The kinetic energy of the object at point *B* is (1) 1000 J (2) 500 J (3) 100 J (4) 50 J

75. What distance did the object travel in moving from point *A* to point *B*? (1) 2.5 m (2) 10. m (3) 20. m (4) 100 m

76. As the mass moves up the incline, its potential energy (1) decreases (2) increases (3) remains the same

77. The object comes to rest at a vertical height of *S* (point *D*) when ∠θ = 30°. If ∠θ were increased to 40°, the object would come to rest at a vertical height (1) less than *S* (2) greater than *S* (3) equal to *S*

Base your answers to questions 78 through 80 on the diagram below, which represents a simple pendulum with a 2.0-kilogram bob and a length of 10. meters. The pendulum is released from rest at position 1 and swings without friction through position 4. At position 3, its lowest point, the speed of the bob is 6.0 meters per second.

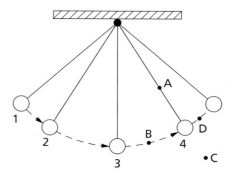

78. At which position does the bob have its maximum kinetic energy? (1) 1 (2) 2 (3) 3 (4) 4

79. What is the potential energy of the bob at position 1 in relation to position 3? (1) 18 J (2) 36 J (3) 72 J (4) 180 J

80. Compared with the sum of the kinetic and potential energies of the bob at position 1, the sum of the kinetic and potential energies of the bob at position 2 is (1) less (2) greater (3) the same

---

Physics in Your Life

Heat Exchange

Some devices we use every day operate on the principles of the laws of thermodynamics, which are discussed in the Enrichment section. Examples of such devices include refrigerators, air conditioners, and heat pumps. Many of these devices transfer heat from a cooler environment to a warmer one. This is much like pumping water uphill. Heat is supposed to flow from hot to cold, just as water is supposed to flow downhill, not the other way around.

Let us consider the operation of a typical electric refrigerator. A refrigerator consists of three basic parts: the evaporator, the compressor, and the condenser. The refrigerator also needs a refrigerant, a substance that absorbs heat quickly and in large amounts. During the refrigeration cycle the liquid refrigerant in the evaporator evaporates (boils), becoming a gas. The liquid absorbs the heat it needs to boil from inside the refrigerator, cooling it. The gas then moves on to the compressor, which compresses the gas and sends it on to the condenser. In the condenser, the compressed gas is returned to the liquid state, releasing the heat it absorbed from the refrigerator to the air in the room. To sum up the process, the refrigerant absorbs heat at a low temperature, and then through the action of mechanical work by the compressor, the gas is compressed and raised to a high enough temperature to allow it to give up the heat it absorbed.

In reality, the mechanical energy used to cool a given area is always greater than the energy lost by the area. The additional energy is used to overcome friction. Thus, you cannot cool your kitchen by opening the refrigerator door—in fact, this will actually heat the kitchen.

An air conditioner operates in much the same way. The main difference is that the air conditioner cools the inside of a room or building, transferring the heat to the air outside the building. This is why, as you are walking outside, you feel a blast of hot air when you pass a working air conditioner. A heat pump, which operates in a similar manner, transfers heat from outside a building inside to heat the building.

Questions

1. How is the action of a refrigerator comparable to pumping water uphill?
2. What happens to the heat energy lost by the refrigerant as it condenses?
3. From where does the liquid refrigerant get the heat it needs to boil?
4. Which part of the refrigerator does work on the refrigerant?
5. In what way is a heat pump different from an air conditioner?

Chapter Review Questions

PART A

1. The amount of work done against friction to slide a box in a straight line across a uniform, horizontal floor depends most on the (1) time taken to move the box (2) distance the box is moved (3) speed of the box (4) direction of the box's motion

2. A 3.0-kilogram block is initially at rest on a frictionless, horizontal surface. The block is moved 8.0 meters in 2.0 seconds by the application of a 12-newton horizontal force, as shown in the diagram below.

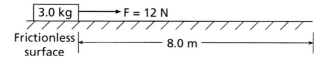

What is the average power developed while moving the block? (1) 24 W (2) 32 W (3) 48 W (4) 96 W

3. One watt is equivalent to one (1) N·m (2) N/m (3) J·s (4) J/s

4. The following diagram shows a 0.1-kilogram apple attached to a branch of a tree 2 meters above a spring on the ground below.

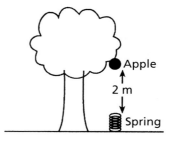

The apple falls and hits the spring, compressing it 0.1 meter from its rest position. If all of the gravitational potential energy of the apple on the tree is transferred to the spring when it is compressed, what is the spring constant of this spring? (1) 10 N/m (2) 40 N/m (3) 100 N/m (4) 400 N/m

5. A 1-kilogram rock is dropped from a cliff 90 meters high. After falling 20 meters, the kinetic energy of the rock is approximately (1) 20 J (2) 200 J (3) 700 J (4) 900 J

6. A unit for kinetic energy is the (1) watt (2) joule (3) newton (4) kilogram-meter/ second

7. A 2.0-newton book falls from a table 1.0 meter high. After falling 0.5 meter, the book's kinetic energy is (1) 1.0 J (2) 2.0 J (3) 10 J (4) 20 J

8. An object is lifted at constant speed a distance h above Earth's surface in a time t. The total

potential energy gained by the object is equal to the (1) average force applied to the object (2) total weight of the object (3) total work done on the object (4) total momentum gained by the object

9. If a 5-kilogram mass is raised vertically 2 meters from the surface of Earth, its gain in potential energy is approximately (1) 0 J (2) 10 J (3) 20 J (4) 100 J

10. If the kinetic energy of a given mass is to be doubled, its speed must be multiplied by (1) 8 (2) 2 (3) $\sqrt{2}$ (4) 4

11. A 2.0-kilogram mass falls freely for 10. meters near Earth's surface. The total kinetic energy gained by the object during its free fall is approximately (1) 400 J (2) 200 J (3) 100 J (4) 50 J

12. If the velocity of a moving object is doubled, the object's kinetic energy is (1) unchanged (2) halved (3) doubled (4) quadrupled

13. Which mass has the greatest potential energy with respect to the floor? (1) 50-kg mass resting on the floor (2) 2-kg mass 10 meters above the floor (3) 10-kg mass 2 meters above the floor (4) 6-kg mass 5 meters above the floor

14. A ball is thrown upward from Earth's surface. While the ball is rising, its gravitational potential energy will (1) decrease (2) increase (3) remain the same

15. Ten joules of work are done in accelerating a 2.0-kilogram mass from rest across a horizontal frictionless table. The total kinetic energy gained by the mass is (1) 3.2 J (2) 5.0 J (3) 10. J (4) 20. J

16. Which graph best represents the relationship between potential energy (*PE*) and height above ground (*h*) for a freely falling object released from rest?

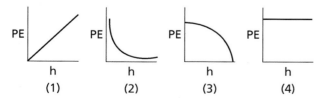

17. A 10.-kilogram object and a 5.0-kilogram object are released simultaneously from a height of 50. meters above the ground. After falling freely for 2.0 seconds, the objects will have different (1) accelerations (2) speeds (3) kinetic energies (4) displacements

18. At what point in its fall does the kinetic energy of a freely falling object equal its potential energy? (1) at the start of the fall (2) halfway between the start and the end (3) at the end of the fall (4) at all points during the fall

19. A 20.-newton block falls freely from rest from a point 3.0 meters above the surface of Earth. With respect to Earth's surface, what is the gravitational potential energy of the block-Earth system after the block has fallen 1.5 meters? (1) 20. J (2) 30. J (3) 60. J (4) 120 J

20. A 24-newton force applied to a spring causes the spring to increase in length by 0.40 meter. What is the spring constant? (1) 0.017 N/m (2) 9.6 N/m (3) 24.4 N/m (4) 60. N/m

21. A 2.0 newton force is applied to a spring with a spring constant of 10. newtons per meter. What is the resultant change in the length of the spring? (1) 0.20 m (2) 0.50 m (3) 5.0 m (4) 20. m

22. An unstretched spring has a length of 0.50 meter and a spring constant of 100. newtons per meter. What force is required to stretch this spring to a length of 0.60 meter? (1) 10. N (2) 60. N (3) 100 N (4) 170 N

23. What is the potential energy stored in a spring with a spring constant of 100. newtons per meter when it is stretched 0.10 meter from the original length? (1) 0.50 J (2) 1.0 J (3) 5.0 J (4) 10. J

24. What work is required to make a spring's length increase by 3.0 meters if its spring constant is 60. newtons per meter? (1) 20. J (2) 180 J (3) 270 J (4) 540 J

PART B-1

25. Which graph best represents the relationship between the gravitational potential energy of a freely falling object and the object's height above the ground near the surface of Earth?

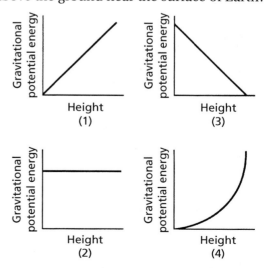

26. When a 1.53-kilogram mass is placed on a spring with a spring constant of 30.0 newtons per meter, the spring is compressed 0.500 meter. How much energy is stored in the spring? (1) 3.75 J (2) 7.50 J (3) 15.0 J (4) 30.0 J

27. Which graph best represents the relationship between the kinetic energy, *KE*, and the velocity of an object accelerating in a straight line?

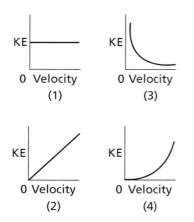

28. When 800 joules of work are done on an object, its potential energy increases by 300 joules and its kinetic energy increases by 400 joules. What was the work done against friction? (1) 100 J (2) 700 J (3) 800 J (4) 1500 J

29. In driving down a hill a car loses 500 joules of potential energy but gains 1500 joules of kinetic energy. If the work done to overcome friction was 200 joules, determine the total work done by the engine to go down the hill. (1) 800 J (2) 1200 J (3) 1800 J (4) 2200 J

30. In the following diagram a cart starts from rest at ground level (point *A*). It is pulled up the inclined plane to the top at a constant speed where its potential energy is 20. joules and its kinetic energy is 30. joules (point *B*). The friction force was 3.0 newtons. The length of the inclined plane is 5.0 meters. What is the total work done in moving the object from *A* to *B*? (1) 10. J (2) 15 J (3) 50. J (4) 65 J

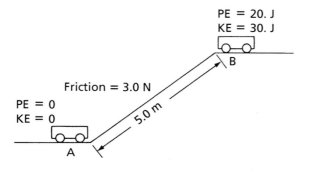

Base your answers to questions 31 through 33 on the following graph taken from an experiment with a spring of original length 1.0 meter.

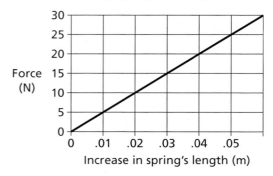

31. What is the spring constant of this spring? (1) 0.05 N/m (2) 5 N/m (3) 50 N/m (4) 500 N/m

32. How much force must be applied to this spring to change its length to 1.025 meters? (1) 1.025 N (2) 6.0 N (3) 12.5 N (4) 513 N

33. What is the potential energy stored in the spring when it has been stretched 0.40 m from its original length? (1) 0.40 J (2) 40 J (3) 80 J (4) 100 J

PART B-2

Base your answers to questions 34 and 35 on the information and graph below.

The graph represents the relationship between the force applied to each of two springs, *A* and *B*, and their elongations.

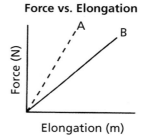

34. What physical quantity is represented by the slope of each line?

35. A 1.0-kilogram mass is suspended from each spring. If each mass is at rest, how does the potential energy stored in spring *A* compare to the potential energy stored in spring *B*?

Base your answers to questions 36 through 39 on the following graph, which represents the relationship between vertical height and gravitational potential energy for an object near Earth's surface.

Gravitational Potential Energy vs. Vertical Height

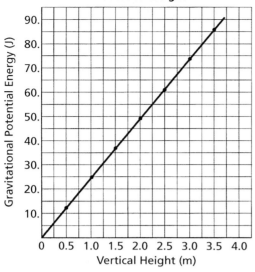

36. Based on the graph, what is the gravitational potential energy of the object when it is 2.25 meters above the surface of Earth?

37. Using the graph, calculate the mass of the object. Show all work, including the equation and substitution with units.

38. What physical quantity does the slope of the graph represent?

39. Copy the graph, then, using a straightedge, draw a line on the graph to represent the relationship between gravitational potential energy and vertical height for an object having a greater mass.

Base your answers to questions 40 and 41 on the information and diagram below.

A 160.-newton box sits on a 10.-meter-long frictionless plane inclined at an angle of 30.° to the horizontal as shown. Force (*F*) applied to a rope attached to the box causes the box to move with a constant speed up the incline.

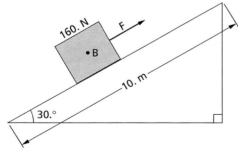

40. Copy the diagram, then construct a vector to represent the weight of the box. Use a metric ruler and a scale of 1.0 centimeter = 40. newtons. Begin the vector at point *B* and label its magnitude in newtons.

41. Calculate the amount of work done in moving the box from the bottom to the top of the inclined plane. Show all work, including the equation and substitution with units.

Base your answers to questions 42 through 45 on the information and table below.

The table lists the kinetic energy of a 4.0-kilogram mass as it travels in a straight line for 12.0 seconds. Using the information in the data table, construct a graph following the directions below.

Time (s)	Kinetic Energy (J)	Time (s)	Kinetic Energy (J)
0.0	0.0	6.0	32
2.0	8.0	10.0	32
4.0	18	12.0	32

42. The title of the graph is Kinetic Energy vs. Time. Label the *x*-axis "Time." Label the *y*-axis Kinetic Energy (J) and mark it with an appropriate scale.

43. Plot the data points for kinetic energy versus time.

44. Calculate the speed of the mass at 10.0 seconds. Show all work, including the equation and substitution with units.

45. Compare the speed of the mass at 6.0 seconds to the speed of the mass at 10.0 seconds.

PART C

Base your answers to questions 46 through 48 on the information and diagram below.

A mass, *M*, is hung from a spring and reaches equilibrium at position *B*. The mass is then raised to position *A* and released. The mass oscillates between positions *A* and *C*. (Neglect friction.)

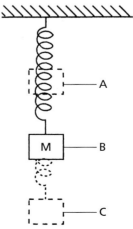

46. At which position, *A*, *B*, or *C*, is mass *M* located when the kinetic energy of the system is at a maximum? Explain your choice.

47. At which position, *A*, *B*, or *C*, is mass *M* located when the gravitational potential energy of the system is at a maximum? Explain your choice.

48. At which position, *A*, *B*, or *C*, is mass *M* located when the elastic potential energy of the system is at a maximum? Explain your choice.

Base your answers to questions 49 through 51 on the information and diagram below.

A 1000.-kilogram empty cart moving with a speed of 6.0 meters per second is about to collide with a stationary loaded cart having a total mass of 5000. kilograms, as shown. After the collision, the carts lock and move together. Assume friction is negligible.

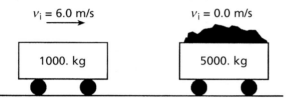

49. Calculate the speed of the combined carts after the collision. Show all work, including the equation and substitution with units.

50. Calculate the kinetic energy of the combined carts after the collision. Show all work, including the equation and substitution with units.

51. How does the kinetic energy of the combined carts after the collision compare to the kinetic energy of the carts before the collision?

Base your answers to questions 52 through 54 on the information and diagram below.

A 250.-kilogram car is initially at rest at point *A* on a roller coaster track. The car carries a 75-kilogram passenger and is 20. meters above the ground at point *A*. (Neglect friction.)

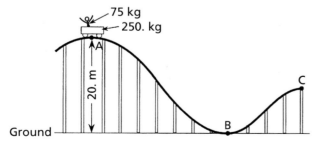

52. Calculate the total gravitational potential energy, relative to the ground, of the car and the passenger at point *A*. Show all work, including the equation and substitution with units.

53. Calculate the speed of the car and passenger at point *B*. Show all work, including the equation and substitution with units.

54. Compare the total mechanical energy of the car and passenger at points *A*, *B*, and *C*.

Base your answers to questions 55 through 57 on the information and data table below.

In an experiment, a student applied various forces to a spring and measured the spring's corresponding elongation. The table below shows his data.

Force (N)	Elongation (m)
0	0
1.0	0.30
3.0	0.67
4.0	1.00
5.0	1.30
6.0	1.50

55. Plot the data points for force versus elongation on a grid.

56. Draw the best-fit line.

57. Using your graph, calculate the spring constant of the spring. Show all work, including the equation and substitution with units.

Base your answers to questions 58 through 61 on the information below.

The driver of a car made an emergency stop on a straight horizontal road. The wheels locked and the car skidded to a stop. The marks made by the rubber tires on the dry asphalt are 16 meters long, and the car's mass is 1200 kilograms.

58. Determine the weight of the car.

59. Calculate the magnitude of the frictional force the road applied to the car in stopping it. Show all work, including the equation and substitution with units.

60. Calculate the work done by the frictional force in stopping the car. Show all work, including the equation and substitution with units.

61. Assuming that energy is conserved, calculate the speed of the car before the brakes were applied. Show all work, including the equation and substitution with units.

Base your answers to questions 62 through 64 on the information below.

A 50.-kilogram child running at 6.0 meters per second jumps onto a stationary 10.-kilogram sled. The sled is on a level frictionless surface.

62. Calculate the speed of the sled with the child after she jumps onto the sled. Show all work, including the equation and substitution with units.

63. Calculate the kinetic energy of the sled with the child after she jumps onto the sled. Show all work, including the equation and substitution with units.

64. After a short time, the moving sled with the child aboard reaches a rough level surface that exerts a constant frictional force of 54 newtons on the sled. How much work must be done by friction to bring the sled with the child to a stop?

Base your answers to questions 65 and 66 on the information below and on your knowledge of physics.

Using a spring toy like the one shown in the diagram, a physics teacher pushes on the toy, compressing the spring, causing the suction cup to stick to the base of the toy.

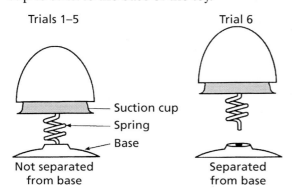

When the teacher removes her hand, the toy pops straight up and just brushes against the ceiling. She does this demonstration five times, always with the same result.

When the teacher repeats the demonstration for the sixth time the toy crashes against the ceiling with considerable force. The students notice that in this trial, the spring and toy separated from the base at the moment the spring released.

The teacher puts the toy back together, repeats the demonstration and the toy once again just brushes against the ceiling.

65. Describe the conversions that take place between pairs of the three forms of mechanical energy, beginning with the work done by the teacher on the toy and ending with the form(s) of energy possessed by the toy as it hits the ceiling. Neglect friction.

66. Explain, in terms of mass and energy, why the spring toy hits the ceiling in the sixth trial and not in the other trials.

Enrichment
Energy

INTERNAL ENERGY

Temperature

The temperature of an object indicates how hot or cold the object is with respect to a chosen standard. It is a property that originates from the motions and vibrations of the molecules in matter. When the average kinetic energy of the molecules increases, the temperature of the object increases. Conversely, the temperature of an object decreases when the average kinetic energy of its molecules decreases.

Internal Energy and Heat

The **internal energy** of an object is the total kinetic and potential energy associated with the motions and relative positions of the molecules of the object, apart from any kinetic and potential energy the object as a whole may possess. The amount of internal energy an object has depends on its temperature, mass, phase, and intermolecular bonds. The internal energy of an object can be changed either by changing the kinetic energy of its molecules, or by changing the potential energy of its molecules, or both.

    Heat energy refers to energy that is transferred from a warm object to a cooler one due to the temperature difference between them. Whenever this occurs, the internal energy of both objects is changed. Heat is a scalar quantity and is measured in joules. It is important to understand the difference between internal energy, temperature, and heat. For example, even though the Arctic Ocean is at a much lower temperature than a cup of boiling water, it has far more internal energy and can release far more heat. This is because temperature is a measure of the *average kinetic energy* of the molecules, while internal energy is the *total kinetic and potential energy* of the molecules. The Arctic Ocean is colder because the average kinetic energy of its molecules is lower than the average kinetic energy of the molecules in the cup of boiling water. However, the Arctic Ocean has so many more molecules that the total kinetic and potential energy of its molecules is much greater than the total kinetic and potential energy of the molecules in the cup of boiling water.

In our discussion of work and energy, we noted that mechanical energy is converted into internal energy when work is done against friction. The increase in internal energy in turn raises the average kinetic energy of the molecules. As a result, whenever we rub our hands together, or drive a nail into wood, or drag an object over a rough surface, we can detect an increase in the temperature of these objects.

Conservation of Internal Energy

Whenever there is a transfer of internal energy without conversion to or from other forms of energy, the total internal energy of a closed system remains constant. For example, if two liquids at different temperatures are mixed, the sum of the internal energies of the liquids before they are mixed is equal to the total internal energy of the mixture. The *law of conservation of internal energy* states that when materials having different temperatures are put into contact, *the heat gained by the cooler material is equal to the heat lost by the warmer material*. Heat energy flows from the warmer material to the cooler material until both materials are at the same temperature and a condition of **thermal equilibrium** is achieved.

THE LAWS OF THERMODYNAMICS

Thermodynamics is the study of heat and its relationship to other forms of energy and to work. The principles of thermodynamics are based on the law of conservation of energy and can be summarized in three fundamental laws.

The First Law

The *first law of thermodynamics* states that the amount of heat energy added to a system is equal to the increase in the internal energy of the system plus the work done by the system. For example, consider the results of adding heat to a gas in a sealed container. If the container does not expand, no work is done by the gas and all of the added heat is used to increase the internal energy of the gas. On the other hand, if the lid of the container is replaced by a movable piston and the gas is allowed to expand as heat is added, the gas will do work as it exerts a force on the outwardly moving piston (Figure 2E-1). In this case in order that energy be conserved, the increase in the internal energy of the gas must equal the amount of heat energy added minus the amount of work done on the piston.

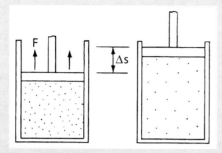

Figure 2-E1. Heat added equals the increase in internal energy plus the work done.

The Second Law

The *second law of thermodynamics* states that unless external work is done to produce the effect, heat cannot flow from a colder to a warmer region. If objects at different temperatures are brought into contact, heat energy is transferred from the warmer to the colder object. The reverse will occur only if external work is done. For example, to remove heat from a cool refrigerator and transfer it to a warm kitchen, an electric motor must expend energy and do work to drive the process.

The fact that heat does not flow on its own from a colder to a warmer region can be explained in terms of **entropy**. *During all natural processes, the entropy, or amount of disorder, of a system tends to increase.* Disorder is simply the absence of organization. If heat were to flow from a colder to a warmer object, all the faster moving molecules would end up in one region and all the slower molecules in another. The molecules would then be organized according to their kinetic energies and disorder would *decrease*. According to the principle of increasing entropy, this can never occur. Instead, heat will always flow from the warmer to the colder object. As the temperature difference between the objects becomes zero, the organization of molecules in terms of kinetic energy is eliminated and the system becomes less organized.

The second law has important consequences. It places limitations on the transfer and conversion of heat energy and, therefore, on the usefulness of all the heat energy around us. For example, a heat engine converts heat energy into useful mechanical energy as heat flows through it from a warmer to a colder region. However, a heat engine can convert only a fraction of the heat energy into mechanical energy. Most of the heat is absorbed by the cold region and cannot be used to do work. If there is no cold region to transfer the heat to, no work can be obtained. This is why an ocean liner cannot be powered by the enormous reservoir of heat energy in the ocean. The extraction of heat from the water, without there being a colder region in contact with it, would necessitate the expenditure of energy (as in the case of the refrigerator). Thus, all of the internal energy stored in the ocean is practically useless.

The Third Law

The *third law of thermodynamics* states that it is impossible to reduce the temperature of a system to absolute zero. In order to lower the temperature of a system, heat must be removed from the system. One way to remove heat from a system is to place it in contact with another system at a lower temperature. In order to lower the temperature of a system to absolute zero, it would have to be placed in contact with a system whose temperature was lower than absolute zero. Since absolute zero is the lowest possible temperature, this is impossible.

Another way to remove heat from a system is to allow the system to do work. This method can be used to lower the temperature of gases. The lowest known condensation temperature of any gas is 3.2 K for the light isotope of helium. Below 3.2 K all substances are in the liquid or solid phase, so absolute zero cannot be achieved in this way either.

Scientists have been able to reach temperatures lower than 1 K by pumping away the vapor emitted by supercold liquids and by using a variety of magnetic effects. However, the closer to 0 K a system gets, the more difficult it is to further reduce its temperature. Absolute zero has never been achieved and is assumed to be unattainable.

Questions

E1. As the temperature of a substance increases, the average kinetic energy of its molecules (1) decreases (2) increases (3) remains the same

E2. The direction of exchange of internal energy between objects is determined by their relative (1) inertias (2) momentums (3) temperatures (4) masses

E3. When an object moves at a constant speed against friction on a horizontal tabletop, there is an increase in the object's (1) temperature (2) momentum (3) potential energy (4) acceleration

E4. Below, is a sketch of a cylinder in an automobile engine. After the fuel is ignited the internal energy of the cylinder (and its contents) increases by 300 J. As the piston is pushed forward, 500 J of work is done by the system. How much heat energy was added to the cylinder when the fuel ignited? (1) 200 J (2) 300 J (3) 500 J (4) 800 J

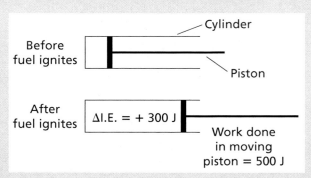

Before fuel ignites

Cylinder

Piston

After fuel ignites — ΔI.E. = + 300 J

Work done in moving piston = 500 J

E5. In the following system, heat energy is flowing from region *A* at 0°C to region *B* at 50°C. Which statement is true about this situation? (1) It is impossible. (2) It occurs naturally. (3) It occurs only if external work is done on the system. (4) It occurs only if the mass of *A* is greater than the mass of *B*.

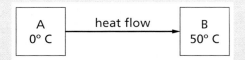

| A 0° C | heat flow | B 50° C |

E6. An increase in the entropy of a system means the system is (1) hotter (2) less ordered (3) more ordered (4) cooler

E7. There is a tendency in nature for systems to proceed toward (1) absolute zero (2) greater energy (3) less energy (4) less order

E8. Which statement describes the thermodynamics involved when water evaporates to a gas? (1) The water absorbs energy, and entropy increases. (2) The water absorbs energy, and entropy decreases. (3) Tthe water releases energy, and entropy increases. (4) The water releases energy, and entropy decreases.

E9. The total energy of the universe is (1) decreasing (2) constant (3) increasing

E10. The total entropy of the universe is (1) decreasing (2) constant (3) increasing

E11 According to the third law of thermodynamics, a substance can be cooled to (1) 0 K (2) below 0 K (3) a little above 0 K (4) −273 K

CHAPTER 3

Electricity and Magnetism

The Fundamentals of Electricity

The following fundamental principles govern all electric phenomena:

1. There are two kinds of electric charge: positive charge and negative charge.

2. Just as gravitational force exists between masses, there is an electrical force between charges. Unlike gravitational force, which is one of attraction only, charges can be attracted to or repelled by other charges. A positive charge will repel another positive charge, and a negative charge will repel another negative charge. A positive charge and a negative charge will attract each other. In other words, *like charges repel and unlike charges attract.*

3. All ordinary objects are electrically neutral, that is, they contain equal amounts of positive and negative charge.

Although both types of charge exist all around us, we are generally unaware of them because the electric forces they exert balance one another. For example, between two neutral objects there are forces of attraction between unlike charges balanced by equally strong forces of repulsion between like types of charge. Thus, the net force exerted on either object is equal to zero.

In all cases of charge transfer, the *law of conservation of charge* applies. Charge is never created or destroyed. A gain of charge in one place must correspond to a loss of that type of charge in another place.

The Atom and Electricity

The basic unit of matter is the **atom** (Figure 3-1). Every atom consists of three types of particles: *protons*, *neutrons*, and **electrons**. Each electron carries a negative charge and each proton carries a positive charge. Neutrons carry no charge. The amount of negative charge on an electron is the same as the amount of positive charge on a proton. Therefore, an atom with an equal number of protons and electrons has no net charge.

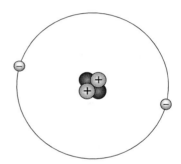

Figure 3-1. Model of an atom.

Protons and neutrons are located at the center of the atom, in the **nucleus**, which occupies a tiny fraction of the volume of the atom. Protons and neutrons are hundreds of times more massive than electrons and are held tightly together by strong nuclear forces. Electrons, bound to the nucleus by electrical forces, orbit the nucleus at various distances. The outermost electrons are often loosely held, and can be removed from the atom.

An object becomes charged if it either loses or gains electrons. An object that loses electrons and is left with fewer electrons than protons has a net positive charge. An object that gains electrons and has more electrons than protons has a net negative charge.

Electrons can only be transferred intact (fragments of electrons do not exist). As a result, any deficiency or excess of charge must consist of a whole-number multiple of the charge on one electron or proton, the **elementary unit of charge**. The charge on one electron is known as one *negative elementary charge*. The charge on one proton is referred to as one *positive elementary charge*.

The unit of charge is the **coulomb** (C). One coulomb is equal to the charge carried by 6.25×10^{18} electrons. Therefore the magnitude of the charge on one electron is equal to $\dfrac{1}{6.25 \times 10^{18}}$ C, or 1.6×10^{-19} C.

Static Electricity

Static electricity is associated with charges at rest. The term "at rest" means that there is no continuous flow of charge in any direction. There are many familiar effects of static electricity. For example, a person shuffling across a carpeted floor and then touching a metal doorknob will often experience a mild shock.

The behavior of objects that have a static charge can be demonstrated with pith balls. When a rubber rod or a glass rod is rubbed against fur or silk fabric and then brought near a suspended pith ball, the ball approaches the rod, makes contact with it, and then moves away from it. After that, the ball remains as far from the rod as possible. One pith ball touched by a stroked rubber rod and another pith ball touched by a stroked glass rod experience a force of attraction and move toward each other. On the other hand, two pith balls touched by the same type of stroked rod experience a force of repulsion and move away from each other.

Charging by Contact

An exchange of electrons, or transfer of charge, causes the phenomenon associated with the rubber and glass rods and the pith balls.

Electrons are bound more tightly in some materials than in others. Two dissimilar, neutral objects may become charged when they are rubbed against each other. For example, rubber tends to hold onto electrons more firmly than fur. When a rubber rod is rubbed against a piece of fur, electrons transfer from the fur to the rubber rod. The fur loses electrons and becomes positively charged. The rubber rod gains electrons and acquires a net negative charge. The magnitude of the charge on the fur and the rod is equal and the signs are opposite.

If the rubber rod is then brought near a suspended, neutral pith ball, the negative charge on the rod repels electrons in the pith ball, forcing them to move to the far side of the ball. As a whole, the ball remains neutral, but the redistrib-

ution of electrons causes the ball to become *electrically polarized*. The side of the ball closest to the rod becomes positively charged and the side of the ball farthest from the rod becomes negatively charged (Figure 3-2). Note that the positive charges do not move, only the electrons move.

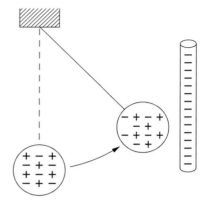

Figure 3-2. A negatively charged rod polarizes the pith ball by the redistribution of electrons.

The rod's excess electrons attract the protons on the near side of the ball and repel the electrons on the far side. The force of attraction is stronger (owing to the shorter distance between the rod and the ball's positively charged near side), and so the ball moves toward the rod. When the ball touches the rod, some of the rod's excess electrons transfer to the pith ball. The ball gains electrons and becomes negatively charged. This method of charge transfer is called *charging by contact*. The ball and the rubber rod are now both negatively charged and they repel each other. The ball moves (and remains) as far away from the rod as possible (Figure 3-3).

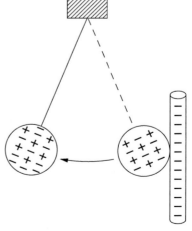

Figure 3-3. When the pith ball touches the rod it becomes negatively charged by contact.

When a glass rod is rubbed against a silk cloth, electrons transfer from the glass rod to the silk. The glass rod loses electrons and becomes positively charged. Like the charged rubber rod, the charged glass rod also polarizes a pith ball, but in the opposite way. The ball's electrons are attracted to the positively charged glass rod and move to the side of the ball closest to it. When the ball touches the rod, some of the ball's electrons transfer to the rod, leaving the ball with more protons than electrons. (The rod remains positively charged because it gains only a few electrons.) At this point, both the ball and the glass rod are positively charged, and the force of repulsion keeps them apart.

Two pith balls that have made contact with a charged rubber rod will repel each other because they have both gained electrons and acquired a negative charge. Two pith balls that have made contact with a charged glass rod will repel each other because they have both lost electrons and become positively charged. Finally, two pith balls that have touched differently charged rods (rubber or glass) attract each other because they become oppositely charged (Figure 3-4).

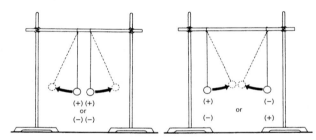

Figure 3-4. Two similarly charged pith balls repel each other, and two oppositely charged pith balls attract each other.

Charging by Induction

A charged object may induce an opposite charge in a neutral object without touching it. This is called **induction**. First, the neutral object must be **grounded**—that is, it must be connected to an object so large that it can either accept or give up a significant number of electrons without becoming noticeably charged. (Earth is often used as a ground.) Once the neutral object is grounded, a charged rod is brought near it, without touching it, as in Figure 3-5. If the rod is negatively charged, it repels the electrons of the neutral object, forcing them to transfer to the ground. If the object is then disconnected from the ground while the negatively charged rod is nearby, the object will be left with a net positive charge. On the other hand, if the rod is positively charged, it attracts electrons, which transfer from the ground into the neutral object. If the object is then disconnected from the ground while the positively charged rod is nearby, the object is left with a net negative charge. In both cases, the charge acquired by the previously neutral object is *opposite* to that of the rod used to charge it.

Detection of Charge The presence of excess charge on an object can be detected by bringing the object near an **electroscope**, a device that consists of a metal knob attached to two light metallic leaves (Figure 3-6 on page 84). If either a positively or a negatively charged object is brought near the knob of the electroscope, the electrons within the electroscope are forced to rearrange themselves, and the electroscope becomes electrically polarized. Each leaf acquires the same type of charge as the charged object, and the leaves diverge.

A neutral electroscope cannot be used to distinguish between a positively charged object and a negatively charged object. However, an electroscope can be given a known charge (either by contact or by induction) and then used to detect both the presence and the type of charge brought near it.

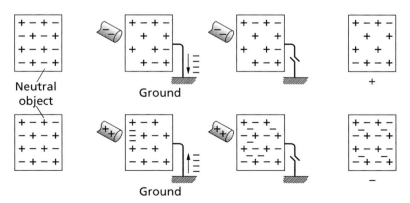

Figure 3-5. Charging by induction.

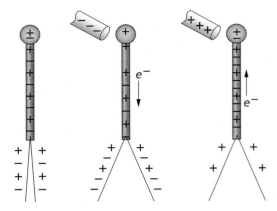

Figure 3-6. Detection of charges using a neutral electroscope.

The leaves of a charged electroscope are separated due to the charge on the leaves (Figure 3-7). If an object brought near the knob of the charged electroscope has the same type of charge as the electroscope, the leaves will diverge even more. If the object brought near the charged electroscope is oppositely charged, the leaves will converge to a vertical position.

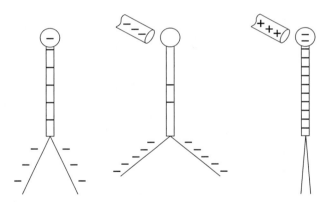

Figure 3-7. Identification of charges using an electroscope with a known charge.

Coulomb's Law

Experiments performed by Charles Coulomb demonstrate that for two charged objects that are much smaller than the distance between them (**point charges**), the electric force between the charges is directly proportional to the product of the amounts of charge and inversely proportional to the square of the distance between them. This relationship is known as *Coulomb's law*. Doubling the distance between two charges results in a force that is one-fourth as strong. Tripling the distance between the charges results in a force that is one-ninth as strong.

As indicated by Newton's third law of motion, the force exerted by one charge on a second charge is equal in magnitude and opposite in direction to the force exerted by the second charge on the first.

Coulomb's law can be expressed as

$$F = \frac{kq_1q_2}{d^2}$$

where F is the force exerted by either charge on the other, in newtons; q_1 and q_2 are the amounts of charge, in coulombs; d is the distance between the charges, in meters; and k, the *electrostatic constant*, is equal to 8.99×10^9 N·m²/C².

1. How strong is the repulsive force exerted on two point charges that each carry 1.0×10^{-6} C of negative charge and are 0.3 m apart?

   Solution:

$$F = \frac{kq_1q_2}{d^2}$$

$$F = \frac{(8.99 \times 10^9 \text{ N·m}^2/\text{C}^2) \times (1.0 \times 10^{-6} \text{ C})(1.0 \times 10^{-6} \text{ C})}{(0.3 \text{ M})^2}$$

$$F = \frac{8.99 \times 10^{-3} \text{ N·m}^2}{9.0 \times 10^{-2} \text{ m}^2} = 0.1 \text{ N}$$

2. Two identical point charges separated by 2.5 m exert a repulsive force on each other of 25 N. What is the magnitude of the charge on each object?

   Solution:

$$F = \frac{kq_1q_2}{d^2} \quad \text{and} \quad q_1 = q_2$$

$$q^2 = \frac{Fd^2}{k}$$

$$q^2 = \frac{(25 \text{ N})(2.5 \text{ m})^2}{8.99 \times 10^9 \text{ N·m}^2/\text{C}^2}$$

$$q^2 = 1.7 \times 10^{-8} \text{ C}^2$$

$$q = 1.3 \times 10^{-4} \text{ C}$$

Questions

1. An electroscope is a device with a metal knob, a metal stem, and freely hanging metal leaves used to detect charges. The following diagram shows a positively charged leaf electroscope.

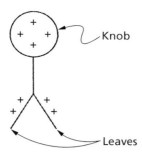

Knob

Leaves

As a positively charged glass rod is brought near the knob of the electroscope, the separation of the electroscope leaves will (1) decrease (2) increase (3) remain the same

2. The ratio of the magnitude of charge on an electron to the magnitude of charge on a coulomb is (1) 1:1 (2) 1:2 (3) 1:1840 (4) 1.6×10^{-19}:1

3. A negatively charged plastic comb is brought close to, but does not touch, a small piece of paper. If the comb and the paper are attracted to each other, the charge on the paper (1) may be negative or neutral (2) may be positive or neutral (3) must be negative (4) must be positive

4. The diagram below shows three neutral metal spheres, x, y, and z, in contact and on insulating stands.

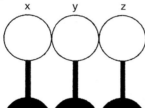

Which diagram best represents the charge distribution on the spheres when a positively charged rod is brought near sphere x, but does not touch it?

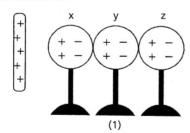

(1)

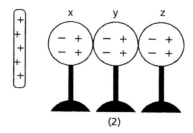

(2)

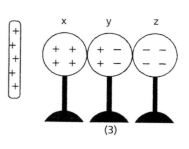

(3)

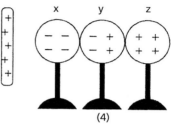

(4)

5. When a neutral metal sphere is charged by contact with a positively charged glass rod, the sphere (1) loses electrons (2) gains electrons (3) loses protons (4) gains protons

6. After two neutral solids, A and B, were rubbed together, solid A acquired a net negative charge. Solid B, therefore, experienced a net (1) loss of protons (2) increase of protons (3) loss of electrons (4) increase of electrons

7. The ratio of the magnitude of charge on an electron to the magnitude of charge on a proton is (1) 1:2 (2) 1:1 (3) $1 : 6.25 \times 10^{18}$ (4) 1:1840

8. How many electrons are contained in a charge of 8.0×10^{-19} coulomb? (1) 5 (2) 2 (3) 8 (4) 4

9. Which is equivalent to three elementary charges? (1) 2.0×10^{-19} C (2) 2.4×10^{-19} C (3) 4.8×10^{-19} C (4) 5.4×10^{-19} C

10. The coulomb is a unit of electrical (1) charge (2) current (3) potential (4) resistance

11. After a neutral object loses 2 electrons, it will have a net charge of (1) -2 elementary charges (2) $+2$ elementary charges (3) -3.2×10^{-19} elementary charge (4) $+3.2 \times 10^{-19}$ elementary charge

12. If a positively charged rod touches a neutral metal sphere, the number of electrons on the rod will (1) decrease (2) increase (3) remain the same

13. An object *cannot* have a charge of (1) 3.2×10^{-19} C (2) 4.5×10^{-19} C (3) 8.0×10^{-19} C (4) 9.6×10^{-19} C

14. The following diagram shows two identical metal spheres, A and B, separated by distance d. Each sphere has mass m and possesses charge q.

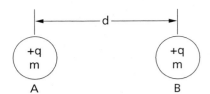

Which diagram best represents the electrostatic force F_e and the gravitational force F_g acting on sphere B due to sphere A?

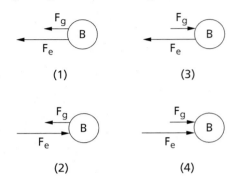

(1)   (3)

(2)   (4)

15. The diagram below shows two small metal spheres, A and B. Each sphere possesses a net charge of 4.0×10^{-6} coulomb. The spheres are separated by a distance of 1.0 meter.

Which combination of charged spheres and separation distance produces an electrostatic force of the same magnitude as the electrostatic force between spheres A and B?

(1)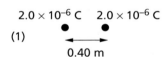

(2)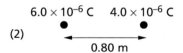

(3)

(4)

16. A positive test charge is placed between an electron, e, and a proton, p, as shown in the following diagram.

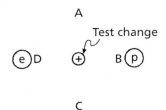

When the test charge is released, it will move toward (1) A (2) B (3) C (4) D

17. Two protons are located one meter apart. Compared to the gravitational force of attraction between the two protons, the electrostatic force between the protons is (1) stronger and repulsive (2) weaker and repulsive (3) stronger and attractive (4) weaker and attractive

18. In the diagram below, proton p, neutron n, and electron e are located as shown between two oppositely charged plates.

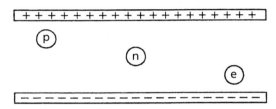

The magnitude of acceleration will be greatest for the (1) neutron, because it has the greatest mass (2) neutron, because it is neutral (3) electron, because it has the smallest mass (4) proton, because it is farthest from the negative plate

19. Oil droplets may gain electrical charges as they are projected through a nozzle. Which quantity of charge is *not* possible on an oil droplet? (1) 8.0×10^{-19} C (2) 4.8×10^{-19} C (3) 3.2×10^{-19} C (4) 2.6×10^{-19} C

20. The diagram below shows two identical metal spheres, A and B, on insulated stands. Each sphere possesses a net charge of -3×10^{-6} coulomb.

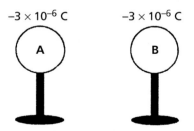

If the spheres are brought into contact with each other and then separated, the charge on sphere A will be (1) 0 C (2) $+3 \times 10^{-6}$ C (3) -3×10^{-6} C (4) -6×10^{-6} C

21. If the charge on each of two small charged metal spheres is doubled and the distance between the spheres remains fixed, the magnitude of the electric force between the spheres will be (1) the same (2) two times as great (3) one-half as great (4) four times as great

22. What is the smallest electric charge that can be put on an object? (1) 9.11×10^{-31} C (2) 1.60×10^{-19} C (3) 9.00×10^9 C (4) 6.25×10^{18} C

23. Two positively charged masses are separated by distance r. Which statement best describes the gravitational and electrostatic forces between the two masses? (1) Both forces are attractive. (2) Both forces are repulsive. (3) The gravitational force is repulsive and the electrostatic force is attractive. (4) The gravitational force is attractive and the electrostatic force is repulsive.

24. A metal sphere has a net negative charge of 1.1×10^{-6} coulomb. Approximately how many more electrons than protons are on the sphere? (1) 1.8×10^{12} (2) 5.7×10^{12} (3) 6.9×10^{12} (4) 9.9×10^{12}

25. The electrostatic force of attraction between two small spheres that are 1.0 meter apart is F. If the distance between the spheres is decreased to 0.5 meter, the electrostatic force will then be (1) $F/2$ (2) $2F$ (3) $F/4$ (4) $4F$

26. If the charge on one of two small charged spheres is doubled while the distance between them remains the same, the electrostatic force between the spheres will be (1) halved (2) doubled (3) tripled (4) unchanged

27. Charge A is $+2.0 \times 10^{-6}$ coulomb and charge B is $+1.0 \times 10^{-6}$ coulomb. If the force that A exerts on B is 1.0×10^{-2} newton, the force that B exerts on A is (1) 1.0×10^{-2} N (2) 2.0×10^{-2} N (3) 3.0×10^{-2} N (4) 5.0×10^{-2} N

PART B-1

28. Which of the following diagrams shows the leaves of the electroscope charged negatively by induction?

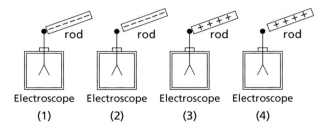

Electroscope (1)   Electroscope (2)   Electroscope (3)   Electroscope (4)

29. An inflated balloon that has been rubbed against a person's hair is touched to a neutral wall and remains attracted to it. Which diagram best represents the charge distribution on the balloon and wall?

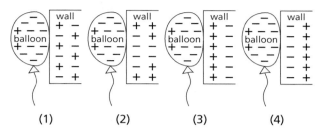

(1)   (2)   (3)   (4)

30. What is the approximate electrostatic force between two protons separated by a distance of 1.0×10^{-6} meter? (1) 2.3×10^{-16} N and repulsive (2) 2.3×10^{-16} N and attractive (3) 9.0×10^{21} N and repulsive (4) 9.0×10^{21} N and attractive

31. The magnitude of the electrostatic force between two point charges is F. If the distance between the charges is doubled, the electrostatic force between the charges will become (1) $F/4$ (2) $2F$ (3) $F/2$ (4) $4F$

32. An object possessing an excess of 6.0×10^6 electrons has a net charge of (1) 2.7×10^{-26} C (2) 5.5×10^{-24} C (3) 3.8×10^{-13} C (4) 9.6×10^{-13} C

33. The charge-to-mass ratio of an electron is (1) 5.69×10^{-12} C/kg (2) 1.76×10^{-11} C/kg (3) 1.76×10^{11} C/kg (4) 5.69×10^{12} C/kg

34. In the diagram below, a positive test charge is located between two charged spheres, A and B. Sphere A has a charge of $+2q$ and is located 0.2 meter from the test charge. Sphere B has a charge of $-2q$ and is located 0.1 meter from the test charge.

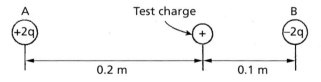

If the magnitude of the force on the test charge due to sphere A is F, what is the magnitude of the force on the test charge due to sphere B? (1) $F/4$ (2) $2F$ (3) $F/2$ (4) $4F$

35. An object that has a net charge of 3 coulombs possesses an excess of (1) 1.9×10^{19} electrons (2) 3.8×10^{19} electrons (3) 4.8×10^{19} electrons (4) 1.9×10^{20} electrons

36. Which graph best represents the relationship between the magnitude of the electrostatic force and the distance between two oppositely charged particles?

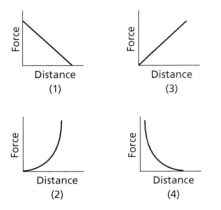

37. In the diagram below, two positively charged spheres, *A* and *B*, of masses m_A and m_B are located a distance *d* apart.

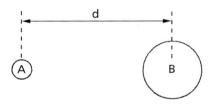

Which diagram best represents the directions of the gravitational force, F_g, and the electrostatic force, F_e, acting on sphere *A* due to the mass and charge of sphere *B*? Vectors are not drawn to scale.

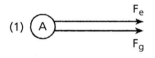

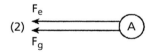

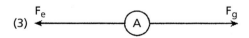

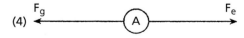

Base your answers to questions 38 through 41 on the following diagram, which represents a system consisting of two charged metal spheres with equal radii.

38. What is the magnitude of the electrostatic force exerted on sphere *A*? (1) 1.1×10^{-9} N (2) 1.3×10^8 N (3) 120 N (4) 10. N

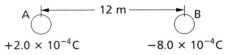

39. Compared with the force exerted on sphere *B* at a separation of 12 meters, the force exerted on sphere *B* at a separation of 6.0 meters would be (1) one-half as great (2) 2 times as great (3) one-quarter as great (4) 4 times as great

40. If the two spheres were touched together and then separated, the charge on sphere *A* would be (1) -6.0×10^{-4} C (2) 2.0×10^{-4} C (3) -3.0×10^{-4} C (4) -8.0×10^{-4} C

41. The following diagram represents two charges at a separation of *d*. Which would produce the greatest increase in the force between the two charges?

(1) doubling charge q_1, only (2) doubling *d*, only (3) doubling charge q_1 and *d*, only (4) doubling both charges and *d*

Base your answers to questions 42 through 44 on the following diagram, which shows two identical metal spheres. Sphere A *has a charge of +12 coulombs and sphere* B *is a neutral sphere.*

42. When spheres *A* and *B* are in contact, the total charge of the system is (1) neutral (2) + 6 C (3) +12 C (4) −24 C

43. After contact, when spheres *A* and *B* are separated, the charge on *A* will be (1) +12 C (2) one-quarter the original amount (3) one-half the original amount (4) 4 times the original amount

44. After spheres *A* and *B* are separated, which graph best represents the relationship of the force between the spheres and their separation?

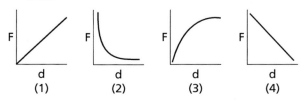

Electric Fields

An electric charge in empty space experiences no electric forces, but an electric charge near another electric charge does. The reason for this difference is that a charge creates an **electric field** that fills the surrounding space and affects other nearby charges.

As long as no other charge is present, an electric field does nothing. When another charge (referred to as a **test charge**) is introduced into the field, the field exerts a force on it. The presence of an electric field can be detected by its effect on a test charge introduced into the region.

Electric Field Intensity

Fields, like forces, are vector quantities: they have magnitude and direction. The magnitude of an electric field is referred to as its **intensity**. The intensity of an electric field at a particular point is defined as the magnitude of the force that the field exerts on a test charge of one coulomb at that point. The inverse square law applies to an electric field around a point charge. The direction of the electric field at any given point is defined as the direction of the force that the field exerts on a positive test charge at that point. The intensity and direction of an electric field are different at different positions in the vicinity of a charge, and therefore, the magnitude and direction of the electric force also vary from point to point.

The electric field intensity E at a particular point is given by the relationship

$$E = \frac{F_e}{q}$$

where F_e is the force, in newtons, exerted on a test charge q, in coulombs. The unit for electric field intensity is the newton/coulomb (N/C).

If the electric field intensity at a particular point is known, the force exerted on any amount of charge at that point is provided by the formula

$$F_e = qE$$

Sample Problems

3. What is the intensity of an electric field at a point where a 0.50-C charge experiences a force of 20. N?

Solution:

$$E = \frac{F_e}{q}$$

$$E = \frac{20.\ \text{N}}{0.50\ \text{C}} = 40.\ \text{N/C}$$

4. If one elementary unit of charge is placed at that point, how strong a force will it experience?

Solution:

$$F_e = qE$$

$$F_e = (1.6 \times 10^{-19}\ \text{C})(40.\ \text{N/C})$$

$$F_e = 6.4 \times 10^{-18}\ \text{N}$$

Electric Field Diagrams

A representation of the field around a single positive point charge is shown in Figure 3-8. The arrows radiating from the charge are called **field lines**. By convention, field lines always point in the direction of the force that would be exerted by the field on a positive test charge.

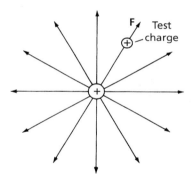

Figure 3-8. The field around a single positive point charge.

The concentration of the field lines at any point in a field diagram indicates the intensity of the field. The intensity of an electric field decreases inversely with the square of the distance from a point charge. Thus, the number of field lines per unit area (perpendicular to the field's direction) also decreases inversely with the square of the distance from the point charge in Figure 3-8. At a point twice as far from the charge, the concentration of field lines per unit area is one-fourth as great, as is the intensity of the field.

The field lines in the vicinity of a negative point charge are shown in Figure 3-9. Except for the fact that the field lines are directed radially inward, the figure is the same as that shown for a positive point charge.

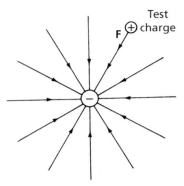

Figure 3-9. The field around a single negative point charge.

The electric field in the vicinity of a spherical shell of positive charge is shown in Figure 3-10. At any point inside the shell, the field intensity is zero. The net force on a test charge inside the shell, found by adding the forces exerted by all the charges on the shell, is zero. Therefore, there is no electric field inside the shell. Outside the shell, the field is the same as though all the charges were concentrated at the center of the shell.

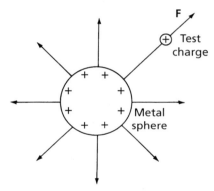

Figure 3-10. The field around a positively charged sphere.

The field created by two equal positive point charges is shown in Figure 3-11. The field around

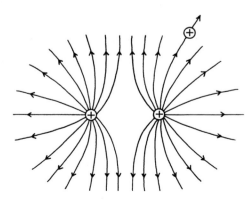

Figure 3-11. The field around two equal positive charges.

two negative charges is the same as the field around two positive charges, except for the fact that the field lines point toward the charges (Figure 3-12).

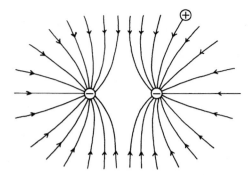

Figure 3-12. The field around two equal negative charges.

The field created by two equal but opposite charges is shown in Figure 3-13. Note that the field lines originate at the positive charge and terminate at the negative charge.

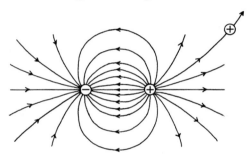

Figure 3-13. The field around equal but opposite charges.

The field between two oppositely charged parallel plates is shown in Figure 3-14. In the case where the distance between the plates is small compared with their size and the charge on the plates is uniformly distributed, the field lines are parallel to one another and perpendicular to the plates (except near the edges). If a positive test charge is released between the plates, it will be pushed in a straight line from the positive to the

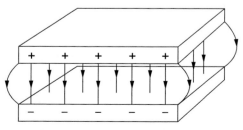

Figure 3-14. The field between oppositely charged parallel plates.

negative plate. Since the field between the plates is uniformly intense, the magnitude of the force exerted on a test charge is the same everywhere in the region between the plates (except near the edges). Outside the plates, the net force is practically zero, so no field lines appear there.

The following features are common to all electric field diagrams:

1. Field lines always begin on positively charged objects and terminate on negatively charged objects.
2. Field lines never intersect each other.
3. Where field lines meet a charged object, they are perpendicular to the surface of the charged object.

Electric Potential Energy

The electric force, like the gravitational force, can push or pull an object through a distance and do work. Just as a mass has gravitational potential energy because of its position in a gravitational field, a charged object has **electric potential energy** due to its position in an electric field.

If an electric field does work on any charged object, such as when the field created by one positive charge repels another positive charge, electric potential energy decreases. Similarly, electric potential energy decreases when the field created by a positive charge attracts a negative charge, pulling the charges closer together. This is analogous to a falling object on Earth. When the gravitational field does work, gravitational potential energy decreases. In accordance with the law of conservation of energy, the lost potential energy is converted to an equal amount of energy of some other type, such as kinetic energy or heat.

If work is done against an electric field, such as when a positive charge is moved closer to another positive charge, electric potential energy increases. Similarly, electric potential energy increases when a negative charge is moved away from a positive charge. This is analogous to lifting an object on Earth. When work is done against the gravitational field, gravitational potential energy increases. Once again, the gain in potential energy must correspond to a loss of energy of some other type, so that energy is conserved. The work done by (or against) an electric field in moving a charge from one point to another is independent of the path taken. In other words, the electric force is a conservative force, as is gravity.

Electric Potential

The **electric potential** at any given point in an electric field is defined as the total amount of work required to bring one coulomb of positive charge from infinity to that point. At a point in a field created by a positive charge, the force is repulsive. Thus, the work done against the field in moving the positive test charge from infinity to the point is positive. The closer the point is to the positive charge that set up the field, the greater the electric potential.

At a point in a field created by a negative charge, the force is attractive. Thus, the work done in moving the positive test charge from infinity to the point is negative (the field does work on the charge). The closer the point is to the negative charge that set up the field, the smaller the electric potential.

In summary, electric potential at any given point in an electric field is positive if work must be done *against* the field to move a positive test charge from infinity to that point, and negative if work is done *by* the field to move a positive test charge from infinity to that point.

Potential Difference

The work required to move a test charge of one coulomb from one point to another in an electric field is equal to the difference in the electric potential between the two points. We refer to this difference in electric potential as the **potential difference** between the two points.

From the amount of work W, in joules, required to move a charge q, in coulombs, between two points in an electric field, you can calculate the potential difference V between those two points by using the formula

$$V = \frac{W}{q}$$

Potential difference is expressed in joules per coulomb. One joule per coulomb is more commonly referred to as one **volt**. Since the potential difference between two points is measured in volts, it is also known as the **voltage**.

If it takes 6 J of work to push a 3-C charge from one point to another in an electric field, the potential difference between the two points is 6 J/3 C or 2 V. If a manufacturer labels a battery 9 V, it means that the electric field created by the two charged terminals will do 9 J of work on every coulomb that is pushed by the battery from one terminal to the other. Since the quantity of work done is independent of the path taken between the terminals, the characteristics of the materials and appliances connected to the battery play no role in determining the voltage.

If the voltage between two points is known, the work done in moving any given amount of charge between those points can be determined by using the formula

$$W = qV$$

Sample Problems

5. What is the potential difference between the terminals of a battery if 60. J of work are done when 3.0 C are pushed through a wire from one terminal to the other?

Solution:

$$V = \frac{W}{q}$$

$$V = \frac{60. \text{ J}}{3.0 \text{ C}} = 20. \text{ J/C} = 20. \text{ V}$$

6. How much work does the same battery do on every electron pushed through the wire?

Solution:

$$W = qV$$

$$W = (1.60 \times 10^{-19} \text{ C})(20. \text{ V})$$

$$W = 3.2 \times 10^{-18} \text{ J}$$

The Electron Volt

A convenient unit of work and energy when working with small charges, such as electrons and protons, is the **electron volt** (eV). One electron volt is defined as the work required to move one elementary charge between two points with a 1-V potential difference between them. Using the equation $W = qV$, this amount of work is

$$W = (1.60 \times 10^{-19} \text{ C})(1 \text{ V})$$

$$W = (1.60 \times 10^{-19} \text{ C})(1 \text{ J/C})$$

$$W = 1.60 \times 10^{-19} \text{ J}$$

One electron volt, therefore, is equivalent to 1.60×10^{-19} J. When this amount of work is done on a charge, the charge gains 1 eV of energy.

Questions

45. The following diagram represents a source of potential difference connected to two large, parallel metal plates separated by a distance of 4.0×10^{-3} meter.

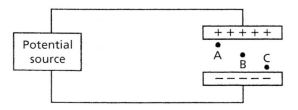

Which statement best describes the electric field strength between the plates? (1) It is zero at point *B*. (2) It is a maximum at point *B*. (3) It is a maximum at point *C*. (4) It is the same at points *A*, *B*, and *C*.

46. An object with a net charge of 4.80×10^{-6} coulomb experiences an electrostatic force having a magnitude of 6.00×10^{-2} newton when placed near a negatively charged metal sphere. What is the electric field strength at this location?
(1) 1.25×10^4 N/C directed away from the sphere
(2) 1.25×10^4 N/C directed toward the sphere
(3) 2.88×10^{-8} N/C directed away from the sphere
(4) 2.88×10^{-8} N/C directed toward the sphere

47. The electric field intensity at a given distance from a point charge is *E*. If the charge is doubled and the distance remains fixed, the electric field intensity will be (1) *E*/2 (2) 2*E* (3) *E*/4 (4) 4*E*

48. Two charged spheres are shown in the diagram. Which polarities will produce the electric field shown? (1) *A* and *B* both negative (2) *A* and *B* both positive (3) *A* positive and *B* negative (4) *A* negative and *B* positive

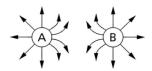

49. Which diagram best represents an electric field?

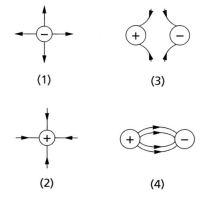

50. Which of the following diagrams best represents the electric field around the two spheres?

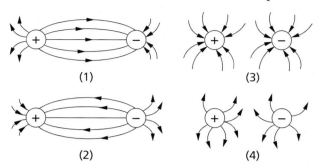

51. Which diagram best represents the electric field between the two spheres?

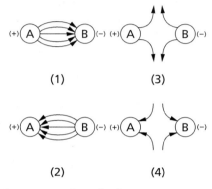

52. An electron placed between oppositely charged parallel plates *A* and *B* moves toward plate *A*, as represented in the diagram below.

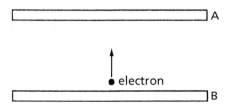

What is the direction of the electric field between the plates? (1) toward plate A (2) toward plate B (3) into the page (4) out of the page

53. How much work is required to move a single electron through a potential difference of 100. volts? (1) 1.6×10^{-21} J (2) 1.6×10^{-19} J (3) 1.6×10^{-17} J (4) 1.0×10^2 J

54. If 4.8×10^{-17} joule of work is required to move an electron between two points in an electric field, what is the electric potential difference between these points? (1) 1.6×10^{-19} V (2) 4.8×10^{-17} V (3) 3.0×10^2 V (4) 4.8×10^2 V

55. In an electric field, 0.90 joule of work is required to bring 0.45 coulomb of charge from point *A* to point *B*. What is the electric potential difference between points *A* and *B*? (1) 5.0 V (2) 2.0 V (3) 0.50 V (4) 0.41 V

56. How much electric energy is required to move a 4.00-microcoulomb charge through a potential difference of 36.0 volts? (1) 9.00×10^6 J (2) 144 J (3) 1.44×10^{-4} J (4) 1.11×10^{-7} J

57. The energy required to move one elementary charge through a potential difference of 5.0 volts is (1) 8.0 J (2) 5.0 J (3) 8.0×10^{-19} J (4) 1.6×10^{-19} J

Base your answers to questions 58 and 59 on the following diagram, which represents two small, charged conducting spheres, identical in size, located 2.0 meters apart.

+5.00 × 10⁻⁶coulomb   −4.00 × 10⁻⁶coulomb

58. What is the net combined charge on both spheres? (1) $+1.0 \times 10^{-6}$ C (2) -1.0×10^{-6} C (3) $+9.0 \times 10^{-6}$ C (4) -9.0×10^{-6} C

59. The force between these spheres is (1) 1.8×10^{-2} N (2) 3.6×10^{-2} N (3) 4.5×10^{-2} N (4) 9.0×10^{-2} N

60. What is the magnitude of the force acting on an electron when it is in the 1.00×10^6 N/C electric field? (1) 1.6×10^{-25} N (2) 1.6×10^{-13} N (3) 1.0×10^6 N (4) 1.6×10^{25} N

61. As an electron moves between two parallel plates from a negatively charged plate to a positively charged plate, the force on the electron due to the electric field (1) decreases (2) increases (3) remains the same

62. The electron above is replaced by a proton. Compared with the magnitude of the force on the electron, the magnitude of the force on the proton will be (1) less (2) greater (3) the same

Base your answer to question 63 on the following diagram, which represents two charged metal spheres.

q = +0.020 coulomb   q = −0.020 coulomb

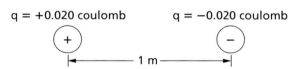

63. What is the magnitude of the force between the two spheres? (1) 3.6×10^6 N (2) 1.8×10^8 N (3) 3.6×10^9 N (4) 9.0×10^9 N

Base your answer to question 64 on the following diagram, which represents an electron projected into the region between two parallel

charged plates that are 10^{-3} meter apart. The electric field intensity between the plates is 10^6 newtons per coulomb.

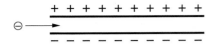

64. In which direction will the electron be deflected? (1) into the page (2) out of the page (3) toward the bottom of the page (4) toward the top of the page

65. Which graph best represents the relationship between the magnitude of the electric field strength, *E*, around a point charge and the distance, *r*, from the point charge?

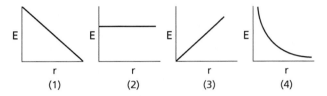

Electric Current

Conduction in Solids

Just as water can be made to flow through a network of pipes by the action of a pump, charged particles (usually electrons) can be made to flow through certain materials by the action of an electric field. For this flow of electric charge, or **electric current**, to occur, the following conditions must be satisfied:

1. The materials must be **conductors**, substances that allow electrons to flow through them. Some materials, called **insulators**, strongly resist such a flow.

2. The conductor must be connected to a **source**, a device that creates an electric field, such that a potential difference exists between the ends of the conductor. Two common sources are the battery and the generator.

3. The source and the conductor must form a closed loop, called a **circuit**, with no gaps across that electrons cannot travel across. If such a gap exists, electrons will gather on one side of the gap, and the flow of charge will not occur.

Before scientists learned that electrons flow in an electric current, they assumed that current consisted of positive charges flowing from the positive to the negative terminal of a battery. This flow of positive charges is today called **conventional current**. We adopt this convention and the

term "current" in this book will henceforth refer to this conventional current, unless otherwise specified. Keep in mind that it is directed opposite to the real flow of electrons in the conductor, which shall be referred to as the **real current**.

Conductors and Insulators

Metals are good conductors because their atoms do not hold firmly onto their outermost electrons. These "free" electrons can easily move through the metal, from atom to atom. The greater the number of free electrons per unit volume in a material, the better a conductor it makes.

Nonmetallic solids are insulators because they hold strongly to their electrons. It is difficult or impossible to get electrons to flow through such materials as rubber, glass, and wood.

The Ampere

The amount of charge flowing past a given point in a conductor is expressed in coulombs per second. If one coulomb passes a point in a conductor each second, the current in the conductor is one **ampere**. The ampere (A) is a fundamental unit.

The current in a conductor can be found by using the formula

$$I = \frac{\Delta q}{t}$$

where Δq is the amount of charge, in coulombs, passing a given point in a conductor in an amount of time *t*, in seconds, and *I* is the current, in amperes. For example, if 8 C pass a point in a conductor in 4 s, the current is 8 C/4 s, or 2 A.

Ohm's Law and Resistance

The amount of current passing through a conductor at a particular temperature is directly proportional to the potential difference across the ends of the conductor. For example, if the potential difference is doubled, the amount of current is doubled. This relationship is known as *Ohm's law*. Ohm's law is generally applicable to metallic conductors.

Ohm's law can be stated mathematically as

$$R = \frac{V}{I}$$

where *V* is the potential difference, in volts; *I* is the current, in amperes; and *R* is a constant. This constant is different for different materials and, for any particular material, is different at different temperatures.

The greater the R value of a material, the smaller the amount of current for a particular voltage. Thus, the greater the R value, the poorer the conductivity. The R value of a material, therefore, is a measure of the material's **resistance** to the flow of electrons through it.

The resistance of a metallic conductor at a given temperature is equal to the slope of a graph of potential difference versus current for that conductor (Figure 3-15). You can use Figure 3-15 to determine resistance by calculating the slope of the line. You can use any two points on the line. Using the points given (3,9) and 9,27) you get:

$$R = \frac{V_2 - V_1}{I_2 - I_1}$$

$$R = \frac{27\ V - 9\ V}{9\ A - 3\ A} = \frac{18\ V}{6\ A}$$

$$R = 3\ \Omega$$

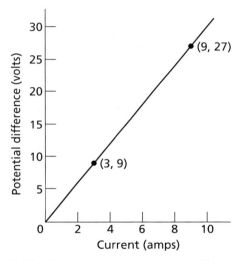

Figure 3-15. The slope of the potential difference-current graph is the resistance.

Resistance

A material that allows a current of one ampere to flow through it when the potential difference across it is one volt is said to have a resistance of one **ohm**. The symbol for the ohm is the Greek letter omega (Ω).

Sample Problems

7. What is the resistance of a metallic conductor that allows a current of 2 A to flow through it when connected to a source that provides a potential difference of 120 V?

Solution:

$$R = \frac{V}{I}$$

$$R = \frac{120\ V}{2\ A}$$

$$R = 60\ \Omega$$

8. If a wire has a resistance of 60 Ω and the potential difference across the wire is 30 V, what is the current flowing through it?

Solution:

$$R = \frac{V}{I}$$

$$I = \frac{V}{R}$$

$$I = \frac{30\ V}{60\ \Omega} = 0.5\ A$$

The resistance of a solid conductor, such as a metallic wire, depends upon the conductor's composition, length, cross-sectional area, and temperature:

1. Resistance is directly proportional to length. A wire that is twice as long as a similar wire has twice as much resistance.

2. Resistance is inversely proportional to cross-sectional area. A wire with twice the cross-sectional area of another wire (all other factors being equal) has one-half as much resistance.

3. The resistance of a metallic conductor increases as temperature rises, but the resistance of some nonmetallic, liquid, and gas conductors decreases as the temperature increases.

Some materials become such good conductors at very low temperatures, near absolute zero (0 K), that their resistance is effectively reduced to zero. Because electrons encounter virtually no resistance in passing through them, these materials are known as **superconductors**. A current of electrons incurs no loss of energy to heat while passing through a superconducting material.

Scientists recently discovered that certain materials act as superconductors at substantially higher temperatures. This discovery may eliminate the need for costly cooling to achieve superconductivity.

4. The resistance of a conductor of given dimensions depends upon the material of which it is made. Copper, for example, is a better conductor and offers less resistance to the flow of charge than aluminum.

Resistivity, represented by the Greek letter ρ (rho), is the measure of the resistance of a wire 1 m long with a uniform cross-sectional area of 1 m^2. You know that the resistance of a conductor is directly proportional to its length and inversely proportional to its cross-sectional area. This is expressed mathematically as $R = \dfrac{\rho L}{A}$, where R is the resistance, ρ is the resistivity of the metal, L is the length of the conductor, and A is the cross-sectional area. If the resistivity of a metal is known, you can calculate the resistance any wire made of that material. Table 3-1 gives the resistivity of several metals.

Table 3-1

Resistivity of Selected Metals at 20°C

Metal	Resistivity ($\Omega \cdot$ m)
Aluminum	2.82×10^{-8}
Copper	1.72×10^{-8}
Gold	2.44×10^{-8}
Nichrome	$150. \times 10^{-8}$
Silver	1.59×10^{-8}
Tungsten	5.6×10^{-8}

Sample Problem

9. What is the resistance of 30.-cm length of copper wire that has a cross-sectional area of 0.50 cm^2?

Solution:

$$R = \frac{\rho L}{A}$$

$$R = \frac{(1.72 \times 10^{-8}\,\Omega \cdot \text{m})(0.30\ \text{m})}{0.50 \times 10^{-4}\ \text{m}^2}$$

$$R = 1.0 \times 10^{-4}\ \Omega$$

Electric Power and Energy

As electrons are pushed through a circuit, work is done on them. Electric potential energy is converted into other forms of energy, such as heat, light, mechanical energy, or a combination of these, depending on the particular circuit. The electrons do not gain kinetic energy as they move through a circuit, because they undergo collisions with the atoms of the conductor. Although the electrons accelerate in between collisions due to the electric force acting on them, the kinetic energy they gain is transferred to the atoms during the collisions. As a result, the average kinetic energy of the atoms increases, and the temperature of the conductor increases.

Electric power is the rate at which electric potential energy is converted into another form of energy. The unit of power is the **watt** (W). One watt is equal to a rate of energy conversion of one joule/second.

The power (P), in watts, can be obtained by using the formula

$$P = VI$$

where V is the potential difference, in volts; and I is the current, in amperes.

For example, if the potential difference between the ends of a conductor is 8 V and the current is 5 A, then the power is 40 W. This means that 40 J of electric potential energy is expended every second and converted into another form of energy. The formula $P = VI$ can be combined with Ohm's law to obtain two other equations for calculating power

$$P = I^2R \quad \text{and} \quad P = \frac{V^2}{R}$$

The total amount of electric energy converted to another form when a current flows through a circuit during a period of time (t) may be obtained from the formula

$$W = Pt$$

where W is the total energy converted, in joules.

For example, if a 40-W circuit operates for 20 s, then 40 J of energy is converted during each of the 20 s of operation. The total amount of energy converted is then 800 J.

The formula $W = Pt$ can be combined with the formulas $P = VI$ and $P = I^2R$ to yield

$$W = VIt, \quad W = I^2Rt, \quad \text{and} \quad W = \frac{V^2t}{R}$$

Sample Problem

10A. A light bulb whose resistance is 240. Ω is connected to a 120.-V source. What is the current through the bulb?

Solution:

$$V = IR$$

$$I = \frac{V}{R}$$

$$I = \frac{120.\ \text{V}}{240.\ \Omega} = 0.500\ \text{A}$$

10B. What is the power of the bulb? (There are two formulas you can use.)

Solution:

$$P = VI$$

$$P = (120.\text{ V})(0.500\text{ A}) = 60.0\text{ W}$$

or $\quad P = I^2R$

$$P = (0.500\text{ A})^2(240.\ \Omega) = 60.0\text{ W}$$

10C. After operating for 10 minutes (600. s), how many joules of heat and light energy are produced? (There are two formulas you can use.)

Solution:

$$W = VIt$$

$$\begin{aligned} W &= (120.\text{ V})(0.500\text{ A})(600.\text{ s}) \\ &= 3.60 \times 10^4\text{ J} \end{aligned}$$

or $\quad W = I^2Rt$

$$\begin{aligned} W &= (0.500\text{ A})^2(240.\ \Omega)(600.\text{ s}) \\ &= 3.60 \times 10^4\text{ J} \end{aligned}$$

QUESTIONS

PART A

66. The table below lists various characteristics of two metallic wires, *A* and *B*.

Wire	Material	Temperature (°C)	Length (m)	Cross-Sectional Area (m²)	Resistance (Ω)
A	Silver	20.	0.10	0.010	R
B	Silver	20.	0.20	0.020	?

If wire *A* has resistance *R*, then wire *B* has resistance (1) *R*/2 (2) *R* (3) 2*R* (4) 4*R*

67. If 10. coulombs of charge are transferred through an electric circuit in 5.0 seconds, then the current in the circuit is (1) 0.50 A (2) 2.0 A (3) 15 A (4) 50. A

68. One watt is equivalent to one (1) N·m (2) N/m (3) J·s (4) J/s

69. In a simple electric circuit, a 110-volt electric heater draws 2.0 amperes of current. The resistance of the heater is (1) 0.018 Ω (2) 28 Ω (3) 55 Ω (4) 220 Ω

70. In a flashlight, a battery provides a total of 3.0 volts to a bulb. If the flashlight bulb has an operating resistance of 5.0 ohms, the current through the bulb is (1) 0.30 A (2) 0.60 A (3) 1.5 A (4) 1.7 A

71. A complete circuit is left on for several minutes, causing the connecting copper wire to become hot. As the temperature of the wire increases, the electrical resistance of the wire (1) decreases (2) increases (3) remains the same

72. The graph below represents the relationship between the potential difference (*V*) across a resistor and the current (*I*) through the resistor.

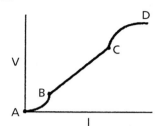

Through which entire interval does the resistor obey Ohm's law? (1) *AB* (2) *BC* (3) *CD* (4) *AD*

73. What must be inserted between points *A* and *B* to establish a steady electric current in the incomplete circuit represented in the diagram below?

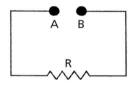

(1) switch (2) voltmeter (3) magnetic field source (4) source of potential difference

74. An incandescent light bulb is supplied with a constant potential difference of 120 volts. As the filament of the bulb heats up, its resistance (1) increases and the current through it decreases (2) increases and the current through it increases (3) decreases and the current through it decreases (4) decreases and the current through it increases

75. During a thunderstorm, a lightning strike transfers 12 coulombs of charge in 2.0×10^{-3} second. What is the average current produced in this strike? (1) 1.7×10^{-4} A (2) 2.4×10^{-2} A (3) 6.0×10^3 A (4) 9.6×10^3 A

76. How much current flows through a 12-ohm flashlight bulb operating at 3.0 volts? (1) 0.25 A (2) 0.75 A (3) 3.0 A (4) 4.0 A

77. Which physical quantity is correctly paired with its unit? (1) power and watt-seconds (2) energy and Newton-seconds (3) electric current and amperes/coulomb (4) electric potential difference and joules/coulomb

78. The current through a lightbulb is 2.0 amperes. How many coulombs of electric charge pass through the lightbulb in one minute? (1) 60. C (2) 2.0 C (3) 120 C (4) 240 C

79. A 330.-ohm resistor is connected to a 5.00-volt battery. The current through the resistor is (1) 0.152 mA (2) 15.2 mA (3) 335 mA (4) 1650 mA

PART B-1

80. What is the total electrical energy used by a 1500-watt hair dryer operating for 6.0 minutes? (1) 4.2 J (2) 250 J (3) 9.0×10^3 J (4) 5.4×10^5 J

81. As the potential difference across a given resistor is increased, the power expended in moving charge through the resistor (1) decreases (2) increases (3) remains the same

82. An electric iron operating at 120 volts draws 10. amperes of current. How much heat energy is delivered by the iron in 30. seconds? (1) 3.0×10^2 J (2) 1.2×10^3 J (3) 3.6×10^3 J (4) 3.6×10^4 J

83. A 1200-watt speaker used 6.3×10^5 joules of electrical energy. How long did it operate? (1) 6.3 minutes (2) 8.75 minutes (3) 525 minutes (4) 875 minutes

84. The current traveling from the cathode to the screen in a television picture tube is 5.0×10^{-5} ampere. How many electrons strike the screen in 5.0 seconds? (1) 3.1×10^{24} (2) 6.3×10^{18} (3) 1.6×10^{15} (4) 1.0×10^5

85. A 10.-meter length of wire with a cross-sectional area of 3.0×10^{-6} square meter has a resistance of 9.4×10^{-2} ohm at 20° Celsius. The wire is most likely made of (1) silver (2) copper (3) aluminum (4) tungsten

86. A 12.0-meter length of copper wire has a resistance of 1.50 ohms. How long must an aluminum wire with the same cross-sectional area be to have the same resistance? (1) 7.32 m (2) 8.00 m (3) 12.0 m (4) 19.7 m

87. Several pieces of copper wire, all having the same length but different diameters, are kept at room temperature. Which graph best represents the resistance, R, of the wires as a function of their cross-sectional areas, A?

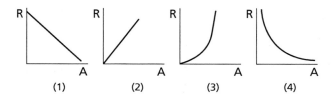

88. A potential drop of 50. volts is measured across a 250-ohm resistor. What is the power developed in the resistor? (1) 0.20 W (2) 5.0 W (3) 10. W (4) 50. W

89. The diagram below represents a lamp, a 10-volt battery, and a length of nichrome wire connected in series.

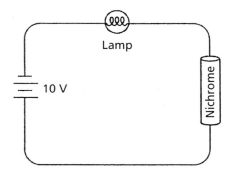

As the temperature of the nichrome is decreased, the brightness of the lamp will (1) decrease (2) increase (3) remain the same

90. The graph below shows the relationship between the potential difference across a metallic conductor and the electric current through the conductor at constant temperature T_1.

Potential Difference vs. Current at Temperature T_1

Which graph best represents the relationship between potential difference and current for the same conductor maintained at a higher constant temperature, T_2?

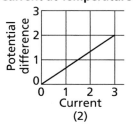

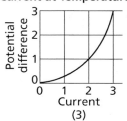

Potential Difference vs. Current at Temperature T$_2$

(3)

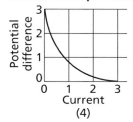

Potential Difference vs. Current at Temperature T$_2$

(4)

91. Which graph best represents the relationship between the electrical power and the current in a resistor that obeys Ohm's law?

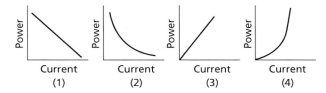

92. A 1.5-volt, AAA cell supplies 750 milliamperes of current through a flashlight bulb for 5.0 minutes, while a 1.5-volt, C cell supplies 750 milliamperes of current through the same flashlight bulb for 20. minutes. Compared to the total charge transferred by the AAA cell through the bulb, the total charge transferred by the C cell through the bulb is (1) half as great (2) twice as great (3) the same (4) four times as great

93. If the potential difference applied to a fixed resistance is doubled, the power dissipated by that resistance (1) remains the same (2) doubles (3) halves (4) quadruples

94. Aluminum, copper, gold, and nichrome wires of equal lengths of 1.0×10^{-1} meter and equal cross-sectional areas of 2.5×10^{-6} meter^2 are at 20.°C. Which wire has the greatest electrical resistance? (1) aluminum (2) copper (3) gold (4) nichrome

95. Which changes would cause the greatest increase in the rate of flow of charge through a conducting wire?
  (1) increasing the applied potential difference and decreasing the length of wire
  (2) increasing the applied potential difference and increasing the length of wire
  (3) decreasing the applied potential difference and decreasing the length of wire
  (4) decreasing the applied potential difference and increasing the length of wire

96. An operating electric heater draws a current of 10. amperes and has a resistance of 12 ohms. How much energy does the heater use in 60. seconds? (1) 120 J (2) 1200 J (3) 7200 J (4) 72,000 J

97. The potential difference applied to a circuit element remains constant as the resistance of the element is varied. Which graph best represents the relationship between power (P) and resistance (R) of this element?

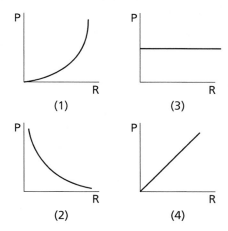

98. What is the resistance at 20°C of a 1.50-meter-long aluminum conductor that has a cross-sectional area of 1.13×10^{-6} meter^2? (1) 1.87×10^{-3} Ω (2) 2.28×10^{-2} Ω (3) 3.74×10^{-2} Ω (4) 1.33×10^{6} Ω

99. An immersion heater has a resistance of 5.0 ohms while drawing a current of 3.0 amperes. How much electrical energy is delivered to the heater during 200. seconds of operation? (1) 3.0×10^{3} J (2) 6.0×10^{3} J (3) 9.0×10^{3} J (4) 1.5×10^{4} J

100. Which graph best represents the relationship between resistance and length of a copper wire of uniform cross-sectional area at constant temperature?

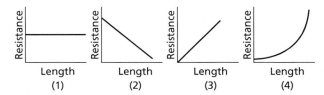

Circuit Combinations

Frequently we connect many appliances to one source so that electrons flow through all of them simultaneously. The appliances may be connected in *series*, in *parallel*, or in combinations of these.

Two fundamental principles apply to the operation of all circuits, no matter how complicated the connections. First, in accordance with the law of conservation of charge, no charge is ever created or destroyed. Thus, for any circuit, the sum of the currents entering any junction is equal to the sum of the currents leaving it. This is known as *Kirchhoff's first law*. Second, in accordance with the law of conservation of energy, the total energy output of the circuit is equal to the total electric energy made available to the circuit by the source. This is referred to as *Kirchhoff's second law*.

Schematic diagrams facilitate the study of electric circuits. Some of the symbols used to represent various circuit elements are shown in Figure 3-16.

Circuit Symbols

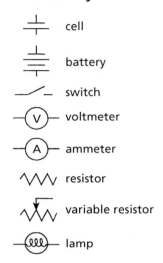

Figure 3-16. Electric circuit symbols.

A **resistor** is any conductor with a measurable resistance. The resistors in a circuit are usually various types of appliances. The connecting wires are assumed to have negligible resistance.

A **cell**, a chemical source of potential difference, is illustrated by a set of parallel lines. The long line represents the positive terminal and the short line represents the negative terminal. A **battery**, or combination of cells, is illustrated with a set of long and short parallel lines for every cell it contains.

An **ammeter** is a device used to measure current and is usually calibrated in amperes. It is connected in series (see below) with the part of the circuit whose current is to be measured. Its insertion into a circuit does not significantly affect the current.

A **voltmeter** is used to measure the potential difference between two points in a circuit. It is usually calibrated in volts. It is connected in parallel with the part of the circuit whose voltage is to be measured. Its insertion into the circuit has no significant effect on the potential difference between any two points in the circuit.

Series Circuits

In a **series circuit** there is only one path through which current can flow. A series circuit with three resistors is shown in Figure 3-17. Since the circuit consists of a single conducting path, the currents in all the resistors must be the same, in accordance with Kirchhoff's first law. This principle is represented by the relationship

$$I = I_1 = I_2 = I_3 = \cdots$$

where I is the total current, and I_1, I_2, and I_3 are the currents through resistors R_1, R_2, and R_3, respectively. (There is no limit to the number of resistors permitted.)

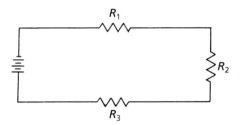

Figure 3-17. Series circuit.

The work done per coulomb of charge as it goes through a particular resistor is equal to the potential difference between the ends of that resistor. This potential difference is referred to as the **voltage drop** across the resistor. Thus, Kirchhoff's second law can be stated as: The sum of the voltage drops across all the resistors in series must equal the potential difference produced by the source. This is represented by the formula

$$V = V_1 + V_2 + V_3 + \cdots$$

where V is the voltage created by the source and V_1, V_2, and V_3 are the voltage drops across resistors R_1, R_2, and R_3, respectively (up to the nth resistor).

The equivalent (total) resistance, R_{eq}, is found by adding all the individual resistances in the series combination

$$R_{eq} = R_1 + R_2 + R_3 + \cdots$$

By combining these principles with Ohm's law and the formulas introduced earlier, the voltage drop, current, and power of all resistors in series can be calculated.

Sample Problem

11. The circuit diagram in Figure 3-18 represents three appliances connected in series to a 90.-V source. If their resistances are 2.0 Ω, 3.0 Ω, and 5.0 Ω, respectively, find the voltage drop across each appliance, the amount of current passing through each appliance, and the power of each appliance.

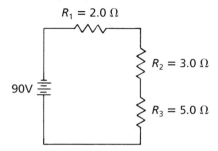

$R_1 = 2.0\ \Omega$

$R_2 = 3.0\ \Omega$

90V

$R_3 = 5.0\ \Omega$

Figure 3-18.

Solution:

First, find the total resistance.

$$R_t = R_1 + R_2 + R_3$$
$$= 2.0\ \Omega + 3.0\ \Omega + 5.0\ \Omega = 10.0\ \Omega$$

Next, determine the current by applying Ohm's law to the total voltage and the total resistance.

$$R = \frac{V}{I}$$

$$I = \frac{V}{R}$$

$$I = \frac{90\ V}{10\ \Omega} = 9.0\ A$$

Since $I = I_1 = I_2 = I_3$, we know that $I_1 = 9.0$ A, $I_2 = 9.0$ A, and $I_3 = 9.0$ A.

Next, find the voltage drop across each appliance by applying Ohm's law to each resistance.

$$V_1 = I_1R_1 = (9.0\ A)(2.0\ \Omega) = 18\ V$$

$$V_2 = I_2R_2 = (9.0\ A)(3.0\ \Omega) = 27\ V$$

$$V_3 = I_3R_3 = (9.0\ A)(5.0\ \Omega) = 45\ V$$

At this point check to make sure that the sum of the individual voltage drops equals the voltage of the source.

$$V = V_1 + V_2 + V_3$$

$$V = 18\ V + 27\ V + 45\ V = 90.\ V$$

Next, determine each appliance's power by applying $P = VI$ to each.

$$P_1 = V_1I_1 = (18\ V)(9.0\ A) = 160\ W$$

$$P_2 = V_2I_2 = (27\ V)(9.0\ A) = 240\ W$$

$$P_3 = V_3I_3 = (45\ V)(9.0\ A) = 410\ W$$

12. Figure 3-19 represents two appliances connected in series to a 100-V source. The resistances of R_1 and R_2 are 3 Ω and 7 Ω, respectively.

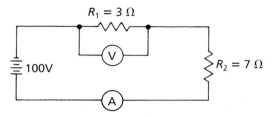

$R_1 = 3\ \Omega$

V

100V

$R_2 = 7\ \Omega$

A

Figure 3-19.

What is the current recorded by the ammeter?
Solution:

$$I = \frac{V}{R_{eq}}$$

$$I_t = \frac{100\ V}{10\ \Omega} = 10\ A$$

What is the voltage drop recorded by the voltmeter?
Solution:

$$V = I_1R_1$$

$$V = (10\ A)(3\ \Omega) = 30\ V$$

How much electric energy is expended in R_1 in 2 seconds?
Solution:

$$W_1 = V_1I_1t$$

$$W_1 = (30\ V)(10\ A)(2\ s) = 600\ J$$

The voltage drop and power of an appliance in series is directly proportional to its resistance. The greater the appliance's resistance, the greater its voltage drop and power.

Parallel Circuits

In a **parallel circuit**, current flows through two or more alternative paths, called **branches**. Every appliance in our homes is a branch in a network of branches. The part of the circuit through which the full amount of current flows is called the **main line**. Each branch of the circuit is independent of the others. A break in one branch only stops the current in that branch, and current continues to flow through the others (Figure 3-20).

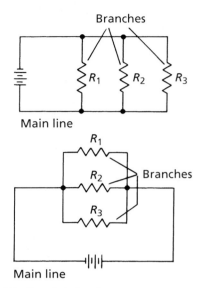

Figure 3-20. Parallel circuits.

Since electric force is a conservative force, the amount of work done per coulomb of charge is equal to the potential difference created by the source irrespective of the particular branch through which the charge passes. It follows that the voltage drop across each resistor connected in parallel is the same. This is expressed mathematically as

$$V = V_1 = V_2 = V_3 = \cdots$$

where V is the voltage of the source, and V_1, V_2, and V_3 are the voltage drops across resistors R_1, R_2, and R_3, respectively (up to the nth resistor).

In accordance with Kirchhoff's first law, the sum of the branch currents must equal the current in the main line. Mathematically this is stated as

$$I = I_1 + I_2 + I_3 + \cdots$$

where I is the current in the main line and I_1, I_2, and I_3 are the currents in branches 1, 2, and 3, respectively (up to the nth branch).

The total current can be found by replacing all the parallel resistors with one whose resistance, R_{eq}, is provided by the formula

$$\frac{1}{R_{eq}} = \frac{1}{R_1} + \frac{1}{R_2} + \frac{1}{R_3} + \mathrm{L}$$

where R_1, R_2, and R_3 are the resistances of branches 1, 2, and 3, respectively (up to the nth branch). The total current I can then be found by using Ohm's law

$$I = \frac{V}{R_{eq}}$$

R_{eq} is referred to as the **combined**, or **equivalent**, **resistance** of the parallel circuit.

By combining these principles with the formulas introduced earlier, the current, voltage drop, and power of each branch in a parallel circuit can be calculated.

Sample Problems

13. What is the current, voltage drop, and power of each of three appliances connected in parallel to a 72-V source in Figure 3-21, if their resistances are 12 Ω, 24 Ω, and 8.0 Ω?

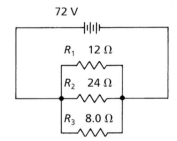

Figure 3-21.

Solution: First, the equivalent resistance of the circuit is determined by applying the formula

$$\frac{1}{R_{eq}} = \frac{1}{R_1} + \frac{1}{R_2} + \frac{1}{R_3}$$

$$\frac{1}{R_{eq}} = \frac{1}{12\ \Omega} + \frac{1}{24\ \Omega} + \frac{1}{8.0\ \Omega} = \frac{1}{4.0\ \Omega}$$

$$R_{eq} = 4.0\ \Omega$$

Note that the equivalent resistance of the circuit is less than the resistance of any one of the branches. The effect of adding more branches to a parallel circuit is to decrease the equivalent resistance of the circuit and to draw a greater amount of current from the source.

Next, Ohm's law is applied to determine the total current through the circuit.

$$R_{eq} = \frac{V}{I}$$

$$I = \frac{V}{R_{eq}}$$

$$I = \frac{72\ \text{V}}{4.0\ \Omega} = 18\ \text{A}$$

Now, the voltage drop across each of the three appliances in parallel is found

$$V = V_1 = V_2 = V_3 = 72 \text{ V}$$

Next, the current in each branch is determined by applying Ohm's law to each one.

$$I_1 = \frac{V_1}{R_1} = \frac{72 \text{ V}}{12 \text{ }\Omega} = 6.0 \text{ A}$$

$$I_2 = \frac{V_2}{R_2} = \frac{72 \text{ V}}{24 \text{ }\Omega} = 3.0 \text{ A}$$

$$I_3 = \frac{V_3}{R_3} = \frac{72 \text{ V}}{8.0 \text{ }\Omega} = 9.0 \text{ A}$$

At this point we check to see if the branch currents add up to the total current.

$$I = I_1 + I_2 + I_3$$

$$I = 6.0 \text{ A} + 3.0 \text{ A} + 9.0 \text{ A} = 18 \text{ A}$$

Note that the amount of current in each branch is inversely proportional to the resistance of the branch. The greater the resistance of the branch, the smaller the current.

The power of each branch can now be determined by applying $P = VI$ to each.

$$P_1 = V_1 I_1 = (72 \text{ V})(6.0 \text{ A}) = 430 \text{ W } (432 \text{ W})$$

$$P_2 = V_2 I_2 = (72 \text{ V})(3.0 \text{ A}) = 220 \text{ W } (216 \text{ W})$$

$$P_3 = V_3 I_3 = (72 \text{ V})(9.0 \text{ A}) = 650 \text{ W } (648 \text{ W})$$

The total power of the parallel circuit (430 W + 220 W + 650 W = 1300 W) would be the same if the branches were replaced by one path with a resistance equal to the equivalent resistance, R_{eq}. The power would then be 72 volts $\times$ 18 amperes, or 1300 watts.

QUESTIONS
PART A

101. What is the total resistance of the circuit segment shown in the diagram below?

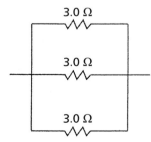

3.0 Ω

3.0 Ω

3.0 Ω

(1) 1.0 Ω (2) 9.0 Ω (3) 3.0 Ω (4) 27 Ω

102. A 10.-ohm resistor and a 20.-ohm resistor are connected in series to a voltage source. When the current through the 10.-ohm resistor is 2.0 amperes, what is the current through the 20.-ohm resistor? (1) 1.0 A (2) 2.0 A (3) 0.50 A (4) 4.0 A

103. The diagram below represents an electric circuit consisting of a 12-volt battery, a 3.0-ohm resistor, R_1, and a variable resistor, R_2.

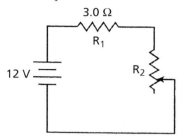

At what value must the variable resistor be set to produce a current of 1.0 ampere through R_1? (1) 6.0 Ω (2) 9.0 Ω (3) 3.0 Ω (4) 12 Ω

104. In which circuit would ammeter A show the greatest current?

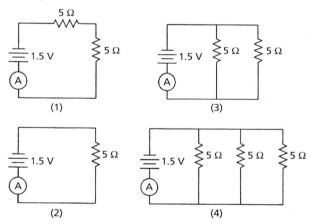

Base your answers to questions 105 and 106 on the following diagram

105. A 9.0-volt battery is connected to a 4.0-ohm resistor and a 5.0-ohm resistor as shown in the diagram below.

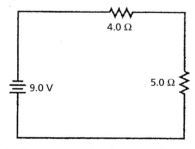

What is the current in the 5.0-ohm resistor?
(1) 1.0 A (2) 1.8 A (3) 2.3 A (4) 4.0A

106. How much electric energy is expended in both resistors per second? (1) 9 J (2) 18 J (3) 45 J (4) 50 J

107. In a series circuit containing two lamps, the battery supplies a potential difference of 1.5 volts. If the current in the circuit is 0.10 ampere, at what rate does the circuit use energy? (1) 0.015 W (2) 0.15 W (3) 1.5 W (4) 15 W

108. A 30.-ohm resistor and a 60.-ohm resistor are connected in an electric circuit as shown below.

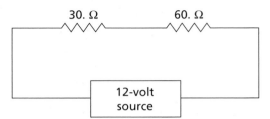

Compared to the electric current through the 30.-ohm resistor, the electric current through the 60.-ohm resistor is (1) smaller (2) larger (3) the same

109. The diagram below represents part of an electric circuit containing three resistors.

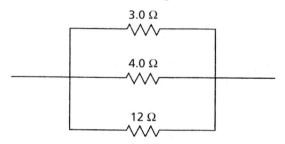

What is the equivalent resistance of this part of the circuit? (1) 0.67 Ω (2) 1.5 Ω (3) 6.3 Ω (4) 19 Ω

110. In the circuit diagram below, what are the correct readings of voltmeters V_1 and V_2?

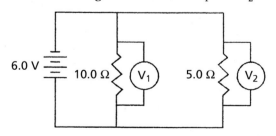

(1) V_1 reads 2.0 V and V_2 reads 4.0 V (2) V_1 reads 4.0 V and V_2 reads 2.0 V (3) V_1 reads 3.0 V and V_2 reads 3.0 V (4) V_1 reads 6.0 V and V_2 reads 6.0 V

111. A unit of electric potential difference is the (1) ampere (2) joule (3) volt (4) coulomb

112. The electron volt is a unit of (1) charge (2) potential difference (3) current (4) energy

Base your answers to questions 113 and 114 on the circuit diagram below.

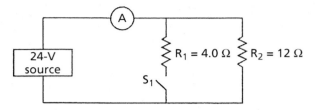

113. If switch S_1 is open, the reading of ammeter A is (1) 0.50 A (2) 2.0 A (3) 1.5 A (4) 6.0 A

114. If switch S_1 is closed, the equivalent resistance of the circuit is (1) 8.0 Ω (2) 2.0 Ω (3) 3.0 Ω (4) 16 Ω

115. In the diagram below, lamps L_1 and L_2 are connected to a constant voltage power supply.

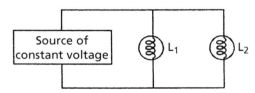

If lamp L_1 burns out, the brightness of L_2 will (1) remain the same (2) increase (3) decreases

116. Which circuit diagram shows voltmeter V and ammeter A correctly positioned to measure the total potential difference of the circuit and the current through each resistor?

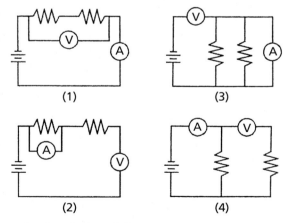

117. Which of the following circuit diagrams correctly shows the connection of ammeter A and voltmeter V to measure the current through and potential difference across resistor R?

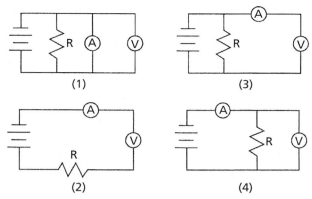

(1)

(3)

(2)

(4)

118. Three resistors of 10 ohms, 20 ohms, and 30 ohms are connected in series to a 120-volt source. The power developed is (1) greatest in the 10-ohm resistor (2) greatest in the 20-ohm resistor (3) greatest in the 30-ohm resistor (4) the same in all three resistors

119. Compared with the potential drop across the 10-ohm resistor shown in the diagram, the potential drop across the 5-ohm resistor is (1) the same (2) twice as great (3) one-half as great (4) four times as great

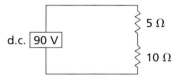

120. The diagram represents a circuit with two resistors in series. If the total resistance of R_1 and R_2 is 24 ohms, the resistance of R_2 is (1) $1.0\,\Omega$ (2) $0.50\,\Omega$ (3) $100.\,\Omega$ (4) $4.0\,\Omega$

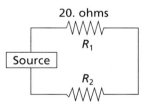

121. In the circuit shown in the diagram below, the rate at which electrical energy is being expended in resistor R_1 is (1) less than in R_2 (2) greater than in R_2 (3) less than in R_3 (4) greater than in R_3

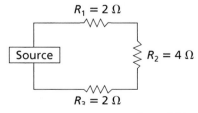

Base your answers to questions 122 and 123 on the diagram, which represents an electrical circuit.

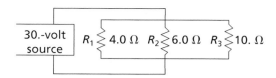

122. The equivalent resistance of R_1, R_2, and R_3 is approximately (1) $10\,\Omega$ (2) $2\,\Omega$ (3) $20\,\Omega$ (4) $7\,\Omega$

123. The current in R_1 is (1) 3.8 A (2) 7.5 A (3) 15 A (4) 60. A

124. Compared with the current in the 10.-ohm resistor in the circuit shown below, the current in the 5.0-ohm resistor is (1) one-half as great (2) one-fourth as great (3) the same (4) twice as great

125. The following diagram represents a segment of a circuit. What is the current in ammeter A? (1) 1 A (2) 0 A (3) 3.5 A (4) 7 A

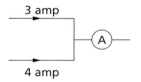

126. Two resistors of 10 ohms and 5 ohms are connected as shown in the diagram. If the current through the 10-ohm resistor is 1.0 ampere, then the current through the 5.0-ohm resistor is (1) 15 A (2) 2.0 A (3) 0.50 A (4) 0.30 A

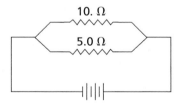

127. Two resistors are connected in parallel to a 12-volt battery as shown in the diagram. If the current in resistance R is 3.0 amperes, the rate at which R consumes electrical energy is

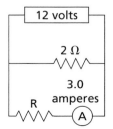

(1) 1.1×10^2 W (2) 36 W (3) 24 W (4) 4.0 W

128. The diagram below represents a segment of an electrical circuit. What is the current in wire *AB*? (1) A (2) 2 A (3) 5 A (4) 6 A

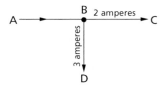

Magnetism

Natural Magnets

Ancient scientists discovered that long slender pieces of certain ores, when free to rotate, would align themselves in the same direction: one end points north (toward Earth's magnetic north pole), the other south (toward the magnetic south pole). Objects that behave this way are called **magnets**. They point toward the earth's magnetic poles so dependably that sailors and navigators rely on them to determine direction. A magnet used in this way is called a **compass**.

The end that points north is referred to as the *north pole (N-pole)* of the magnet, and the end that points due south is the *south pole (S-pole)*. When two or more magnets are brought together, they exert forces on each other; like poles repel, unlike poles attract. Earth itself is a giant magnet that affects the orientation of all magnetic compasses on Earth.

Some substances, such as iron, become *magnetized* when placed near a magnet. When this happens, the magnetic poles created in the iron are so arranged that an attraction to the magnet results. Thus, a magnet is attracted to a nonmagnet, because being close to the magnet has at least temporarily magnetize the nonmagnet. Different substances have varying degrees of response to this type of magnetization.

Breaking a magnet into pieces does not separate the poles from each other. Instead, all the pieces become magnets, each with its own two poles. Objects that are magnets always have both poles; single poles do not exist. (See Figure 3-22.)

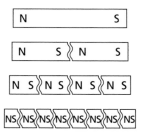

Figure 3-22. Breaking a magnet results in new, complete magnets.

Heating a magnet weakens its magnetism, but cooling does not ordinarily strengthen it. Dropping or hammering a magnet also weakens its magnetism.

Magnetic Fields

Just as charges create electric fields that act on other charges, magnets create **magnetic fields** that act on other magnets. Magnetic fields, like electric and gravitational fields, are *vector quantities*, with magnitude (intensity) and direction.

A magnetic field's direction at a given point is defined as the direction in which the N-pole of a test magnet points due to the action of the field at that point. The field's magnitude, or *intensity*, is determined by the force it exerts on a test magnet at a given point. A more precise definition of intensity will be presented in the enrichment section (pp. 122–123).

Magnetic fields, like electric fields, can be represented by field lines if certain rules are adopted. First, at any point in the field, the direction of the field is tangent to the field line at that point, with the arrow on the line indicating the direction in which the N-pole of the test magnet would point. Second, the concentration of field lines in a region indicates the intensity of the field in the region. Where the field is more intense, the field lines are more crowded together. Based on these rules, the field lines in the vicinity of a bar-shaped magnet are as shown in Figure 3-23.

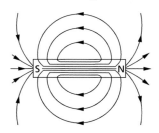

Figure 3-23. Magnetic field around a bar magnet.

A simple and useful field picture exists between two magnets with opposite poles facing each other, as shown in Figure 3-24. Note that the

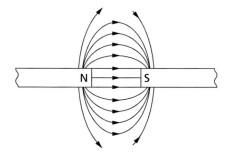

Figure 3-24. Magnetic field between opposite poles.

field between the poles is uniformly intense and the field lines are parallel to one another. Magnet field lines always form closed loops and never intersect one another.

Currents Act on Magnets

With the development of electric currents came the discovery that a flow of charge affects any magnet in the vicinity of the current. This means that currents create magnetic fields, just as magnets do.

The magnetic field created by a current is best described by field lines. Magnetic field lines in the vicinity of a straight current-carrying wire can be represented as concentric circles around the wire, with each circle forming a plane perpendicular to the current.

A test magnet placed on the right side of the downward-moving current in Figure 3-25 experiences a force that makes its N-pole point out of the page, toward the reader. A test magnet placed on the left side, will have its N-pole point into the page, away from the reader. Behind the current, the N-pole points to the reader's right; in front, it points to the left. These directions are reversed for an upward-moving current. (Recall that the term "current" refers to conventional current.)

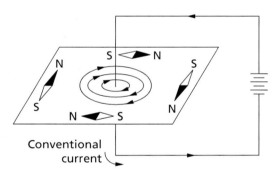

Figure 3-25. The magnetic field around a straight current-carrying conductor.

In the vicinity of a coil-shaped current-carrying wire (called a **solenoid**) or a single loop of current-carrying wire, the magnetic field lines appear as shown in Figure 3-26. Inside the solenoid or loop, the field lines are parallel to one another and perpendicular to the circles of current.

Field lines emerge from one end of the solenoid. At some distance from the solenoid they begin to curve back, until the arrows point in the opposite direction. The lines curve back again and complete the loop by reentering the other side of the solenoid.

A test magnet placed inside the coil in Figure 3-26 experiences a force that makes its N-pole

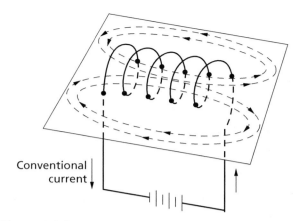

Figure 3-26. The magnetic field around a solenoid.

point to the right. The N-pole of a magnet situated outside, but near either end of the solenoid, is also forced to point to the right. At other locations, a magnet's N-pole is forced to point in the direction indicated by the local field line.

Note: The field created by a solenoid looks very much like that of a bar magnet. We will return to this important fact in a later section (p. 108).

When certain materials, such as iron, are inserted inside a solenoid they become magnetized in such a way that the magnetic field of the solenoid becomes stronger. A solenoid whose magnetic field is in this way intensified is referred to as an **electromagnet**. Materials that behave this way are said to be **ferromagnetic**.

Magnets Act on Currents

Further experimentation with magnets and currents revealed that not only do currents affect magnets, but magnets also exert forces on currents. A straight, current-carrying wire situated in a magnetic field experiences a force if the charges move perpendicularly to the field, or at least a component of their motion is perpendicular to the field. No force appears if the direction of the current is parallel to the field. The force is greatest when the angle between the field lines and the current is exactly 90°.

The force exerted on the wire is perpendicular to both the direction of the current and the direction of the magnetic field.

Currents Act on Currents

Since currents create magnetic fields and magnetic fields exert forces on currents, it follows that one current exerts a force on another. (No electric force exists between current-carrying wires, since they are electrically neutral.)

Two straight, parallel wires carrying current in the same direction attract each other. If the currents move in opposite directions, the wires repel each other (see Figure 3-27). The force on either current is always equal in magnitude and opposite in direction to the force on the other current. If the currents are perpendicular to each other, no forces appear.

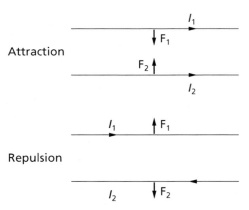

Figure 3-27. The direction of current in the wires determines the directions of the forces exerted by the wires.

The Magnetic Force

All the various forces described in this section—those exerted by magnets on magnets, magnets on currents, currents on magnets, and currents on currents—are different manifestations of one and the same principle of nature: *moving charges exert forces on other moving charges*. This force is independent of the electric force between charges; we call it the **magnetic force**. Each moving charge creates a magnetic field. The intensity of the field created by a moving charge is proportional to the amount of charge and to the velocity of the charge. Intensity weakens with distance from the moving charge.

Magnets behave as they do because the electrons in their atoms spin in the same direction, as illustrated in Figure 3-28. As we look at the N-pole of a magnet, the electron rotations are clockwise; as we look at the S-pole, they are counterclockwise. Magnets are, in effect, solenoids, with each atom supplying one of the "coils" of current. This is why a bar magnet's field lines look like those of a solenoid.

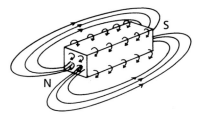

Figure 3-28. Magnetic field around a bar magnet.

In nonmagnets the atomic currents (as the electron rotations are called) are not all oriented in the same direction. The magnetic field created by one atom is cancelled by the oppositely directed field created by another atom. The net result is no magnetic field. The object neither exerts a magnetic force nor responds to the presence of a magnetic field.

Bringing two magnets together is the same as bringing two solenoids together. The magnetic field of each exerts a twisting force (torque) on the other. Each attempts to realign the other's atomic currents. As a result, like poles repel and unlike poles attract.

A magnet tries to align the atomic currents within a nonmagnet so that they all rotate in the same direction. When this realignment occurs, the nonmagnet becomes magnetized. Many substances, however, resist the twisting of their atoms, and it is difficult to magnetize them.

Heating a magnet weakens its magnetism because the increased kinetic energy of the atoms and the more violent collision between them disrupt the uniform orientation of the atomic currents in one direction. Breaking a magnet produces many new magnets, each with two poles, because each piece is still a solenoid, albeit a smaller one.

QUESTIONS

129. Which diagram best represents magnetic flux lines around a bar magnet?

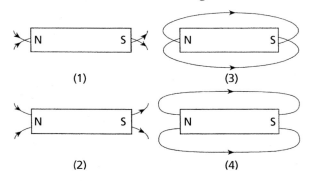

130. The diagram below represents the magnetic field near point *P*.

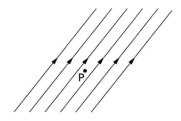

If a compass is placed at point *P* in the same plane as the magnetic field, which arrow represents the direction the north end of the compass needle will point?

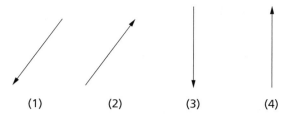

(1)   (2)   (3)   (4)

131. Which type of field is present near a moving electric charge? (1) an electric field, only (2) a magnetic field, only (3) both an electric field and a magnetic field (4) neither an electric field nor a magnetic field

132. The diagram below represents magnetic lines of force within a region of space.

Lines of Force

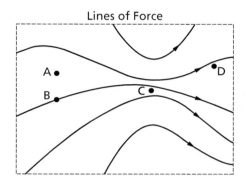

The magnitude of the field is strongest at point (1) *A* (2) *B* (3) *C* (4) *D*

133. The diagram below shows the lines of magnetic force between two north magnetic poles.

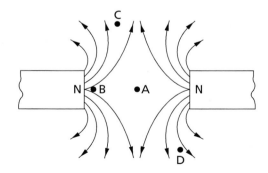

At which point is the magnetic field strength the greatest? (1) *A* (2) *B* (3) *C* (4) *D*

134. As two parallel conductors with currents in the same direction are moved apart, their force of (1) attraction increases (2) attraction decreases (3) repulsion increases (4) repulsion decreases

135. The diagram below shows a bar magnet.

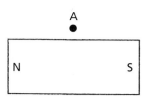

Which arrow best represents the direction of the needle of a compass placed at point *A*? (1) ↑ (2) ↓ (3) → (4) ←

136. In the following diagram, what is the direction of the magnetic field at point *A*? (1) to the left (2) to the right (3) toward the top of the page (4) toward the bottom of the page

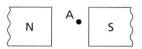

137. Which diagram correctly shows a magnetic field configuration?

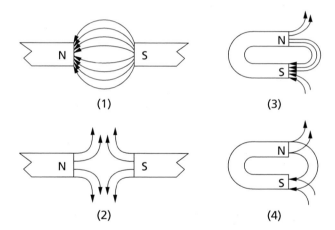

(1)   (3)

(2)   (4)

Electromagnetism

Electromagnetic Induction

If a straight piece of metal wire (or some other conductor) is moved through a magnetic field perpendicularly to the field lines, the field exerts a force on the electrons in the wire. The electrons in the wire move with the wire, and magnetic fields act on moving charges. The direction of the force on the moving electrons is perpendicular to both the magnetic field lines and the wire's direction of motion (see Figure 3-29), and may produce a current in the wire.

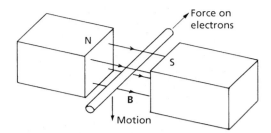

Figure 3-29. A current is induced in a wire moving across a magnetic field.

A current—one that can make a bulb glow, for example—is created if the wire moving in the magnetic field is connected to a complete circuit. This current continues as long as the wire moves across the field lines.

Electromagnetic induction also occurs when the intensity (or strength) of a magnetic field varies over time. As the field becomes stronger or weaker, the number of magnetic field lines in the region enclosed by a loop of wire varies. This gives rise to current in the loop, even if the wire is not moving.

The current created by a changing magnetic field or by moving a wire across a magnetic field is referred to as **induced current**, and the potential difference (voltage) created is referred to as **induced EMF** (electromotive force). The magnitude of the induced voltage is proportional to the rate of change in the number of field lines enclosed by the loop per unit time. The faster the number of field lines changes, the greater the induced voltage (and current) in the loop.

Since the changing magnetic field leads to a force on electrons even when the wire is at rest, we say *changing magnetic fields create electric fields.* Electric fields exert forces on charges whether or not the charges are in motion; magnetic fields act only on moving charges.

The physicist James Clerk Maxwell proposed that the converse statement should also be true, that *changing electric fields create magnetic fields.* This has been found to be the case.

Electromagnetic Applications

Current-Carrying Loops in Magnetic Fields

If a solenoid or a single loop of current-carrying wire is placed in a magnetic field, the wire experiences a torque (twisting force) that rotates the plane of the loop (and every individual loop in the case of a solenoid) until the plane of the loop is perpendicular to the magnetic field lines (see Figure 3-30). When this occurs, the magnetic field created by the loop of current inside the loop be-comes *parallel* to and points in the same direction as the external magnetic field. The strength of the torque is proportional to the amount of current in the loop and to the intensity of the external field.

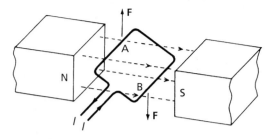

Figure 3-30. Forces on a current-carrying loop of wire in a magnetic field.

This twisting action exerted on coils of current in magnetic fields is the basis of the operation of many useful devices, including the ammeter, the voltmeter, and the electric motor.

Ammeter (–Ⓐ–) Current is measured with an ammeter. The ammeter is placed in series with the circuit (or part of the circuit) whose current is to be measured.

Voltmeter (–Ⓥ–) Potential difference, or voltage, is measured with a voltmeter. The voltmeter is connected in parallel with the part of the circuit whose voltage is to be measured.

Motors An electric **motor** is a device that converts electrical energy into rotational kinetic energy. Its operation is based on the torque exerted by a magnetic field on a loop of current in the field. A wire loop situated between opposite magnetic poles is connected to a battery, one end to the positive terminal and the other end to the negative terminal (see Figure 3-31).

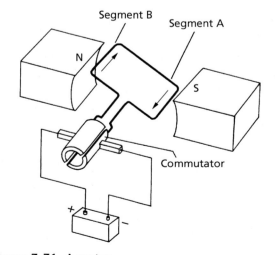

Figure 3-31. A motor.

The Generator The process of inducing current and EMF by moving a conducting wire across magnetic field lines is used today to provide millions of homes with electric current. The most common device for this purpose is the **generator**. A simple generator consists of a loop of wire that is made to rotate between two opposite magnetic poles. As the loop rotates, the current and potential difference generated vary in magnitude and alternate in direction. This is called **alternating current** (AC).

Transformers (⫴) A **transformer** is a device inserted into a circuit carrying alternating current (AC) in order to change the voltage to some higher or lower value. It is used with fluorescent and neon lights, thermostats, bells, and many other appliances that require voltages greater or smaller than the 120 volts supplied by all electric utility companies in the U.S.

Questions

138. The following diagram shows a copper wire located between the poles of a magnet. Maximum electric potential will be induced in the wire when it is moved at a constant speed toward which point? (1) *A* (2) *B* (3) *C* (4) *D*

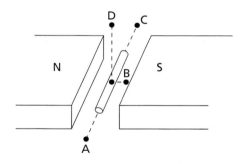

139. A conductor is moving perpendicularly to a uniform magnetic field. Increasing the speed of the conductor will cause the potential difference induced across the ends of the conductor to (1) decrease (2) increase (3) remain the same

140. The following diagram shows the cross section of a wire that is perpendicular to the page and a uniform magnetic field directed to the right. Toward which point should the wire be moved to induce the maximum electric potential? (Assume the same speed would be used in each direction.)

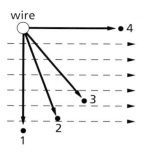

(1) 1 (2) 2 (3) 3 (4) 4

141. In the circuit represented by the diagram below, what is the reading of voltmeter *V*?

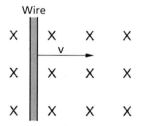

(1) 20. V (2) 2.0 V (3) 30. V (4) 40. V

142. The diagram below shows a wire moving to the right at speed *v* through a uniform magnetic field that is directed into the page.

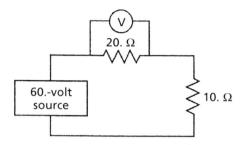

Magnetic field directed into page

As the speed of the wire is increased, the induced potential difference will (1) decrease (2) increase (3) remain the same

143. A motor is to rotational mechanical energy as a generator is to (1) chemical potential energy (2) induced electrical energy (3) thermal internal energy (4) elastic potential energy

144. A student uses a voltmeter to measure the potential difference across a circuit resistor. To obtain a correct reading, the student must connect the voltmeter (1) in parallel with the circuit resistor (2) in series with the circuit resistor (3) before connecting the other circuit components (4) after connecting the other circuit components

145. Which statement best describes the torque experienced by a current-carrying loop of wire in an external magnetic field?
(1) It is due to the current in the loop of wire, only.
(2) It is due to the interaction of the external magnetic field and the magnetic field produced by current in the loop.
(3) It is inversely proportional to the length of the conducting loop in the magnetic field.
(4) It is inversely proportional to strength of the permanent magnetic field.

146. The turning force on the armature of an operating electric motor may be increased by
(1) decreasing the current in the armature
(2) decreasing the magnetic field strength of the field poles
(3) increasing the potential difference applied to the armature
(4) increasing the distance between the armature and the field poles

Physics in Your Life

Storing Information Magnetically

If you have ever recorded your voice on tape, used a videotape recorder, or a computer, you have taken advantage of magnetism as a means of storing information. The tape used in audio and videotape records consists of a thin layer of magnetic oxide on a thin plastic tape. When a sound is recorded, the sound is first converted into an electronic current. That current is then sent to the recording head, which acts as a tiny electromagnet.

The electromagnet produces a magnetic field that varies with the electric current. As the tape moves through the recording head, the varying magnetic field magnetizes the tiny section of the tape passing over the narrow gap at each instant. When the tape is played back, the changing magnetism of the tape causes changes in a magnetic field within the player, which then induces a current. This induced current is the output signal that is amplified and sent to a speaker and converted back into sound.

Computer disks, which include hard disks and floppy disks, and magnetic tape are read and written in a similar manner to audiotapes and videotapes. The hard disk inside a computer, for example, is made up of a stack of metal disks that have magnetic particles on one surface. When information is saved to the hard disk, a device changes the orientation of the magnetic particles on the disk's surface. Orientation in one direction represents 0 and orientation in the other direction represents 1. When the disk is read, the 0s and 1s are converted to pulses of electric current. The current is then converted into a visible image that can be observed on a computer screen or printed page.

Questions

1. What roles do electricity and magnetism play in the recording and playback of a movie on a videotape?

2. Describe the process in which digital information is stored on a hard disk.

3. Why could bringing a computer disk near a strong magnet destroy the information stored on the disk?

Chapter Review Questions

1. How much work is done in moving 6 electrons through a potential difference of 2.0 volts? (1) 6.0 eV (2) 2.0 eV (3) 3.0 eV (4) 12 eV

2. If 6.0 joules of work is done to move 2.0 coulombs of charge from point A to point B, what is the electric potential difference between points A and B? (1) 6.0 V (2) .33 V (3) 3.0 V (4) 12 V

3. The work required to move a charge of 0.04 coulomb from one point to another point in an electric field is 200 joules. What is the potential difference between the two points? (1) 0.0002 V (2) 8 V (3) 200 V (4) 5000 V

4. If 4 joules of work are required to move 2 coulombs of charge through a 6-ohm resistor, the potential difference across the resistor is (1) 1 V (2) 2 V (3) 6 V (4) 8 V

5. When 20. coulombs of charge pass a given point in a conductor in 4.0 seconds, the current in the conductor is (1) 80. A (2) 0.20 A (3) 16 A (4) 5.0 A

6. Which graph best represents the relationship between the current in a metallic conductor and the applied potential difference?

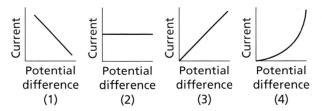

7. A potential difference of 12 volts is applied across a circuit which has a 4.0-ohm resistance. What is the magnitude of the current in the circuit? (1) 0.33 A (2) 48 A (3) 3.0 A (4) 4.0 A

8. The ratio of the potential difference across a metallic conductor to the current in the conductor is known as (1) potential drop (2) conductivity (3) resistance (4) electromagnetic force

9. If the current in a wire is 2.0 amperes and the potential difference across the wire is 10. volts, what is the resistance of the wire? (1) 5.0 Ω (2) 8.0 Ω (3) 12 Ω (4) 20. Ω

10. Which condition must exist between two points in a conductor in order to maintain a flow of charge? (1) a potential difference (2) a magnetic field (3) a low resistance (4) a high resistance

11. As the temperature of a metal conductor is reduced, the resistance of the conductor will (1) decrease (2) increase (3) remain the same

12. The resistance of a metallic wire conductor is inversely proportional to its (1) tensile strength (2) cross-sectional area (3) length (4) temperature

13. If the cross-sectional area of a fixed length of wire were decreased, the resistance of the wire would (1) decrease (2) increase (3) remain the same

14. A piece of wire has a resistance of 8 ohms. A second piece of wire of the same composition, diameter, and temperature, but one-half as long as the first wire, has a resistance of (1) 8 Ω (2) 2 Ω (3) 16 Ω (4) 4 Ω

15. If energy is used in an electric circuit at the rate of 20. joules per second, then the power supplied to the circuit is (1) 5.0 W (2) 20. W (3) 25 W (4) 100. W

16. What is the current in a 1200-watt heater operating on 120 volts? (1) 0.10 A (2) 5.0 A (3) 10. A (4) 20. A

17. The potential difference across a 100.-ohm resistor is 4.0 volts. What is the power dissipated in the resistor? (1) 0.16 W (2) 25 W (3) 4.0×10^2 W (4) 4.0 W

18. Identical resistors (R) are connected across the same 12-volt battery. Which circuit uses the greatest power?

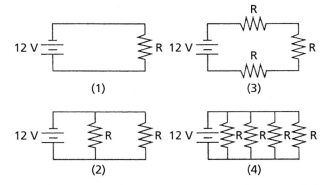

19. Two identical resistors connected in parallel have an equivalent resistance of 40. ohms. What is the resistance of each resistor? (1) 20. Ω (2) 40. Ω (3) 80. Ω (4) 160 Ω

Base your answers to questions 20 through 22 on the information and diagram below.

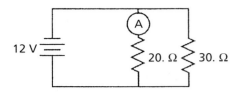

A 20.-ohm resistor and a 30.-ohm resistor are connected in parallel to a 12-volt battery as shown. An ammeter is connected as shown.

20. What is the equivalent resistance of the circuit? (1) 10. Ω (2) 12 Ω (3) 25 Ω (4) 50. Ω

21. What is the current reading of the ammeter? (1) 1.0 A (2) 0.60 A (3) 0.40 A (4) 0.20 A

22. What is the power of the 30.-ohm resistor? (1) 4.8 W (2) 12 W (3) 30. W (4) 75 W

23. The diagram below shows a circuit with two resistors.

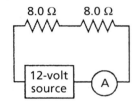

What is the reading on ammeter A? (1) 1.3 A (2) 1.5 A (3) 3.0 A (4) 0.75 A

24. The diagram below represents currents in a segment of an electric circuit.

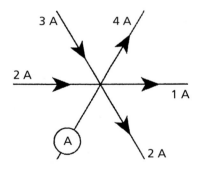

What is the reading of ammeter A? (1) 1 A (2) 2 A (3) 3 A (4) 4 A

25. A 100.-ohm resistor and an unknown resistor are connected in series to a 10.0-volt battery. If the potential drop across the 100.-ohm resistor is 4.00 volts, the resistance of the unknown resistor is (1) 50.0 Ω (2) 100. Ω (3) 150. Ω (4) 200. Ω

26. In the following circuit diagram, ammeter A₁ reads 10. amperes.

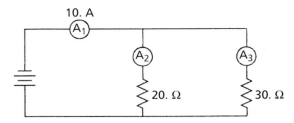

What is the reading of ammeter A₂? (1) 6.0 A (2) 10. A (3) 20. A (4) 4.0 A

Base your answers to questions 27 and 28 on the circuit diagram below, which shows two resistors connected to a 24-volt source of potential difference.

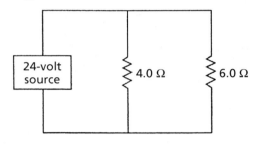

27. Copy the diagram then, using the appropriate circuit symbol, indicate a correct placement of a voltmeter to determine the potential difference across the circuit.

28. What is the total resistance of the circuit? (1) 0.42 Ω (2) 2.4 Ω (3) 5.0 Ω (4) 10. Ω

29. A long copper wire was connected to a voltage source. The voltage was varied and the current through the wire measured, while temperature was held constant. The collected data are represented by the graph below.

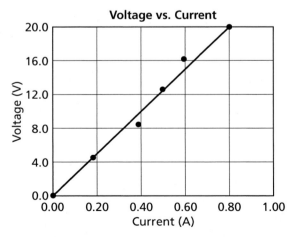

Using the graph, determine the resistance of the copper wire.

30. The diagram below shows two compasses located near the ends of a bar magnet. The north pole of compass *X* points toward end *A* of the magnet.

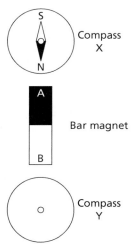

On your own paper, draw the correct orientation of the needle of compass *Y* and label its polarity.

Base your answers to questions 31 through 34 on the information, circuit diagram, and data table below.

In a physics lab, a student used the circuit shown to measure the current through and the potential drop across a resistor of unknown resistance, *R*. The instructor told the student to use the switch to operate the circuit only long enough to take each reading. The student's measurements are recorded in the data table.

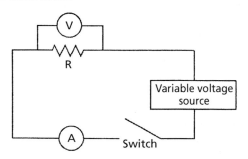

Data Table

Current (A)	Potential Drop (V)
0.80	21.4
1.20	35.8
1.90	56.0
2.30	72.4
3.20	98.4

Using the information in the data table, construct a graph as follows:

31. Mark appropriate scale on the axis labeled "Potential Drop (V)" and "Current (I)."

32. Plot the data points for potential drop versus current.

33. Draw the line or curve of best fit.

34. Calculate the slope of the line or curve of best fit. Show all work, including the equation and substitution with units.

Base your answers to questions 35 through 37 on the information and graph below.

A student conducted an experiment to determine the resistance of a lightbulb. As she applied various potential differences to the bulb, she recorded the voltages and corresponding currents and constructed the graph below.

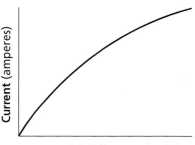

Current vs. Potential Difference

35. The student concluded that the resistance of the lightbulb was not constant. What evidence from the graph supports the student's conclusion?

36. According to the graph, as the potential difference increased, the resistance of the lightbulb (1) decreased (2) increased (3) changed, but there is not enough information to know which way

37. While performing the experiment the student noticed that the lightbulb began to glow and became brighter as she increased the voltage. Of the factors affecting resistance, which factor caused the greatest change in the resistance of the bulb during her experiment?

38. The diagram below represents a wire conductor, *RS*, positioned perpendicular to a uniform magnetic field directed into the page.

```
              R
  X   X  ⌂  X   X  Magnetic
  X   X  │  X   X  field
  X   X  │  X   X  directed
  X   X  ⌄  X   X  into the page
              S
```

Describe the direction in which the wire could be moved to produce the maximum potential difference across its ends, R and S.

39. What is the magnitude of the charge, in coulombs, of a lithium nucleus containing three protons and four neutrons?

40. A light bulb attached to a 120.-volt source of potential difference draws a current of 1.25 amperes for 35.0 seconds. Calculate how much electrical energy is used by the bulb.

41. A student is given two pieces of iron and told to determine if one or both of the pieces are magnets. First, the student touches an end of one piece to one end of the other. The two pieces of iron attract. Next, the student reverses one of the pieces and again touches the ends together. The two pieces attract again. What does the student definitely know about the initial magnetic properties of the two pieces of iron?

Base your answers to questions 42 through 46 on the information and data table below.

A variable resistor was connected to a battery. As the resistance was adjusted, the current and power in the circuit were determined. The data are recorded in the table below.

Current (A)	Power (W)
0.75	2.27
1.25	3.72
2.25	6.75
3.00	9.05
4.00	11.9

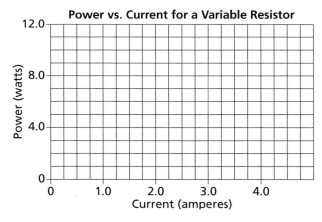

Power vs. Current for a Variable Resistor

42. Plot the data points as a graph for power versus current.

43. Draw the best-fit line.

44. Using your graph, determine the power delivered to the circuit at a current of 3.5 amperes.

45. Calculate the slope of the graph. Show all calculations, including the equation and substitution with units.

46. What is the physical significance of the slope of the graph?

Base your answers to questions 47 and 48 on the following diagram, which represents a source connected to two large, parallel metal plates. The electric field intensity between the plates is 3.75×10^4 newtons per coulomb; the distance between the plates is 4.00×10^{-3} m. The relationship between electric field intensity (E), voltage (V), and distance between the plates (d) is given by $E = \dfrac{V}{d}$.

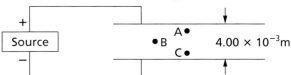

47. What is the potential difference of the source?

48. What would be the magnitude of the electric force on a proton at point A?

Base your answers to questions 49 and 50 on the information below.

A proton starts from rest and gains 8.35×10^{-14} joule of kinetic energy as it accelerates between points A and B in an electric field.

49. What is the final speed of the proton?
(1) 7.07×10^6 m/s   (2) 1.00×10^7 m/s
(3) 4.28×10^8 m/s (4) 5.00×10^{13} m/s

50. Calculate the potential difference between points A and B in the electric field. Show all work, including the equation and substitution with units.

Base your answers to questions 51 and 52 on the information below.

A lightweight sphere hangs by an insulating thread. A student wishes to determine if the sphere is neutral or electrostatically charged. She has a negatively charged hard rubber rod and a positively charged glass rod. She does not touch the sphere with the rods, but runs tests by bringing them near the sphere one at a time.

51. Describe the test result that would prove the sphere is neutral.

52. Describe the test result that would prove the sphere is positively charged.

Base your answers to questions 53 through 56 on the information and data table below.

An experiment was performed using various lengths of a conductor of uniform cross-sectional area. The resistance of each length was measured and the data recorded in the table below.

Length (m)	Resistance (Ω)
5.1	1.6
11.0	3.8
16.0	4.6
18.0	5.9
23.0	7.5

Using the information in the data table, construct a graph following the directions below.

53. Mark appropriate scales on the axis labeled "Length (m)" and "Resistance (Ω)".

54. Plot the data points for resistance versus length.

55. Draw the best-fit line.

56. Calculate the slope of the best-fit line. Show all work, including the equation and substitution with units.

Base your answers to questions 57 through 59 on the information below.

An 18-ohm resistor and a 36-ohm resistor are connected in parallel with a 24-volt battery. A single ammeter is placed in the circuit to read its total current.

57. Draw a diagram of this circuit using appropriate symbols

58. Calculate the equivalent resistance of the circuit. Show all work, including the equation and substitution with units.

59. Calculate the total power dissipated in the circuit. Show all work, including the equation and substitution with units.

Base your answers to questions 60 and 61 on the information below.

A 1.00-meter length of nichrome wire with a cross-sectional area of 7.85×10^{-7} meter^2 is connected to a 1.50-volt battery.

60. Calculate the resistance of the wire. Show all work, including the equation and substitution with units.

61. Determine the current in the wire.

Base your answers to questions 62 and 63 on the following information.

An electric circuit contains two 3.0-ohm resistors connected in parallel with a battery. The circuit also contains a voltmeter that reads the potential difference across one of the resistors.

62. On your own paper draw a diagram of this circuit, using appropriate symbols.

63. Calculate the total resistance of the circuit. Show all work, including the equation and substitution with units.

Base your answers to questions 64 and 65 on the information below.

A toaster having a power rating of 1050 watts is operated at 120. volts.

64. Calculate the resistance of the toaster. Show all work, including the equation and substitution with units.

65. The toaster is connected in a circuit protected by a 15-ampere fuse. (The fuse will shut down the circuit if it carries more than 15 amperes.) Is it possible to simultaneously operate the toaster and a microwave oven that requires a current of 10.0 amperes on this circuit? Justify your answer mathematically.

66. An electron is accelerated through a potential difference of 2.5×10^4 volts in the cathode ray tube of a computer monitor. Calculate the work, in joules, done on the electron. Show all work, including the equation and substitution with units.

Base your answers to questions 67 through 71 on the information and data table below.

Three lamps were connected in a circuit with a battery of constant potential. The current, potential difference, and resistance for each lamp are listed in the data table below. There is negligible resistance in the wires and the battery.

Lamp	Current (A)	Potential Difference (V)	Resistance (Ω)
1	0.45	40.1	89
2	0.11	40.1	365
3	0.28	40.1	143

67. Using appropriate circuit symbols, draw a circuit showing how the lamps and battery are connected.

68. What is the potential difference supplied by the battery?

69. Calculate the equivalent resistance of the circuit. Show all work, including the equation and substitution with units.

70. If lamp 3 is removed from the circuit, what would be the value of the potential difference across lamp 1 after lamp 3 is removed?

71. If lamp 3 is removed from the circuit, what would be the value of the current in lamp 2 after lamp 3 is removed?

72. Your school's physics laboratory has the following equipment available for conducting experiments:

accelerometers	batteries
ammeters	electromagnets
bar magnets	lasers
light bulbs	stopwatches
meter sticks	thermometers
power supplies	voltmeters
spark timers	wires

Explain how you would find the resistance of an unknown resistor in the laboratory. Your explanation must include:

(a) measurements required

(b) equipment needed

(c) complete circuit diagram

(d) any equation(s) needed to calculate the resistance

Enrichment
Electricity and Magnetism

CURRENTS AND MAGNETS

Currents Act on Magnets

The magnetic field created by a current is best described by field lines. Magnetic field lines in the vicinity of a straight current-carrying wire can be represented as concentric circles around the wire, with each circle forming a plane perpendicular to the current. The direction of the arrows on these circles is found by applying *hand rule #1*. Imagine that you grasp the wire with your right hand, with the thumb pointing in the direction of the current (that is, the conventional current). The four fingers wrapped around the wire then point in the direction of the arrows to be placed on the circular field lines. (See Figure 3E-1.)

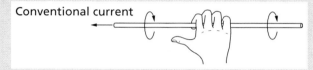

Figure 3E-1. Hand rule #1.

Inside a solenoid or loop, the field lines are parallel to each other and perpendicular to the circles of current. The direction of the magnetic field lines there is found by applying *hand rule #2*. The four fingers of the right hand are made to follow the conventional current flow; the thumb then points in the direction of the field lines *inside* the coil (Figure 3E-2).

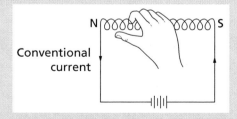

Figure 3E-2. Hand rule #2.

The magnetic field strength of a solenoid is directly proportional to the number loops, or turns, of wire and to the current.

Magnets Act on Currents

A straight current-carrying wire situated in a magnetic field experiences a force if the charges move perpendicularly to the field, or at least a component of their motion is perpendicular to the field. No force appears if the direction of the current is parallel to the field. The force is greatest when the angle between the field lines and the current is exactly 90°.

The force exerted on the wire is perpendicular to both the direction of the current and the direction of the magnetic field. The direction of this force is found by using *hand rule #3*. Hold your right hand flat with the fingers pointing in the direction of the field lines and the thumb pointing in the direction of the conventional current. The palm of your hand pushes in the direction of the force on the current-carrying wire. (See Figure 3E-3.)

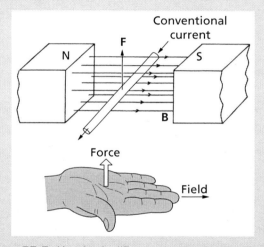

Figure 3E-3. Hand rule #3.

For example, to find the direction of the force on the downward-moving current between two magnetic poles in Figure 3E-4, proceed as follows:

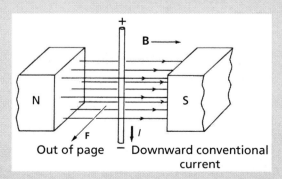

Figure 3E-4.

Point the thumb of the right hand downward in the direction of current; then point your fingers in the direction of the magnetic field, to the right. The palm of the hand now indicates that the force is directed out of the page, toward the reader.

Sample Problems

E1. In what direction will the N-pole of a magnet point if it is placed behind a current directed to the right as shown in Figure 3E-5?

Figure 3E-5.

Solution: Using hand rule #1, the thumb of the right hand points to the right and the fingers curve around the wire such that they point *upward* behind the current.

E2. In what direction will the N-pole of a magnet point if it is placed at point *A* in Figure 3E-6?

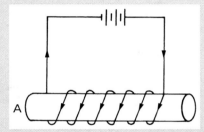

Figure 3E-6.

Solution: Using hand rule #2, the fingers of the right hand curve in the direction of the current and the thumb points *to the right*.

E3. In what direction will the current-carrying wire in Figure 3E-7 be pushed?

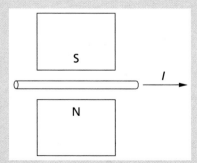

Figure 3E-7.

Solution: Using hand rule #3, the thumb of the right hand points to the right, the fingers point up, toward the top of the page, the palm of the hand pushes *out of the page, toward the reader*.

Currents Act on Currents

Since currents create magnetic fields and magnetic fields exert forces on currents, it follows that one current exerts a force on another. (No electric force exists between current-carrying wires, since they are electrically neutral.)

Let us analyze the case of two long parallel wires carrying current to the left, segments of which are shown in Figure 3E-8. The magnetic field created by the upper current is found via hand rule #1. With the thumb of the right hand pointing to the left, the four fingers wrap around the wire in such a way that they point into the page and away from the reader above the wire, out of the page and toward the reader below the wire.

The lower current is thus situated in a magnetic field that is directed out of the page, toward the reader, as shown in the diagram.

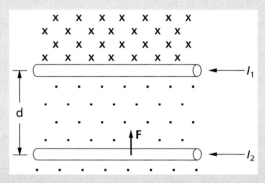

Figure 3E-8. The field created by the upper current-carrying wire is into the page (indicated by x's) above the wire and out of the page (indicated by dots) below the wire. Hand rule #3 is used to find the direction of the force exerted on the lower wire by this field.

Hand rule #3 provides the direction of the force exerted by the field created by the upper wire on the lower wire. With the thumb of the right hand pointing to the left and the fingers perpendicular to the page, pointing toward the reader, the palm of the hand pushes upward, in the plane of the page. The lower current thus experiences a force upward, toward the upper current.

A similar analysis of the force exerted on the upper wire by the magnetic field created by the lower wire would reveal that there is a downward force on the upper wire (toward the lower wire). The net result is that the two wires are attracted to each other.

Each moving charge creates a magnetic field whose direction can be found by applying either hand rule #1 or #2, and the field so created exerts a force on the other moving charge. The direction of this force can be found by using hand rule #3. The intensity of the field created by a moving charge is proportional to the amount of charge and to the velocity of the charge. Intensity weakens with distance from the moving charge.

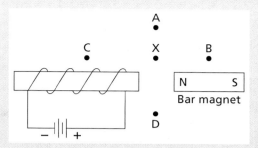

Questions

E1. Which diagram best represents the direction of the magnetic field around a wire conductor in which the current is moving as indicated? (The X's indicate that the field is directed into the paper and the dots indicate that the field is directed out of the page.)

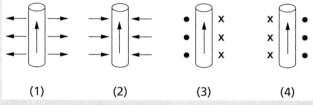

(1)   (2)   (3)   (4)

E2. Which diagram best represents the magnetic field around a current-carrying conductor?

E3. The diagram represents a current-carrying loop of wire. The direction of the magnetic field at point *P* is (1) toward the bottom of the page (2) to the right (3) into the page (4) out of the page

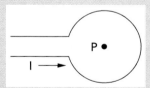

E4. Current flows in a loop of wire as shown in the diagram. What is the direction of the magnetic field at point *A*? (1) into the paper (2) out of the paper (3) toward the left (4) toward the right

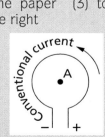

E5. In the diagram, in which direction is the magnetic field at point *X*? (1) toward *A* (2) toward *B* (3) toward *C* (4) toward *D*

Base your answers to questions E6 through E8 on the following diagram, which represents a cross section of an operating solenoid. A compass is located at point C.

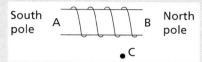

E6. Which diagram best represents the shape of the magnetic field around the solenoid?

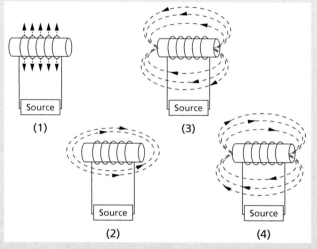

(1)   (3)

(2)   (4)

E7. Which shows the direction of the compass needle at point *C*?

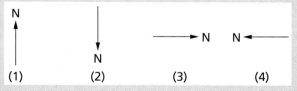

(1)   (2)   (3)   (4)

E8. If *B* is the north pole of the solenoid, which diagram best represents the direction of curren in one of the wire loops?

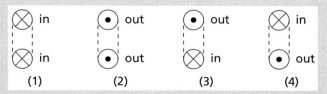

(1)   (2)   (3)   (4)

Base your answers to questions E9 through E12 on the following diagram, which represents a circuit containing a solenoid on a *cardboard tube, a variable resistor R, and a source of potential difference.*

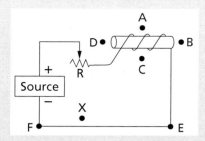

E9. The north pole of the solenoid is nearest to point (1) *A* (2) *B* (3) *C* (4) *D*

E10. Due to the current in the *FE* section of the circuit, the direction of the magnetic field at point *X* is (1) into the page (2) out of the page (3) to the left (4) to the right

E11. If the resistance of resistor *R* is increased, the magnetic field strength of the solenoid will (1) decrease (2) increase (3) remain the same

E12. If the number of turns in the solenoid is increased and the current is kept constant, the magnetic field strength of the solenoid will (1) decrease (2) increase (3) remain the same

E13. Two long, straight parallel conductors carry equal currents and are spaced 1 meter apart. If the current in each conductor is doubled, the magnitude of the magnetic force acting between the conductors will be (1) unchanged (2) doubled (3) halved (4) quadrupled

E14. A rectangular loop of wire is moving perpendicularly to a magnetic field directed out of the page, as illustrated below. For each of the following steps, state whether or not current is induced in the loop and in what direction the current, if any, flows. (a) The loop has not yet entered the field. (b) The loop has partially entered the field. (c) The loop is entirely inside the field, moving across its field lines. (d) The loop has partially exited the field. (e) The loop has entirely exited the field.

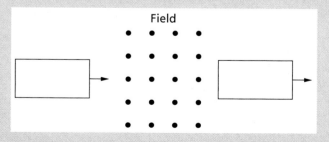

Field

E15. The diagram represents a rectangular conducting loop *ABCD* being moved to the right through a magnetic field, which is directed into the page. Which path will the induced current follow?

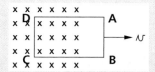

(1) DCBAD   (2) DABCD   (3) CBDAC
(4) ADBCA

Base your answers to questions E16 through E19 on the following diagram, which shows a cross section of a wire (A) moving down through a uniform magnetic field (B). The flux density of the field is 5.0 newtons per ampere-meter. The wire is 1.0 meter long and has a velocity of 2.0 meters per second perpendicular to the magnetic field.

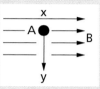

E16. What is the direction of the magnetic force on the electrons in the wire? (1) toward *x* (2) toward *y* (3) into the page (4) out of the page

E17. What is the direction of the magnetic force on the wire due to the induced current in the wire? (1) toward *x* (2) toward *y* (3) into the page (4) out of the page

E18. What is the potential difference across the ends of the wire? (1) 1.6×10^{-18} V (2) 2.5 V (3) 70. V (4) 10. V

E19. The maximum potential difference will be induced across the wire when the angle between the direction of the motion of the wire and the direction of the magnetic field is (1) 0° (2) 45° (3) 90° (4) 180°

Intensity of a Magnetic Field

The magnitude of the force exerted on a straight current-carrying wire situated in a particular magnetic field is proportional to the current in the wire (assuming it is perpendicular to the field lines) and the length of the wire in the field. This is stated mathematically as

$$F = BIL$$

The constant *B* is different for different magnetic fields. Its value is equal to the force that a particular magnetic field exerts on a 1-meter-long wire carrying one amp of current (if the wire is perpendicular to the magnetic field). One magnetic field will exert a stronger force on such a wire than another field only if the field is more intense. Thus the magnitude of *B* represents the intensity of the field. The unit for B, or **magnetic field intensity**, is the newton per ampere-meter (N/A•m).

By convention, the concentration of field lines (the number of lines per unit surface area drawn perpendicularly to the lines) in a region in a magnetic field is made equal to the intensity of the field (the magnitude of B) in that region. Magnetic field lines are sometimes called **flux lines** or webers. The number of flux lines per unit area is the **flux density**, expressed in units of webers per square meter (Wb/m²). The magnetic field intensity in N/A•m is thus identical to the concentration of field lines in Wb/m². Both units can be used for the magnitude of B (1 Wb/m² = 1 N/A•m). Recently the N/A•m and the Wb/m² have been renamed the tesla, symbolized by T.

Sample Problem

E4. An 8-cm segment of a current-carrying wire is situated in a magnetic field that is directed out of the page, toward the reader, as illustrated in Figure 3E-9. The direction of the current is to the left and the rate of flow is 20.0 A. A force of 6.4 newtons is found to act on the wire. What is the direction of the force on the wire and the intensity of the magnetic field?

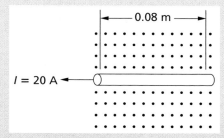

Figure 3E-9. The dots indicate that the magnetic field is directed out of the page.

Solution: The direction of the force is found by applying hand rule #3. Point the thumb of the right hand to the left in the direction of the current and the fingers in the direction of

the magnetic field, perpendicular to the page and toward you. The palm of the hand then indicates that the force is directed upward, in the plane of the page.

The intensity of the field is found by applying the formula

$$F = BIL$$

$$B = \frac{F}{IL}$$

Change 8 cm to 0.08 m.

$$B = \frac{6.4 \text{ N}}{(20.0 \text{ A})(0.08 \text{ m})} = 4 \text{ N/A} \cdot \text{m}$$

or 4 Wb/m²

or 4 Tesla

Electromagnetic Waves

The electric and magnetic fields in the vicinity of an accelerating charge change in magnitude and direction as the position and velocity of the charge continue to change. Since a changing magnetic field creates an electric field and a changing electric field creates a magnetic field, an endless chain of fields is generated by the accelerating charge.

The net result is that electric and magnetic fields propagate away from the vicinity of an accelerating charge. This propagation continues even after the charge stops accelerating because the fields continue to generate each other. We refer to this as **electromagnetic radiation**.

If a charge oscillates back and forth between two fixed points (in the process accelerating and decelerating) and the frequency of oscillation is constant, a periodic *electromagnetic wave* of the same frequency is radiated outward from the vicinity of the charge. The magnitude and direction of the electric and magnetic fields vary from point to point in wavelike fashion. At any particular point in the wave, the electric and magnetic fields are perpendicular to each other and to the direction of propagation of the wave. (Figure 3E-10).

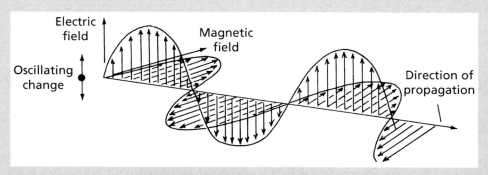

Figure 3E-10. Electromagnetic wave.

Electromagnetic waves propagate through empty space at the speed of light, c, or 3×10^8 meters per second.

Questions

Base your answers to questions E20 through E23 on the following diagram, which represents a U-shaped wire conductor positioned perpendicular to a uniform magnetic field that is directed into the page. AB represents a second wire, which is free to slide along the U-shaped wire. The length of wire AB is 1 meter, and the magnitude of the magnetic field is 8.0 webers/meter².

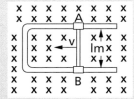

E20. If wire *AB* is moved to the left at a constant speed, the direction of the induced electron motion in wire *AB* will be (1) toward *A*, only (2) toward *B*, only (3) first toward *A* and then toward *B* (4) first toward *B* and then toward *A*

E21. If wire *AB* is moved to the left with a constant speed of 10. meters per second, the potential difference induced across wire *AB* will be (1) 0.8 V (2) 8.0 V (3) 10 V (4) 80 V

E22. Wire *AB* is moved at a constant speed to the left. The current induced in the conducting loop will produce a force on wire *AB* which acts (1) to the right (2) to the left (3) into the page (4) out of the page

E23. The resistance of wire *AB* is increased, and the wire is moved to the left at a constant speed of 10 meters per second. Compared to the induced potential difference before the resistance was increased, the new potential difference will be (1) less (2) greater (3) the same

E24. Which is the unit of magnetic flux in the MKS system? (1) Weber (2) joule (3) coulomb (4) newton per ampere-meter

E25. Magnetic flux density may be measured in (1) N/m² (2) Wb/m² (3) C/m² (4) J/m²

E26. If the current in the solenoid is doubled and the number of turns halved, the magnetic field strength of the solenoid will (1) decrease (2) increase (3) remain the same

E27. Electromagnetic radiation can be produced by charged particles that are (1) held sta-tionary in a uniform magnetic field (2) held stationary in an electric field (3) moving at a constant velocity (4) being accelerated

E28. Electromagnetic radiations such as radio, light, and gamma are propagated by the in-terchange of energy between (1) magnetic fields, only (2) electric fields, only (3) elec-tric and gravitational fields (4) electric and magnetic fields

MORE ELECTROMAGNETIC APPLICATIONS

The Galvanometer (–Ⓖ–)

A galvanometer is used to measure small amounts of current. It consists of a coil-shaped wire placed between the opposite poles of a permanent mag-net. When the current flows through the coil, the field between the poles exerts a torque, forcing the coil to rotate against a spring (see Figure 3E-11). If the spring were not there, the coil would rotate until the plane of the wire was perpendicular to the magnetic field. The spring prevents this from hap-pening. It counteracts the torque from the mag-netic field with a torque of its own. The coil rotates until the opposing torque exerted by the spring be-comes equal to the torque exerted by the field.

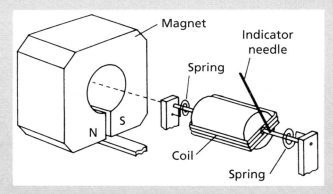

Figure 3E-11. A galvanometer.

The resultant deflection of the coil depends on the amount of current; the greater the current in the coil, the greater the deflection. Typically, an in-dicator needle is attached to the coil. By reading the amount of deflection of the needle one can de-termine the amount of current in the coil.

When the current ceases to flow, the coil is forced back to its starting position by the action of the spring, with the plane of the coil parallel to the field. The galvanometer is then ready to be used once again.

The Ammeter (–Ⓐ–)

An ammeter is a modified galvanometer and is used to measure larger amounts of current. In an ammeter the coil is connected in parallel with a *shunt*, a material whose resistance is much smaller than that of the galvanometer coil (see Figure 3E-12). Most of the current into the ammeter thus bypasses the coil and flows through the shunt. Since the resistance of the coil and shunt are known, the total current through the ammeter can be determined from the current in the coil, which in turn is determined from the deflection of the needle.

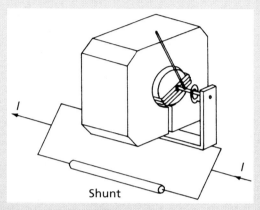

Figure 3E-12. An ammeter.

To measure current, the ammeter is placed in series with the circuit (or part of the circuit) whose current is to be measured. Since the total resistance of the ammeter is very small, its insertion into the circuit changes the current only negligibly (see Figure 3E-13).

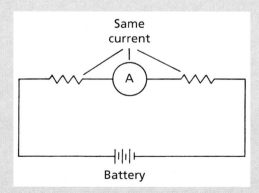

Figure 3E-13. Because of its low resistance, the ammeter does not noticeably affect the current in the circuit.

Though an ammeter can measure more current than a galvanometer, there is still a limit to the amount of current an ammeter can measure. Once the plane of the galvanometer coil has been twisted 90° and is oriented perpendicularly to the external field, increasing the current no longer produces more rotation. The only way to raise this maximum for any particular ammeter is to decrease the resistance of the shunt. Doing so diverts more of the current to the shunt and away from the coil.

The Voltmeter (–Ⓥ–)

A voltmeter is also a modified galvanometer and is used to measure potential difference. It consists of a galvanometer coil connected *in series* with a high-resistance material (see Figure 3E-14). Since the total resistance of the voltmeter is known, the voltage across voltmeter can be determined from the current through the galvanometer (see "Series Circuits," pp. 100–102), which in turn is determined from the deflection of the needle.

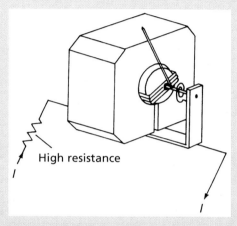

Figure 3E-14. A voltmeter.

To measure potential difference, a voltmeter is connected in parallel with the part of the circuit whose voltage is to be measured. Since the total resistance of the voltmeter is very high, it draws little current, and its insertion in parallel has a negligible effect on the circuit (see Figure 3E-15).

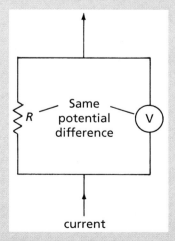

Figure 3E-15. Because of its high resistance, the voltmeter does not noticeably affect the current in the circuit.

Just as there is a limit to the current an ammeter can read, so every voltmeter has a maximum potential difference it can measure. This maximum can be raised for any particular voltmeter by increasing the resistance of the material in series with the coil. The greater that resistance, the smaller the deflection of the coil in the magnetic field.

Questions

E29. Doubling the current in a loop of wire situated in a magnetic field (1) leaves the torque unchanged (2) halves the torque (3) doubles the torque (4) quadruples the torque

E30. A galvanometer measures (1) small amounts of current (2) large amounts of resistance (3) large amounts of current (4) small amounts of power

E31. The torque produced by the magnetic field in a galvanometer on the coil is opposed by (1) gravity (2) tension in the spring (3) an electric force (4) another magnet

E32. When no current exists in a galvanometer, the plane of the coil should be (1) parallel to the field (2) turned 45 degrees from the field (3) perpendicular to the field (4) any one of the above

E33. Ammeters are used to measure (1) small amounts of current (2) large potential differences (3) larger amounts of current (4) small amounts of power

E34. To perform its mission, an ammeter is connected (1) in series with the circuit (2) either in series or in parallel with the circuit (3) in parallel with the circuit

E35. The total resistance of an ammeter (1) must be very large (2) must be very small (3) could be any amount

E36. To raise the maximum current an ammeter can read, the shunt resistance should be (1) decreased (2) made equal to that of the coil (3) increased

E37. An ammeter is a galvanometer with a (1) low resistance in series (2) low resistance in parallel (3) high resistance in series (4) high resistance in parallel

E38. A voltmeter is a galvanometer with a (1) low resistance in series (2) low resistance in parallel (3) high resistance in series (4) high resistance in parallel

E39. The purpose of the shunt in an ammeter is to provide (1) electrostatic deflection of the coil (2) magnetic deflection of the coil (3) resistance to current flow (4) a path for some current to bypass the coil.

E40. To function properly, a voltmeter is connected (1) in series with the circuit (2) in parallel with the circuit (3) either in series or in parallel with the circuit

E41. The total resistance of a voltmeter (1) must be very large (2) must be very small (3) could be any amount

E42. The reading of the ammeter in the diagram should be recorded as (1) 1 A (2) 0.76 A (3) 0.55 A (4) 0.5 A

E43. One milliampere produces a full-scale deflection in a galvanometer whose internal resistance is 50 ohms. To convert this instrument into an ammeter whose full-scale deflection is 1 ampere, it should be shunted with a resistance of approximately (1) 0.005 Ω (2) 0.05 Ω (3) 0.5 Ω (4) 5.0 Ω

Motors

Motors operate on the principle that a current-carrying loop in a magnetic field experiences a torque, or turning force. In its most simple form, a motor is composed of a wire loop situated between opposite magnetic poles. One end of the loop is connected to the positive terminal of a battery, and the other end is connected to the negative terminal. (See Figure 3E-16.)

Current flows into the page on the left side of the loop and out of the page on the right side. Since the magnetic field lines point to the right, the right side of the loop (segment A) is pushed up and the left side (segment B) is pushed down—as dictated by hand rule #3. The loop is thus made to rotate counterclockwise.

To keep the loop rotating in one direction, the current must be reversed every half-turn. Otherwise, the current in segment A would continue to flow out of the page and the force exerted on it would continue to act in the upward direction. When segment A arrives on the left side, this force would act to reverse the rotation of the loop by turning it clockwise.

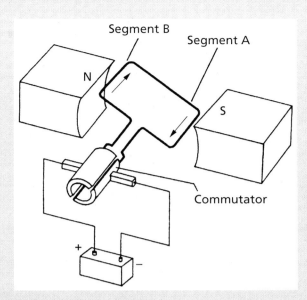

Figure 3E-16.

To prevent this from happening, each end of the wire loop is connected to a conducting material in the shape of a half-ring. Each half-ring makes contact with a wire leading to one of the battery's terminals by rubbing against a brush. This way, when wire segment A gets to the left side, its half-ring makes contact with the other brush, the one that leads to the opposite terminal of the battery. As a result, the current through segment A is reversed and now flows into the page. The force on it now points downward, and the rotation of the loop continues to be counterclockwise. A similar reversal in the current's direction occurs in wire segment B when it gets to the right side. This arrangement is known as a **split-ring commutator**.

Motors usually contain many wire loops that together constitute the *armature* of the motor. Typically, the wire loops are wrapped around a core of "permeable" material, such as soft iron, that becomes magnetized when current flows around it. The magnetic field of the magnetized iron adds to the flux density (or intensity) of the external field. This strengthens the torque acting to rotate the coils.

Back EMF Once the current flows through the armature of a motor and the coil begins rotating, the motion of the wires across the field lines leads to an induced current. (Recall electromagnetic induction, page 109). This induced current is opposite in direction to the current that makes the armature rotate in the first place.

For example, consider wire segment A in Figure 3E-16 as it pushed upward because the current through it is directed out of the page. As soon as it starts moving upward, carrying its load of charged particles along with it, a magnetic force appears that acts to create current *into the page*. To see why this is so, use hand rule #3 again. Point the fingers to the right, in the direction of the field, and the thumb upward, in the direction of the moving protons as they are carried upward by the moving wire. The palm of the hand then pushes into the page, indicating the direction of the force exerted on the protons by the field as a result of the wire's motion.

The force responsible for this oppositely directed induced current is referred to as **back EMF**. It is an example of *Lenz's law*, which states that *all magnetic effects lead to forces that oppose the change that produced the effect*. You will learn more about Lenz's law in the next section. The magnetic effect in this case is the rotation of the wire loop. The magnetically produced rotation leads to a force that opposes the current, the "change" that led to the rotation. As a result of the oppositely-directed induced current, the actual rate of flow of charge in the wire (the conventional current) is reduced.

The existence of the back EMF also guarantees that the Law of Conservation of Energy is obeyed. Had there been no back EMF, the rotational kinetic energy of the rotating armature would have been "free"—there would have been no equivalent loss of energy elsewhere in the system. Once the current is reduced as a result of the motion of the armature, less heat is generated in the coils. Less of the battery's chemical energy is converted into heat, and more is converted into rotational kinetic energy. The gain in kinetic energy is balanced by a reduction in heat energy generated by the current.

Questions

E44. As a torque causes the current-carrying loop in an electric motor to begin rotating, the current in that loop (1) decreases (2) increases (3) remains the same

E45. A current-carrying loop of wire in a magnetic field is forced to (1) move perpendicularly to the field (2) rotate (3) move parallel to the field

E46. The current in the armature of a dc motor (1) flows steadily in one direction (2) varies in magnitude in one direction (3) alternates in direction

E47. The current in the armature of an electric motor switches direction with each rotation. Which motor part produces this phenomenon? (1) magnet (2) split-ring commutator (3) armature (4) stator

E48. The coil in an electric motor is made to rotate by (1) gravity (2) a nuclear force (3) an electric force (4) a magnetic force

E49. The wire loops in a motor are wrapped around soft iron to (1) increase the intensity of the magnetic field (2) increase the current in the loops (3) decrease the intensity of the magnetic field (4) decrease the current in the loops

E50. If the rotation of the coil in an electric motor is stopped while the motor is still connected to the battery, the current in the coil will (1) decrease (2) increase (3) remain the same

E51. The back EMF in an electric motor guarantees that which of the following laws is not violated by the operation of the motor? (1) action-reaction (2) conservation of momentum (3) universal gravitation (4) conservation of energy

Charged Particles in Magnetic Fields

If instead of a continuous train of moving charges in a conducting wire, a single charged particle moves across a magnetic field, the field exerts a force on the particle if its motion, or at least a component of its motion, is perpendicular to the field lines. No force is exerted if the particle travels parallel to the field. Hand rule #3 is still applicable, provided that the thumb points in the direction of the motion of the particle. If the particle is positively charged, as in the case of an alpha particle, the right hand is used. If the particle is negatively charged, as in the case of an electron, the left hand is used. In either case, the fingers point in the direction of the field, the thumb points in the direction of the motion of the charged particle (perpendicular to the field), and the push of the palm indicates the direction of the magnetic force (perpendicular to both the field and the motion of the particle.)

The magnitude of the force on an individual charged particle is provided by the formula

$$F = qvB$$

where F is the force, in newtons; q is the amount of charge in coulombs; v is the component of the velocity of the particle perpendicular to the magnetic field, in meters per second; and B is the intensity of the field in N/A•m, Wb/m², or tesla.

Sample Problems

E5. What is the direction and magnitude of the force exerted on an electron that moves to the left at the rate of 2×10^4 m/s across a 4.0-tesla magnetic field whose lines are perpendicular to the page and point away from the reader? (See Figure 3E-17.)

Figure 3E-17. An electron moving at 2×10^4 m/s to the left across a magnetic field with intensity 4.0 tesla. The field is directed into the page.

Solution: The direction of the force is found by applying hand rule #3 and using the left hand since the particle (an electron) is negatively charged. Point the four fingers of your left hand into the page (the direction of the field) and the thumb to the left (the direction of motion of the electron). The palm of your left hand now pushes upward, indicating that the direction of the magnetic force is upward, in the plane of the page, toward the top of the page.

The magnitude of the force is found by applying the formula

$$F = qvB$$

$$F = (1.6 \times 10^{-19} \text{ C})(2 \times 10^4 \text{ m/s})(4.0 \text{ T})$$

$$F = 1.28 \times 10^{-14} \text{ N}$$

E6. What is the magnitude and direction of the force exerted by the same magnetic field on a proton that travels perpendicularly to the page, toward the reader, at the same speed?

Solution: No force is exerted, since the charged particle is moving parallel to the field.

Thermionic Emission

When metallic substances are heated to incandescence, they emit electrons. This happens because, at high temperatures, the atoms are too energetic to be able to hold on to their outermost electrons. We refer to this phenomenon as **thermionic emission**—*therm* for heat, *ionic* for charged. It is also known as the "Edison effect" after its discoverer, Thomas Edison. As the temperature is increased, the rate of electron emission increases.

As electrons are emitted, however, there is a buildup of negative charge in the space around the incandescent material. This growing negative "space charge" makes it increasingly difficult for other electrons to escape from the material, since they are repelled back into the material. This places a limit on the number of electrons that can be emitted.

Electron Beams The limit can be overcome by placing a positively charged plate, or **anode**, near the negatively charged electron emitter (known as the **cathode**). The positive anode acts to pull emitted electrons toward itself and away from the cathode, thereby eliminating the building of negative charge and enabling other electrons to leave the cathode without difficulty. This results in a continuous beam of electrons traveling from the cathode to the anode.

As the electrons in the beam travel from the cathode to the anode, the electric field between the plates causes them to accelerate. The greater the potential difference between the cathode and anode, the greater the acceleration rate of the electrons toward the anode. Vacuum tubes in electronic devices usually consist of such cathode-anode arrangements.

The cathode-anode arrangement is but one of many devices used in laboratories today to accelerate charged particles to speeds approaching that of light. Such devices are called **particle accelerators**. The fast-moving particles are then aimed at various targets, and they bombard the target nuclei with great force. A great deal of information about the atomic and subatomic worlds has been obtained in this way.

Control of Electron Beams The path of an electron beam can be manipulated by placing an electric or magnetic field in the vicinity of the beam. In an electric field the beam is deflected by a force that is directed *parallel* to the field lines; in a magnetic field the beam is deflected by a force that is directed *perpendicularly* to the field lines and to the beam (as dictated by hand rule #3). The magnitude of the force exerted on each electron by an electric field is provided by the formula $F = Eq$ and that exerted by a magnetic field is provided by $F = qvB$.

Cathode Ray Tubes An evacuated tube that contains a source of electrons at one end (the cathode) and a fluorescent screen (coated with material that glows when electrons strike it) at the other end is referred to as a **cathode ray tube**. The "ray" emitted by the cathode is, of course, a beam of electrons. Between the ends of the tube are placed one or more pairs of plates arranged parallel to each other (see Figure 3E-18). The plates are connected to an outside source of charge, and varying amounts of charge can be imparted to them.

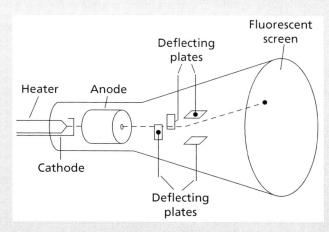

Figure 3E-18. Cathode ray tube.

As the electrons in the beam pass through the field between the plates, they are deflected. By varying the charge on the plates, the shape of the path of the beam can be manipulated and the glowing spot on the fluorescent screen (where the electrons strike) can be moved up, down, or sideways. The brightness of the spot can be varied by controlling the intensity of the electron beam (the rate of emission of electrons from the cathode). The television picture tube and the oscilloscope tube are advanced forms of the cathode ray tube.

Mass Spectroscopy If a beam of electrons is aimed at a gas, the ensuing collisions between the electrons and the molecules of the gas sometimes cause electrons to be knocked out of the gas molecules. The molecules thereby lose their electric neutrality and acquire a positive charge. When the positive ions are then subjected to electric and magnetic fields that act on them and the resultant deflection in their path is measured, their charge-to-mass ratio can be determined. If the charge on the ions is known (they can only be some whole number multiple of the elementary unit of charge), their mass can be calculated from the charge-to-mass ratio. In this manner, the mass of many different atomic species have been determined. The device used in this process is known as a **mass spectrometer**.

This procedure is also used to separate atoms of the same element with different masses (called "isotopes"). Their differing masses make them take divergent paths when charged and placed in electric and magnetic fields.

Mass of the Electron If a beam of electrons is accelerated by an electric field of known intensity, the kinetic energy of the electrons in the beam can be determined. If such a beam is then passed through a magnetic field of known intensity and the deflection of the beam is measured, the charge-to-mass ratio of the electrons can be determined.

Since the charge on an electron is known (1.6×10^{-19} C), knowing the electron's charge-to-mass ratio leads to knowledge of its mass. In this way it was determined that the mass of an electron is 9.11×10^{-31} kg.

The Laser If electrons accelerated by a large potential difference are made to collide with certain types of atoms, the electron-atom collisions stimulate the atoms to emit light. Light produced this way is of one color and usually very intense. A device that emits this kind of light is referred to as a **laser**, an acronym for Light Amplification by Stimulated Emission of Radiation.

In the helium-neon laser most commonly found in classrooms, the electrons are made to collide with helium atoms, and the excited helium atoms then collide with neon atoms. The neon atoms respond by emitting red light.

Induced Voltage

We already know that when a wire is moved perpendicularly across magnetic field lines, the field exerts a force on the charged particles in the wire, and if the conductor is part of a complete circuit, a current is induced in the wire as a result.

Since this force pushes the electrons through a distance (the length of the wire), the field does work on the electrons. The amount of work done, in joules, on every coulomb of charge between the ends of the wire segment in the field (the only place work is done) is known as the **induced potential difference**, or the **induced voltage** of the circuit. Sometimes it is referred to as the induced EMF (for electromotive force, though it represents work, not force). The symbol for induced potential difference is V.

The induced voltage, V, is provided by the formula

$$V = BLv$$

where B is the *magnetic field intensity*, in tesla; L is the length of the wire segment in the field, in meters, and v is the velocity of the wire (perpendicular to the field), in meters per second. The units for V are joules/coulomb, or volts.

If the wire is part of a closed circuit, the induced current is related to the induced voltage by Ohm's law, $V = IR$, where R is the resistance, in ohms, of the *entire* circuit. The power of the circuit in watts, is provided by the relationship $P = VI$.

Sample Problems

E7. What is the voltage and current induced in a 4.0 ohm circuit if a 0.5 meter segment of the circuit is moved perpendicularly to a 2 N/A•m field at the rate of 8 meters/second?

Solution:

$$V = BLv$$

$$V = \left(2.0 \frac{N}{A \cdot m}\right)(0.5 \text{ m})\left(8.0 \frac{m}{s}\right)$$

$$V = \frac{8.0 \text{ N} \cdot \text{m}}{A \cdot s} = \frac{8.0 \text{ N} \cdot \text{m}}{\frac{C}{s}} \cdot s = \frac{8.0 \text{ N} \cdot \text{m}}{C}$$

$$V = 8.0 \frac{\text{joules}}{\text{coulomb}} \quad \text{or} \quad 8.0 \text{ volts}$$

$$V = IR$$

$$I = \frac{V}{R}$$

$$I = \frac{8.0 \text{ V}}{4.0 \text{ } \Omega} = 2.0 \text{ A}$$

E8. What is the power of the circuit?
Solution:

$$P = VI$$

$$P = \left(8.0 \frac{J}{C}\right)\left(2.0 \frac{C}{s}\right) = 16 \text{ watts}$$

Generator

A simple generator consists of a wire loop that rotates between the two opposite poles of a magnet. As the loop of wire rotates, the induced current and voltage vary. Twice during every rotation of the loop, the induced voltage and current are zero; this happens when the plane of the loop is *perpendicular* to the field lines. At those two points, the wire segments are moving parallel to the field, and no force is exerted on the electrons. Also, twice during every rotation, the plane of the loop is *parallel* to the field. At these two points the motion of the wire is perpendicular to the field lines and the induced current and voltage are at a maximum.

In between the perpendicular and parallel orientations, the induced current and voltage are greater than zero but less than the maximum, since only a component of the motion of the wire is then perpendicular to the field. In addition, the direction of the induced voltage and current in the loop is reversed every half turn, since the wire's direction of motion is reversed. The wire segment that is on the right side and moving downward in Figure 3E-19 will be moving upward when it gets to the left side after one-half of a rotation.

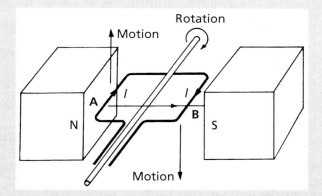

Figure 3E-19. Loop of wire rotating in a magnetic field.

The rotating loop is connected to an external circuit that carries the current to a variety of appliances. As the loop rotates, the current and potential difference generated vary in magnitude and alternate in direction. This is referred to as *alternating current* (ac). The maximum current and voltage values are proportional to the intensity of the magnetic field and to the rotational speed of the loop.

Lenz's Law

All magnetic effects lead to forces that oppose the change that produced the effect. We saw earlier how this rule is applied to motors (p. 127). The current induced in a wire as it moves across magnetic field lines and in the ac generator is also a magnetic effect. As such, it is also subject to the rule known as Lenz's law.

The magnetic effect in this case is the induced current. The "change" that produces it is the motion of the wire across the field lines. As soon as the induced current appears, a force (recall that magnetic fields act on currents) manifests itself to oppose the motion of the wire. For example, the current (directed out of the page) induced by the downward motion of the wire in Figure 3E-20 leads to an upward-acting force that resists the wire's motion. This can be verified by hand rule #3: the thumb points in the direction of current, out of the page; the fingers in the direction of the magnetic field, to the right; and the palm of the hand pushes (representing the force) upward, parallel to the page.

Since an induced current can exist only as long as the wire moves across the field, and a force opposing the motion of the wire appears at the same time as the current, the only way the current can be maintained is by forcing the wire to continue to move against this opposing force. The work done to

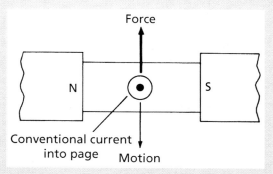

Figure 3E-20.

keep the wire moving is equal to the energy provided by the current. The energy created by the induced current is therefore not "free." Lenz's law, thus, provides yet another example of the Law of Conservation of Energy.

In the case of the ac generator, work must be done to keep the wire loop rotating. This is usually accomplished by a steam turbine that derives its energy from the heat produced by burning coal, oil, or from nuclear reactions.

The "change" that induces the current in the second case of electromagnetic induction (see pp. 109–110) is the change over time in the number of flux lines enclosed by a wire loop. In this case Lenz's Law operates as follows: when the number of external field lines increases, the magnetic field created by the induced current opposes the external field. This causes the number of flux lines through the loops to decrease. When the number of external field lines decreases, the magnetic field created by the induced current supplements the external field. This causes the number of field lines through the loops to increase.

Sample Problem

E9. The plane of a loop of wire is oriented perpendicularly to a magnetic field whose flux lines point into the page, away from the reader (see Figure 3E-21). The intensity of the field weakens over time, and the number of flux lines enclosed by the loop decreases. How will the current induced in the loop be directed?

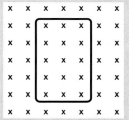

Figure 3E-21.

Solution: According to Lenz's Law, the induced current creates a magnetic field that *increases* the number of flux lines enclosed by the loop. This means that the field lines created by the current inside the loop point in the *same direction* as the external field, in this case into the page and away from the reader. Use of hand rule #2 shows that the induced current has to move clockwise. (Start with the thumb of the right hand pointing into the page, in the direction of the magnetic field. The four fingers then point in the direction of the induced conventional current, clockwise.)

Questions

E52. The space charge near an incandescent cathode is (1) negative (2) positive (3) either positive or negative

E53. The cathode-anode arrangement is designed to (1) decelerate electrons (2) accelerate protons (3) accelerate electrons (4) maintain constant speed of the electrons

E54. An electron traveling at a speed v in the plane of this paper enters a uniform magnetic field. Which diagram best represents the condition under which the electron will experience the greatest magnitude force as it enters the magnetic field?

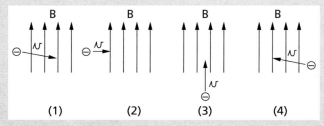

E55. If a charged particle moving through a magnetic field experiences a magnetic force, the angle between the magnetic field and the force exerted on the particle is (1) 0° (2) 45° (3) 90° (4) 180°

E56. If a charged particle moving perpendicularly to a uniform magnetic field increases in velocity, the magnetic force on the charge (1) decreases (2) increases (3) remains the same

Base your answers to questions E57 through E61 on the following diagram, which represents a helium ion with a charge of +2 elementary charges moving toward point A with a constant speed v of 2.0 meters per second perpendicular to a uniform magnetic field between the poles of a magnet. The strength of the magnetic field is 0.10 weber per square meter.

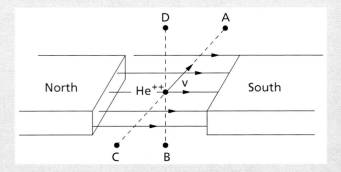

E57. The direction of the magnetic force on the helium ion is toward point (1) A (2) B (3) C (4) D

E58. The magnitude of the magnetic force exerted on the helium ion is (1) 3.2×10^{-20} N (2) 6.4×10^{-20} N (3) 0.10 N (4) 0.20 N

E59. If the strength of the magnetic field and the speed of the helium ion are both doubled, the force on the helium ion will be (1) halved (2) doubled (3) the same (4) quadrupled

E60. If the polarity of the magnet is reversed, the magnitude of the magnetic force on the helium ion will (1) decrease (2) increase (3) remain the same

E61. The helium ion is replaced by an electron moving at the same speed. Compared to the magnitude of the force on the helium ion, the magnitude of the force on the electron is (1) less (2) greater (3) the same

Base your answers to questions E62 through E65 on the following diagram, which represents an electron beam entering the space between two parallel, oppositely charged plates. A uniform magnetic field, directed out of the page, exists between the plates.

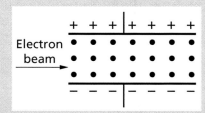

E62. If the magnitude of the electric force on each electron and the magnetic force on each electron are the same, which diagram best represents the direction of the *vector sum* of the forces acting on one of the electrons?

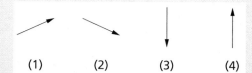

(1)	(2)	(3)	(4)

E63. In which direction would the magnetic field have to point in order for the magnetic force on the electrons to be opposite in direction from the electric force on the electrons? (1) toward the bottom of the page (2) toward the top of the page (3) out of the page (4) into the page

E64. If the electric force were equal and opposite to the magnetic force on the electrons, which diagram would best represent the path of the electrons as they travel in the space between the plates?

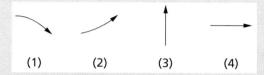

(1)	(2)	(3)	(4)

E65. If only the potential difference between the plates is increased, the force on the electron will (1) decrease (2) increase (3) remain the same

Base your answers to questions E66 through E70 on the following diagram, which represents an electron moving at 2.0×10^6 meters per second into a magnetic field that is directed into the paper. The magnetic field has a strength of 2.0 newtons per ampere-meter.

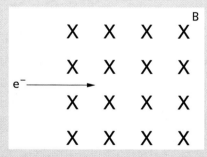

E66. Which vector best indicates the direction of the force on the electron?

(1)	(2)	(3)	(4)

E67. What is the magnitude of the force on the electron? (1) 6.4×10^{-13} N (2) 4.0×10^6 N (3) 6.4×10^6 N (4) 8.0×10^6 N

E68. If the strength of the magnetic field were increased, the force on the electron would

(1) decrease (2) increase (3) remain the same

E69. If the velocity of the electron were increased, the force on the electron would (1) decrease (2) increase (3) remain the same

E70. The electron is replaced with a proton moving with the same velocity. Compared to the magnitude of the force on the electron, the magnitude of the force on the proton would be (1) less (2) greater (3) the same

Base your answers to questions E71 through E74 on the following diagram, which represents an electron beam in a vacuum. The beam is emitted by the cathode C, accelerated by anode A, and passes through electric and magnetic fields.

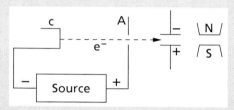

E71. If an electron in the beam is accelerated to a kinetic energy of 4.8×10^{-16} joule, the potential difference between the cathode and the anode is (1) 7.7×10^3 V (2) 4.8×10^{-3} V (3) 3.0×10^3 V (4) 3.0×10^{-3} V

E72. In which direction will the electron beam be deflected by the electric field? (1) into the page (2) out of the page (3) toward the top of the page (4) toward the bottom of the page

E73. In which direction will the force of the magnetic field act on the electron beam? (1) into the page (2) out of the page (3) toward the top of the page (4) toward the bottom of the page

E74. If an electron in the beam moves at 2.0×10^8 meters per second between the magnetic poles where the flux density is 0.20 weber per square meter, the force on the electron is (1) 6.4×10^{-12} N (2) 6.4×10^{-10} (3) 4.0×10^7 N (4) 4.0×10^9 N

Transformers (▦)

A typical transformer is constructed as follows: wire carrying alternating current, usually powered by a generator, is wound around one arm of a rectangular-shaped "core" made of soft iron (see Figure 3E-22). This coil is referred to as the **primary coil.** Another wire is wound around a second arm of the core and is known as the **secondary coil.** The secondary coil leads to the appliance we wish

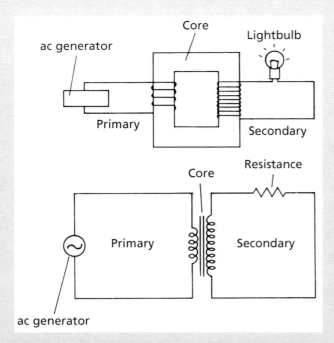

Figure 3E-22. Transformer and schematic equivalent.

to operate, but it is *not* connected to any power source nor to the primary coil. Indeed the two coils usually consist of insulated wires that cannot make contact with each other.

As the current in the primary coil continues to alternate in magnitude and direction, it creates a magnetic field that alternates in intensity and direction (recall "Currents Act on Magnets," pp. 107 and 119). The highly permeable soft iron core becomes alternately magnetized and demagnetized. Consequently, the intensity and direction of the magnetic field created by the core in the region of the secondary coil also alternates. The changing magnetic field induces current (ac) and EMF in the secondary circuit (see p. 000). This induced current powers the appliance connected to that coil.

The voltage induced in the secondary coil is usually different from that in the primary coil. The ratio of the voltage in the primary, V_p, to the voltage in the secondary, V_s, is equal to the ratio of the number of loops of wire around the primary arm, N_p, to the number of loops around the secondary arm, N_s. This is stated mathematically as

$$\frac{V_p}{V_s} = \frac{N_p}{N_s}$$

If the secondary arm contains more loops than the primary, the voltage induced in the secondary coil is greater and the transformer is called a **step-up transformer**. If the secondary arm contains fewer loops than the primary, its voltage is smaller and

the transformer is referred to as a **step-down transformer** (Figure 3E-23).

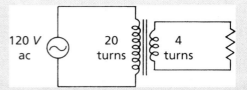

Figure 3E-23. Step-down transformer.

However, as a consequence of the Law of Conservation of Energy, the power output from the secondary (to the appliances connected to it) cannot exceed the power input of the primary (supplied by the generator). At best, the power output of the secondary will equal the power input to the primary, and

$$V_p I_p = V_s I_s$$

Such transformers are said to be 100% efficient; no heat energy is lost between the primary and the secondary.

Most transformers are not 100% efficient, and the power output from the secondary is usually less than the power input to the primary. The efficiency of a transformer is defined as the ratio of the two rates of doing work, expressed in percentage form:

$$\% \text{ efficiency} = \frac{V_s I_s}{V_p I_p} \times 100$$

Sample Problems

E10. A step-down transformer consists of a primary coil containing 20 turns and a secondary coil with 4 turns. The generator provides the primary with 120. volts of potential difference. What is the voltage induced in the secondary?

Solution:

$$\frac{V_p}{V_s} = \frac{N_p}{N_s}$$

$$V_s = \frac{V_p N_s}{N_p}$$

$$V_s = \frac{120.\ \text{V} \times 4}{20} = 24.0\ \text{V}$$

E11. If the transformer is 80% efficient and 1000 watts of power are supplied to the primary, how much current flows through the secondary?

Solution:

$$\% \text{ efficiency} = \frac{V_s I_s}{V_p I_p} \times 100$$

$$I_s = \frac{\% \text{ efficiency} \times V_p \times I_p}{V_s \times 100}$$

$$I_s = \frac{80 \times 1000 \text{ W}}{24.0 \text{ V} \times 100}$$

$$I_s = 33.3 \text{ amps}$$

Questions

E75. Compared with the voltage in the coil of a transformer with more turns of wire, the voltage in the coil with fewer turns is (1) smaller (2) greater (3) the same

E76. An ideal transformer cannot (1) increase the current (2) increase the voltage (3) increase the power (4) decrease the current

E77. A transformer is connected to a source of alternating current. The only factor always common to both primary and secondary windings is the (1) current (2) voltage (3) frequency (4) resistance

E78. The current flowing in the primary of a transformer depends upon (1) the resistance of the secondary coil (2) the resistance of the primary coil (3) the current taken by the secondary (4) the resistance between the primary and secondary coils

E79. A transformer has 50 turns on the primary and 100 turns on the secondary. If the pri-

mary is connected to a 6-volt battery the voltage on the secondary will be (1) zero (2) less than zero (3) twice the voltage on the primary (4) half the voltage on the primary

E80. A transformer changes (1) electrical energy into mechanical energy (2) mechanical energy into electrical energy (3) high-voltage dc to low-voltage dc or vice versa (4) low-voltage ac to high-voltage ac or vice versa

Base your answers to questions E81 through E84 on the following information: A transformer consists of a primary connected to a 120-volt ac source that drives 4 A of current through the primary, and a secondary that contains one-third as many turns of wire as the primary. The transformer's efficiency is 80%.

E81. The voltage of the secondary is (1) 360 V (2) 90 V (3) 40 V (4) 480 V

E82. The current through the secondary is closest to (1) 80 A (2) 8 A (3) 3 A (4) 10 A

E83. The current in the secondary (1) is of the ac type (2) could be either ac or dc (3) is of the dc type (4) is of the rectified ac type

E84. The power output of the secondary is (1) equal to the power input to the primary (2) greater than the power input to the primary (3) less than the power input to the primary.

E85. To step up 6 V to 30 V with a transformer whose primary contains 10 coils of wire, the secondary should consist of (1) 5 coils (2) 3 coils (3) 50 coils (4) 15 coils

E86. What is the power output of a transformer whose efficiency is 60%, if the power input to the primary is 240 W? (1) 144 W (2) 300 W (3) 400 W (4) 180 W

CHAPTER  Wave Phenomena

Introduction to Waves

A disturbance is a change in a body of matter. This change is opposed by a force of nature. The body of matter becomes the **medium** through which the disturbance travels. Traveling disturbances begin at a point of origin and transmit energy through the medium. They can propagate through such material media as solids, liquids, and gases.

A single vibratory disturbance is called a **pulse**. One example of a pulse is a traveling crest of water. When the disturbance reaches a particular part of the pool, the water molecules there are pushed upward, forming a crest. When the disturbance passes that point, the water molecules come down. In this way, the up-and-down vibration travels from place to place.

Disturbances in Solids

When a person holds one end of a rope (the other end attached to, say, a door knob) and moves his or her hand upward, a pulse in the form of a crest is formed in the rope. From its point of origin at one end, the crest moves toward the other end of the rope (Figure 4-1), transmitting the energy gained when the person moved the rope. The

force of tension in the rope acts to pull the crest down. As the crest comes down in response to these forces, it pushes against the adjacent part of the rope, causing the crest to reappear farther along the rope. This happens repeatedly as the crest travels the length of the rope.

Disturbances in Liquids

The circular ripples formed by a pebble dropped into a pond are disturbances traveling in the body of water. A crest is formed by the downward thrust of the pebble into the water (similar to the flick of the wrist with the rope). The forces of tension and gravity then pull this circular crest down. As the crest collapses in one spot, it reemerges farther away from the point of origin, turning the ripple into an ever-expanding circle whose center is the point where the pebble entered the water.

Disturbances in Gases

When a tuning fork is struck, it oscillates back and forth. As one prong of the fork swings outward, it pushes air molecules in that same direction, creating a pocket of compressed air (Figure 4-2). As this compressed air expands and returns to normal, it pushes against the adjacent volume of air, compressing it. As with the crest on the rope and the ripple in the pond, the disturbance (compressed air) travels through a medium (air), away from its source (the tuning fork).

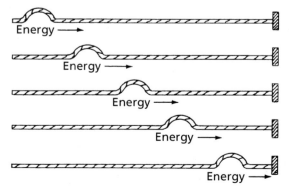

Figure 4-1. A pulse traveling in a rope.

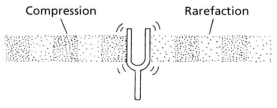

Figure 4-2. A series of compressions and rarefactions form a sound wave.

Periodic Waves

A regularly repeating series of pulses is called a **periodic wave**. Here the vibratory disturbances are evenly spaced, one pulse following the other like cars in a train—thus the term *wave train*. Periodic waves are usually referred to simply as *waves*. Returning to the rope example, moving your hand up and down continuously at a regular rate forms periodic waves. Each upward movement sends a crest along the rope, and each downward movement sends a trough.

The same effect can be produced in water by repeatedly inserting a finger into the water and pulling it out. Each time the finger is inserted, a circular crest forms. Each time the finger is pulled out, a circular trough, or depression, forms. A regularly repeating pattern of crests following troughs travels across the water. The water molecules move up and down while the wave moves forward.

Likewise, the vibration of the tuning fork sends out a series of pulses. As the tuning fork oscillates back and forth, compressions and **rarefactions** (pockets of expanded air) follow each other into the surrounding air (Figure 4-2).

One complete repetition of the pattern in a periodic wave is referred to as a **cycle**. For example, a crest followed by a trough, or a compression followed by a rarefaction, make up one cycle.

Sound and light are two special types of waves. Sound waves consist of compression-rarefaction cycles produced by an oscillating material. Sound needs a medium in which to travel; it cannot travel in a vacuum. Light is an electromagnetic wave. Light waves consist of electromagnetic disturbances that need no material medium to move from place to place; light can travel in a vacuum.

Transverse Waves

Waves in which the disturbances are perpendicular to the direction of wave motion are called **transverse waves**. The vertical disturbances in a rope or body of water produce waves that move horizontally, and so are examples of transverse waves. Light is also a transverse wave.

Longitudinal Waves

Waves in which disturbances are parallel to the direction of wave motion are called **longitudinal waves**. The compressions and rarefactions of sound waves, for example, consist of molecules vibrating parallel to the motion of the wave. Sound waves, therefore, are longitudinal waves.

Some waves, such as large ocean waves, consist of combinations of longitudinal and transverse vibrations.

Wave Characteristics

Waves cause the particles of a medium to vibrate about their rest position; but it is the energy of the wave, not the particles of the medium, that travels through the medium. When a crest, for example, moves over the surface of water, no water is actually moving with the crest. The water moves vertically (up and down) but not horizontally. Only the energy needed to cause the up-and-down vibration moves through the water.

Frequency The number of cycles produced by a vibrating source per second is the **frequency** of the wave, symbolized by the letter *f*. Frequency also represents the number of cycles that pass by a fixed point per second. "Cycles per second" are expressed in **hertz** (Hz), a derived unit. A radio station that transmits at 600 kilohertz (kHz) sends 600,000 cycles per second toward receiving antennas.

The frequency of a wave is determined by its source. The nature of the medium through which the wave travels does not affect its frequency. For example, the frequency of a sound wave produced by a tuning fork is determined by how many times per second the tuning fork oscillates and does not depend upon the temperature or quality of the surrounding air. If the tuning fork completes five full oscillations every second, it produces five compression-rarefaction cycles per second, and five such cycles per second will be seen following each other past any fixed point in the path of the wave.

Period The time required for a complete cycle to be produced or to pass a given point is the **period** of the wave, symbolized by *T*. Like frequency, period is determined by the source of the wave.

The period of a wave is the reciprocal of its frequency. If the frequency is 10 Hz, the period is $\frac{1}{10}$ second. The relationship between period and frequency can be summarized by the formulas

$$T = \frac{1}{f} \quad \text{and} \quad f = \frac{1}{T}$$

Sample Problem

1. If the time necessary for one cycle of a wave to pass a given point is $\frac{1}{8}$ second, what is the frequency of the wave?

Solution:

$$f = \frac{1}{T} = \frac{1}{\frac{1}{8}} = 8 \text{ Hz}$$

Wavelength The length of one complete cycle is called the **wavelength** of the wave, symbolized by the Greek letter lambda (λ). Wavelength can be measured from crest to crest, trough to trough, or between corresponding points on adjacent pulses (Figure 4-3). Both the source and the medium affect the wavelength.

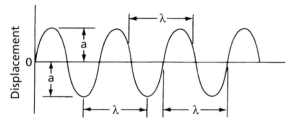

Figure 4-3. Amplitude (*a*) and wavelength (λ).

Velocity The *velocity* of a wave is the distance traveled per unit time by any part of the wave. Velocity is generally expressed in meters per second and is symbolized by *v*. The velocity of a wave depends on the nature of the medium through which it travels. Sound waves, for example, travel faster in warm air than in cold air. The speed of sound in air at 0° C and one atmosphere of pressure is 3.31×10^2 m/s. Light travels fastest in a vacuum, slower in gases, liquids, and solids. In a vacuum, the speed of all electromagnetic waves, including light, is 3.00×10^8 m/s.

Some materials allow waves of different frequencies to travel through them at different velocities. Such a medium is called a **dispersive medium**. Glass is a dispersive medium for light waves.

The velocity of a wave is equal to the product of its frequency and wavelength.

$$v = f\lambda$$

If 5 cycles pass a given point in one second and each cycle is 2 meters long, then 10 meters of "wave" pass the point per second. The frequency (*f*) is 5 Hz, the wavelength (λ) is 2 meters, and the speed ($f\lambda$) is 10 meters per second.

Sample Problem

2. The speed of a radio wave is 3×10^8 m/s. What is the wavelength of a radio wave whose frequency is 600 kHz?

Solution:

$$v = f\lambda$$

$$\lambda = \frac{v}{f}$$

$$= \frac{3 \times 10^8 \text{ m/s}}{6 \times 10^5 \text{ s}^{-1}} = 5 \times 10^2 \text{ m}$$

(*Note*: Hz represents cycles/second but a "cycle" is not a unit since it is not a measured quantity. The unit for frequency is therefore $1/s$, or s^{-1}.)

Amplitude The **amplitude** of a wave traveling through a material is equal to the maximum displacement of particles of the medium from their rest, or equilibrium, position. For waves in water, the amplitude is the maximum height of a crest or the maximum depth of a trough (Figure 4-3). For the compressions and rarefactions of sound waves, the amplitude is the greatest deviation in density or pressure from the norm of the surrounding air. The amount of energy a wave transmits is determined by the amplitude of the wave.

The amplitude of a wave is determined by its source. A tuning fork that goes through wider oscillations produces more compressed compressions and more rarefied rarefactions. A larger pebble dropped into water pushes up taller crests. The greater the amplitude of a wave, the more energy it carries. As the amplitude of a sound wave increases, the loudness of the sound increases. Similarly, as the amplitude of a light wave increases, the brightness of the light increases

Phase Points on a periodic wave that are identically displaced from the equilibrium position and are moving in the same direction away from the equilibrium position are said to be **in phase**. In other words, the phase difference between them is 0°. Points that are in phase are always a whole number of wavelengths apart. Points *A*, *I*, and *Q* in Figure 4-4 meet all of these conditions and are therefore in phase. The same is true of points *B*, *J*, and *R*; and *E* and *M*. Points *A* and *B*, *B* and *D*, or *B* and *H*, on the other hand, do not meet all of these requirements, and are said to *be out of phase*.

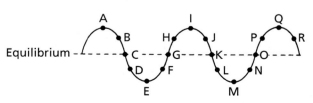

Figure 4-4.

Points on a periodic wave that are equally displaced from the equilibrium position but are moving in opposite directions are said to be 180° out of phase. Points that are 180° out of phase are always an odd number of half wavelengths apart. Points *A* and *E*, *B* and *F*, and *F* and *J* (Figure 4-4) are all pairs of points with a phase difference of 180° between them.

QUESTIONS

PART A

1. A periodic wave transfers (1) energy, only (2) mass, only (3) both energy and mass (4) neither energy nor mass

2. A surfacing whale in an aquarium produces water wave crests having an amplitude of 1.2 meters every 0.40 second. If the water wave travels at 4.5 meters per second, the wavelength of the wave is (1) 1.8 m (2) 2.4 m (3) 3.0 m (4) 11 m

3. A student strikes the top rope of a volleyball net, sending a single vibratory disturbance along the length of the net, as shown in the diagram below.

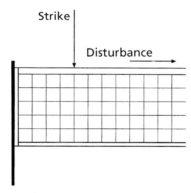

This disturbance is best described as (1) a pulse (2) a periodic wave (3) a longitudinal wave (4) an electromagnetic wave

4. Which form(s) of energy can be transmitted through a vacuum? (1) light, only (2) sound, only (3) both light and sound (4) neither light nor sound

5. A transverse wave passes through a uniform material medium from left to right, as shown in the diagram below.

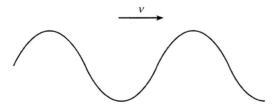

Which diagram best represents the direction of vibration of the particles of the medium?

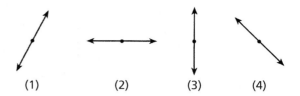

6. An electric bell connected to a battery is sealed inside a large jar. What happens as the air is removed from the jar?
   (1) The electric circuit stops working because electromagnetic radiation can *not* travel through a vacuum.
   (2) The bell's pitch decreases because the frequency of the sound waves is lower in a vacuum than in air.
   (3) The bell's loudness increases because of decreased air resistance.
   (4) The bell's loudness decreases because sound waves *cannot* travel through a vacuum.

7. A tuning fork oscillates with a frequency of 256 hertz after being struck by a rubber hammer. Which phrase best describes the sound waves produced by this oscillating tuning fork?
   (1) electromagnetic waves that require no medium for transmission
   (2) electromagnetic waves that require a medium for transmission
   (3) mechanical waves that require no medium for transmission
   (4) mechanical waves that require a medium for transmission

8. A physics student notices that 4.0 waves arrive at the beach every 20. seconds. The frequency of these waves is (1) 0.20 Hz (2) 5.0 Hz (3) 16 Hz (4) 80. Hz

9. The diagram below shows two points, *A* and *B*, on a wave train.

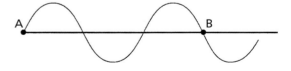

How many wavelengths separate point *A* and point *B*? (1) 1.0 (2) 1.5 (3) 3.0 (4) 0.75

10. The energy of a water wave is most closely related to its (1) frequency (2) wavelength (3) period (4) amplitude

11. A tuning fork vibrating in air produces sound waves. These waves are best classified as

(1) transverse, because the air molecules are vibrating parallel to the direction of wave motion

(2) transverse, because the air molecules are vibrating perpendicular to the direction of wave motion

(3) longitudinal, because the air molecules are vibrating parallel to the direction of wave motion

(4) longitudinal, because the air molecules are vibrating perpendicular to the direction of wave motion

12. The diagram below represents a transverse wave traveling in a string.

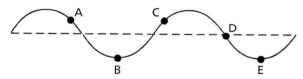

Which two labeled points are 180° out of phase? (1) A and D (2) B and F (3) D and F (4) D and H

13. The diagram below represents a periodic wave.

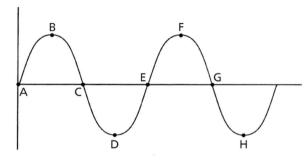

Which two points on the wave are in phase? (1) A and C (2) B and D (3) A and D (4) B and E

14. If the speed of a wave doubles, its wavelength will be (1) unchanged (2) doubled (3) halved (4) quadrupled

15. Which wave diagram has *both* wavelength (λ) and amplitude (A) labeled correctly?

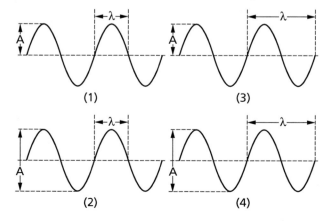

16. An electric guitar is generating a sound of constant frequency. An increase in which sound wave characteristic would result in an increase in loudness? (1) speed (2) period (3) wavelength (4) amplitude

17. A student notices the frequency of a periodic wave is 3.0 hertz. What is the wave's period? (1) 3 seconds (2) 1 second (3) 0.33 second (4) 0.11 second

18. A single vibratory disturbance that moves from point to point in a material medium is known as a (1) phase (2) pulse (3) distortion (4) wavelet

19. In which type of wave is the disturbance parallel to the direction of wave travel? (1) torsional (2) longitudinal (3) transverse (4) circular

20. If the frequency of a sound wave in air remains constant, its energy can be varied by changing its (1) amplitude (2) speed (3) wavelength (4) period

21. The reciprocal of the frequency of a periodic wave is the wave's (1) period (2) amplitude (3) propagation (4) velocity

22. As the frequency of the wave generated by a radio transmitter is increased, the wavelength (1) decreases (2) increases (3) remains the same

23. If the period of a radio wave is doubled, its wavelength will be (1) halved (2) doubled (3) unchanged (4) quartered

24. As the frequency of a wave increases, the period of that wave (1) decreases (2) increases (3) remains the same

25. Periodic waves are produced by a wave generator at the rate of one wave every 0.50 second. The period of the wave is (1) 1.0 s (2) 2.0 s (3) 0.25 s (4) 0.50 s

PART B-1

26. Which diagram below does *not* represent a periodic wave?

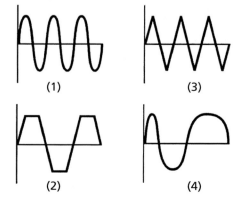

Base your answers to questions 27 through 29 on the following diagram, which represents a segment of a periodic wave traveling to the right in a steel spring.

27. What is the amplitude of the wave? (1) 2.5 m (2) 2.0 m (3) 0.2 m (4) 0.4 m

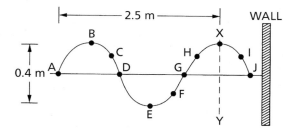

28. What is the wavelength of the wave? (1) 1.0 m (2) 2.0 m (3) 2.5 m (4) 0.4 m

29. If a wave crest passes line *XY* every 0.40 second, the frequency of the wave is (1) 1.0 Hz (2) 2.5 Hz (3) 5.0 Hz (4) 0.4 Hz

Base your answers to questions 30 and 31 on the following information.

The frequency of a wave is 2.0 cycles per second, and its speed is 0.04 meter per second.

30. The period of the wave is (1) 0.005 s (2) 2.0 s (3) 0.50 s (4) 0.02 s

31. The wavelength of the wave is (1) 1.0 m (2) 0.02 m (3) 0.08 m (4) 4.0 m

32. If the frequency of a sound wave is 440. cycles per second, its period is closest to (1) 2.27×10^{-3} s/cycle (2) 0.75 s/ cycle (3) 1.33 s/cycle (4) 3.31×10^2 s/cycle

33. Periodic waves with a wavelength of 0.05 meter move with a speed of 0.30 meter per second. When the waves enter a dispersive medium, they travel at 0.15 meter per second. What is the wavelength of the waves in the dispersive medium? (1) 20. m (2) 1.8 m (3) 0.05 m (4) 0.025 m

Boundary Behavior

When a wave encounters a boundary between two different media, part of the wave is reflected back into the first medium and part is transmitted into the second medium. The fraction of the wave's energy that is reflected and the fraction transmitted depend on the type of wave and on the nature of the two media. For example, when a light wave traveling through the air hits glass, most of the wave's energy is transmitted through the glass and only a small fraction is reflected. On the other hand, when a water wave reaches the end of a pool or lake, or when a sound wave en-

counters a smooth rigid wall, most of the wave's energy is reflected back into the first medium.

The part of the wave that is reflected at the boundary between two media retains the same speed, frequency, and wavelength as the original wave. The part of the wave that is transmitted into the second medium, however, experiences a change in speed, since the speed of a wave is determined by the medium. The transmitted wave's frequency, on the other hand, is determined by its source and does not change. Since wavelength, frequency, and speed are interrelated ($v = f\lambda$), a change in speed must result in a change in at least one of the other two factors. If the frequency remains constant there must be a change in wavelength when the wave enters the second medium. For example, the speed of light in glass is less than it is in air; this means that the wavelength of light becomes shorter as it passes from air to glass. This concept will be discussed further in the section called Refraction.

The Doppler Effect

Although it is generally true that the number of cycles passing by a given point per second (the **observed frequency** of a wave) is equal to the number of cycles produced by the source per second (the **transmitted frequency**), this is not the case if the source and the observer are moving relative to each other.

When the source and the observer approach each other, the number of cycles observed passing by per second is greater than would have been the case had the source and observer been at rest. In this case, the observed frequency is greater than the transmitted frequency.

When the source and observer move away from each other, the number of cycles observed passing by per second is less than would have been the case had the source and observer been at rest. In this case, the observed frequency is less than the transmitted frequency.

These phenomena are known as the *Doppler effect*, and important consequences result from them. The pitch of a sound wave, a quality that is related to frequency, changes noticeably as the source of a sound, such as a siren or radio, first approaches the observer, passes by, then moves away. In the case of light waves, the Doppler effect leads to changes in color. White light from an approaching source, for example, takes on a bluish appearance. From a receding source, the light shifts toward the red.

The amount of increase or decrease in frequency that results from relative motion between a source and an observer depends on the speed of the source or the observer or both. Indeed, it is possible to determine the unknown speed of a source of sound or light, toward or away from the observer, from the change in pitch of the sound or change in color of the light.

QUESTIONS

34. A radar gun can determine the speed of a moving automobile by measuring the difference in frequency between emitted and reflected radar waves. This process illustrates (1) resonance (2) the Doppler effect (3) diffraction (4) refraction

35. A train sounds a whistle of constant frequency as it leaves the train station. Compared to the sound emitted by the whistle, the sound that the passengers standing on the platform hear has a frequency that is
(1) lower, because the sound-wave fronts reach the platform at a frequency lower than the frequency at which they are produced
(2) lower, because the sound waves travel more slowly in the still air above the platform than in the rushing air near the train
(3) higher, because the sound-wave fronts reach the platform at a frequency higher than the frequency at which they are produced
(4) higher, because the sound waves travel faster in the still air above the platform than in the rushing air near the train

36. A source of sound waves approaches a stationary observer through a uniform medium. Compared to the frequency and wavelength of the emitted sound, the observer would detect waves with a
(1) higher frequency and shorter wavelength
(2) higher frequency and longer wavelength
(3) lower frequency and shorter wavelength
(4) lower frequency and longer wavelength

37. A source of waves and an observer are moving relative to each other. The observer will detect a steadily increasing frequency if (1) he moves toward the source at a constant speed (2) the source moves away from him at a constant speed (3) he accelerates toward the source (4) the source accelerates away from him

38. A change in the speed of a wave as it enters a new medium produces a change in (1) frequency (2) period (3) wavelength (4) phase

39. A sound of constant frequency is produced by the siren on top of a firehouse. Compared to the frequency produced by the siren, the frequency observed by a firefighter approaching the firehouse is (1) lower (2) higher (3) the same

40. A 2.00×10^6 hertz radio signal is sent a distance of 7.30×10^{10} meters from Earth to a spaceship orbiting Mars. Approximately how much time does it take the radio signal to travel from Earth to the spaceship? (1) 4.11×10^{-3} s (2) 2.43×10^2 s (3) 2.19×10^8 s (4) 1.46×10^{17} s

Base your answers to questions 41 through 43 on the information and diagram below.

A system consists of an oscillator and a speaker that emits a 1000.-hertz sound wave. A microphone detects the sound wave 1.00 meter from the speaker.

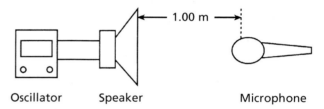

Oscillator   Speaker   Microphone

41. Which type of wave is emitted by the speaker? (1) transverse (2) longitudinal (3) circular (4) electromagnetic

42. The microphone is moved to a new fixed location 0.50 meter in front of the speaker. Compared to the sound waves detected at the 1.00-meter position, the sound waves detected at the 0.50-meter position have a different (1) wave speed (2) frequency (3) wavelength (4) amplitude

43. The microphone is moved at constant speed from the 0.50-meter position back to its original position 1.00 meter from the speaker. Compared to the 1000.-hertz frequency emitted by the speaker, the frequency detected by the moving microphone is (1) lower (2) higher (3) the same

Base your answers to questions 44 through 46 on the following diagram, which represents waves around a sound source that is moving with a constant velocity through air. The source produces waves of a constant frequency.

44. The diagram illustrates (1) interference (2) diffraction (3) reflection (4) the Doppler effect

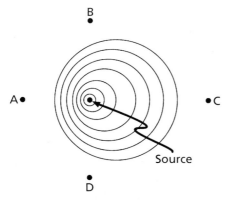

45. The source is moving toward point (1) *A* (2) *B* (3) *C* (4) *D*

46. Compared with the frequency of the waves observed at *C*, the frequency of the waves observed at *A* is (1) less (2) greater (3) the same

When Waves Meet— Interference

Two or more waves may pass through a medium at the same time. When this occurs two rules apply. First, the total displacement experienced at any point where waves meet is equal to the sum of the displacements of the individual waves at that point. This is known as the *principle of superposition*. Second, waves pass through each other, with each wave unaffected by the passage of the others. After meeting, the individual waves continue traveling in their original directions and with the same characteristics they had before they met.

Constructive Interference

As an example, consider two crests traveling toward each other in a rope. Assume one crest is 3 cm tall and the other is 2 cm tall before they meet. At the moment they meet, the rope is displaced 5 cm (3 cm + 2 cm) upward from its normal (equilibrium) position. Then the 3-cm- and 2-cm-tall crests reappear and continue onward, one to the right and the other to the left, in the same directions they were traveling before they met (Figure 4-5).

Since these two crests "interfered" with each other when they met and the result was a larger disturbance, we call this **constructive interference**.

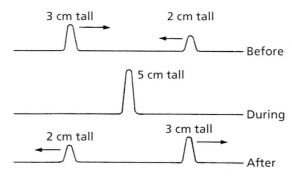

Figure 4-5. Constructive interference.

Destructive Interference

As another example, consider a crest and a trough heading toward each other in a rope. The crest is 2 cm tall (displacement of +2) and the trough is 2 cm deep (displacement of −2). When they meet, the net displacement is zero and the rope straightens. Then the crest and trough reappear, with the crest continuing on its way to the left and the trough continuing on its way to the right. In effect, they passed through each other (Figure 4-6).

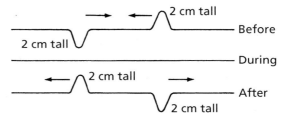

Figure 4-6. Destructive interference.

Since these two pulses interfered with each other only to negate each other's disturbances, this is an example of **destructive interference**.

Maximum constructive interference occurs at points where the phase difference between the waves that meet is 0° and maximum destructive interference occurs where the waves are 180° out of phase. The total destruction of two overlapping periodic waves occurs if they have equal amplitudes and frequencies and are everywhere 180° out of phase.

Two Sources in Phase Two wave sources vibrating at the same frequency and producing waves of equal amplitude are said to be **sources in phase**. In Figure 4-7, two such sources, located at s_1 and s_2, are operating in the same medium, such as a body of water. Each source produces crests and troughs that propagate outward in all directions. The solid semicircles in the figure represent the crests and the dotted semicircles represent

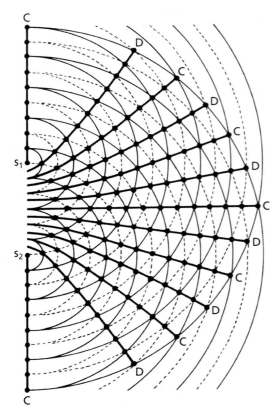

Figure 4-7. Two sources in phase and their interference pattern. Lines labeled "D" are nodal lines; lines labeled "C" are antinodal lines.

Standing Waves

Another example of interference is the guitar string that is fixed at both ends and plucked in the middle. Two waves travel from the point where the string is plucked, in opposite directions, toward the ends of the string. At the ends of the string the waves are reflected and they pass through each other as they travel to the opposite ends. Then they are reflected again, pass through each other again, and so on until the waves die out. For certain combinations of string length and wavelength, the waves repeatedly pass through each other in such a way that at some points on the string they always meet to interfere constructively, while at other points they always interfere destructively. At points where a crest of one wave always meets a crest of the other and a trough always meets a trough (constructive interference), antinodes are formed and the string vibrates visibly. At points where a crest of one wave always meets a trough of the other (destructive interference), nodes are formed and the string remains motionless.

This leads to the familiar appearance of plucked guitar strings (Figure 4-8). Since the eye sees no horizontal movement of crests or troughs along the string, only up-and-down vibrations in place, we call these **standing waves**. Actually, two interfering waves are continuously moving across the string; it is only the locations of the nodes and antinodes that remain stationary.

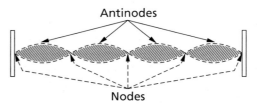

Figure 4-8. A plucked guitar string showing standing waves.

Standing waves are produced whenever two waves of the same frequency and amplitude travel in opposite directions in the same medium. As in the case of the guitar string, they are most often produced by the reflection of a wave at a fixed boundary.

Resonance

Most objects have a natural frequency of vibration. If struck, they respond by vibrating at a particular frequency. When an object is disturbed by a wave whose frequency is the same as its natural vibration frequency, the amplitude of vibration of the object continues to increase. This phenome-

the troughs. The lines labeled C connect points of constructive interference—along these lines the crests of one wave meet the crests of the other wave (solid semicircles meet solid semicircles) and the troughs meet troughs (dotted semicircles meet dotted semicircles). The path distances from any point on these lines to the two sources are either equal or differ by an *even* number of half-wavelengths (the same as a whole number of wavelengths). The waves meet in phase and constructive interference occurs. Points of constructive interference are called **antinodes**, and lines that connect them, along which the water is wavy, are called *antinodal lines*.

The lines labeled D, on the other hand, connect points of destructive interference. Along these lines the crests of one wave meet the troughs of the other (solid semicircles meet dotted semicircles). The path distances from any point on these lines to the two sources differ by an *odd* number of half-wavelengths ($\frac{1}{2}\lambda$, $\frac{3}{2}\lambda$, $\frac{5}{2}\lambda$, . . .). The waves meet out of phase and destroy each other. These points are called **nodes**, and the lines that connect them, along which the water is calm, are called *nodal lines*.

non is called **resonance**. For example, a tuning fork vibrating at its natural frequency in the vicinity of an *identical* tuning fork induces the second tuning fork to vibrate and produce sound waves of the same frequency. The amplitude of the induced sound waves increases until the waves from both tuning forks have equal amplitudes.

QUESTIONS

47. The diagram below shows a standing wave.

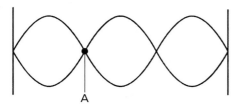

Point *A* on the standing wave is (1) a node resulting from constructive interference (2) a node resulting from destructive interference (3) an antinode resulting from constructive interference (4) an antinode resulting from destructive interference

48. The superposition of two waves traveling in the same medium produces a standing wave pattern if the two waves have
   (1) the same frequency, the same amplitude, and travel in the same direction
   (2) the same frequency, the same amplitude, and travel in opposite directions
   (3) the same frequency, different amplitudes, and travel in the same direction
   (4) the same frequency, different amplitudes, and travel in opposite directions

49. The diagram below shows two pulses of equal amplitude, *A*, approaching point *P* along a uniform string.

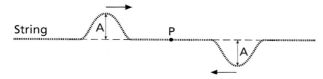

When the two pulses meet at *P*, the vertical displacement of the string at *P* will be (1) *A* (2) 2*A* (3) 0 (4) *A*/2

50. The diagram below represents a wave moving toward the right side of this page.

Which wave shown below could produce a standing wave with the original wave?

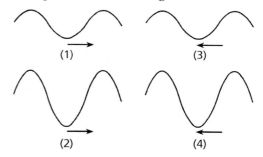

51. The diagram below represents shallow water waves of constant wavelength passing through two small openings, *A* and *B*, in a barrier.

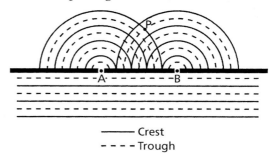

——— Crest
- - - - Trough

Which statement best describes the interference at point *P*? (1) It is constructive, and causes a longer wavelength. (2) It is constructive, and causes an increase in amplitude. (3) It is destructive, and causes a shorter wavelength. (4) It is destructive, and causes a decrease in amplitude.

52. A standing wave pattern is produced when a guitar string is plucked. Which characteristic of the standing wave immediately begins to decrease? (1) speed (2) wavelength (3) frequency (4) amplitude

53. Which phenomenon occurs when an object absorbs wave energy that matches the object's natural frequency? (1) reflection (2) amplitude (3) resonance (4) standing wave

54. The diagram below represents two waves of equal amplitude and frequency approaching point *P* as they move through the same medium.

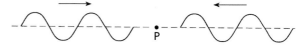

As the two waves pass through each other, the medium at point *P* will (1) vibrate up and down (2) vibrate left and right (3) vibrate into and out of the page (4) remain stationary

55. Standing waves in water are produced most often by periodic water waves (1) being absorbed at the boundary with a new medium (2) refracting at a boundary with a new medium (3) diffracting around a barrier (4) reflecting from a barrier

56. In a demonstration, a vibrating tuning fork causes a nearby second tuning fork to begin to vibrate with the same frequency. Which wave phenomenon is illustrated by this demonstration? (1) the Doppler effect (2) nodes (3) resonance (4) interference

Base your answers to questions 57 through 61 on the following diagram which represents a transverse wave.

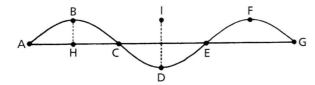

57. Which two points are in phase? (1) A and C (2) B and D (3) C and E (4) B and F

58. The amplitude of the wave is the distance between points (1) A and C (2) A and E (3) B and H (4) I and D

59. How many cycles are shown in the diagram? (1) 1 (2) 2 (3) 3 (4) 1.5

60. A wavelength is the distance between points (1) A and C (2) A and E (3) B and H (4) I and D

61. If the period of the wave is 2 seconds, its frequency is (1) 0.5 cycle/s (2) 2.5 cycles/s (3) 3.0 cycles/s (4) 1.5 cycles/s

Base your answers to questions 62 through 66 on the following diagram, which represents a vibrating string with a periodic wave originating at A and moving to G, a distance of 6.0 meters.

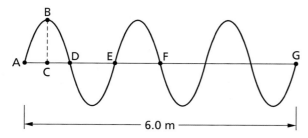

62. What type of wave is represented by the diagram? (1) elliptical (2) longitudinal (3) torsional (4) transverse

63. What is the wavelength of this wave? (1) 1.0 m (2) 2.0 m (3) 3.0 m (4) 6.0 m

64. Which phenomenon would occur if the waves were reflected at G and returned back to A through the oncoming waves? (1) diffraction (2) dispersion (3) standing waves (4) Doppler effect

65. As the wave moves toward G, point E on the string will move (1) to the left and then to the right (2) vertically down and then vertically up (3) diagonally down and then diagonally up (4) diagonally up and then diagonally down

66. If the waves were produced at a faster rate, the distance between points D and E would (1) decrease (2) increase (3) remain the same

67. A tuning fork that vibrates at a frequency of 100 hertz can produce resonance in a fork having which of the following frequencies? (1) 100 Hz (2) 200 Hz (3) 300 Hz (4) 1000 Hz

68. The diagram below shows two pulses traveling toward each other in a uniform medium.

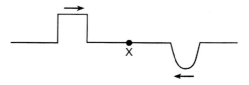

Which diagram best represents the medium when the pulses meet at point X?

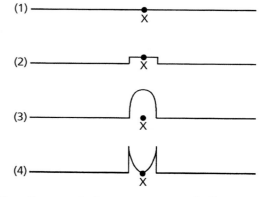

69. The diagram below represents shallow water waves of wavelength λ passing through two small openings, A and B, in a barrier.

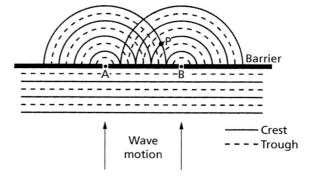

How much longer is the length of path *AP* than the length of path *BP*? (1) 1λ (2) 2λ (3) 3λ (4) 4λ

70. Which pair of moving pulses in a rope will produce destructive interference?

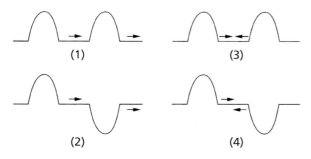

(1) (3)

(2) (4)

71. Which pair of waves will produce a resultant wave with the smallest amplitude?

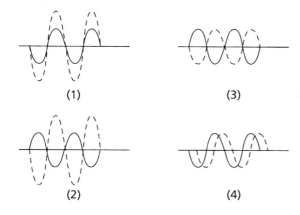

(1) (3)

(2) (4)

The Wave Nature of Light

When a charge, such as an electron, oscillates between two points (in the process accelerating and decelerating) and the frequency of oscillation is constant, a periodic *electromagnetic wave* of the same frequency radiates outward from the vicinity of the charge. The magnitude and direction of the electric and magnetic fields vary from point to point in wavelike fashion. At any particular point in the wave, the electric and magnetic fields are perpendicular to each other and to the direction of propagation of the wave. (See Figure 3E-10 on p. 123.) Light is an electromagnetic wave that can produce the sensation of sight. Much of the behavior of light can be interpreted in terms of wave phenomena.

The Double-Slit Experiment In a famous experiment, Thomas Young passed light through two slits cut into an opaque screen (Figure 4-9). The light produced a pattern of many bright bands, separated by dark bands, on another screen behind the slits. How could as many as twenty bands of light result from only two slits?

Interference Pattern Young explained this phenomenon by postulating that light is a wave. When the light wave arrives at the wall with the slits, the part of every successive wave front that passes through the slits act as a new source of waves. This produces a whole set of semicircles of crests and troughs that spread out behind each slit (Figure 4-9). This behavior is characteristic of waves of all types and is summarized by *Huygen's principle—every point on a wavefront acts as a source of waves with the same speed.* A water wave passing through a slit in a wall, for example, also produces a set of crests and troughs that expand in all directions behind the slit.

Since the light that approaches the two slits comes from one source, the slits act as sources of **coherent light waves**—waves of the same frequency that are produced by sources in phase. When a crest arrives at one slit, a crest arrives at the

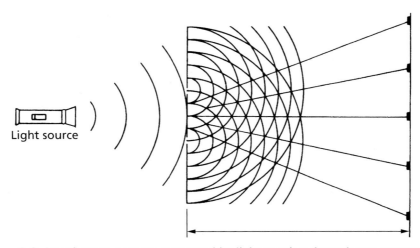

Figure 4-9. Interference pattern generated by light passing through two narrow slits.

other. Thus, behind the wall with the slits, there are two overlapping coherent waves. This leads to the alternating columns of constructive and destructive interference, described earlier for two overlapping water waves. Where a column of constructive interference meets the screen, a band of light appears. Where a column of destructive interference meets the screen a dark spot appears because the two light waves destroy each other.

The pattern of bright and dark bands on the screen reveals the arrangement of the alternating lines of constructive and destructive interference. It is thus referred to as an *interference pattern*. The bands are of equal width and are equally spaced from each other.

Speed of Light Experiments have shown that the speed of light in a vacuum is 3.00×10^8 meters per second. This number is an important physical constant and is used in a wide variety of applications. It is the speed of all colors of light in a vacuum regardless of brightness or source. The speed of light is equal to the product of frequency and wavelength (as is the speed of any periodic wave). This relationship is represented by the formula $c = f\lambda$, where the letter c represents the speed of light, 3.00×10^8 m/s.

The speed of light in a material medium depends on the frequency of the light and the nature of the medium. In media other than a vacuum, light travels at speeds slower than 3.00×10^8 m/s. The exact value of the speed of light differs from material to material, and within the same material it may be different for different colors. In glass,

light travels at a rate equal to about two-thirds its speed in a vacuum (with slight variations from color to color). Through air, on the other hand, the speed of light is only slightly slower than through a vacuum.

The Electromagnetic Spectrum Light is but one member of a large family of waves that are electromagnetic in nature. The members of the family (in order of increasing wavelength) include: gamma rays, x-rays, ultraviolet, visable light, infrared, microwaves, and radio waves. All consist of electric and magnetic fields that propagate from place to place, with the strength and direction of these fields alternating in wavelike fashion. All electromagnetic waves are generated by accelerated charged particles.

The different types of electromagnetic waves make up the **electromagnetic spectrum**. They all travel through empty space at the speed of light (3.00×10^8 meters per second). They differ, however, in frequency and wavelength. As all waves must, they obey the formula $\lambda f = v$. Since they all travel at the same rate, we know that their wavelengths (λ) and frequencies (f) are inversely proportional.

Figure 4-10 shows the different types of electromagnetic waves, in order of increasing wavelength and decreasing frequency. Each wave type represents a range of frequencies and wavelengths.

The pure colors of light—red, orange, yellow, green, blue, and violet—differ from each other in frequency and wavelength. Red light has the

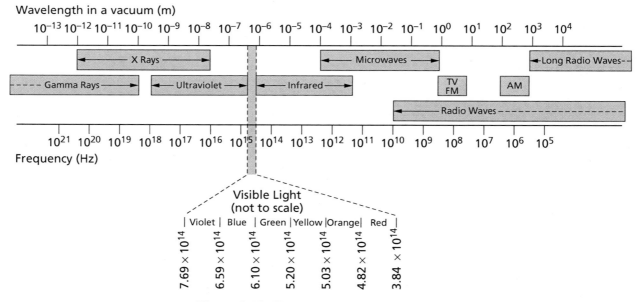

Figure 4-10. The electromagnetic spectrum.

longest wavelength and smallest frequency; violet light has the shortest wavelength and greatest frequency. Wavelengths shorter than that of violet and longer than that of red cannot be sensed by the human eye and are invisible. This is why we cannot see x-rays or radio waves. Mixtures of light waves of different wavelengths (the pure colors) produce colors other than those listed (including white). Each mixture produces a unique sensation in the eye and a unique color.

Polarization of Light Transverse waves, whose disturbances are perpendicular to the direction of motion of the wave, can be **polarized**. That is, a particular plane may be selected for the back-and-forth vibration, to the exclusion of all other possible planes. The crests and troughs in a rope, for example, can be made to point "up and down" in the plane of this page, "in or out" perpendicularly to the page, or in some in-between plane. Longitudinal waves, such as sound waves, on the other hand, cannot be polarized since their disturbances are parallel to the direction of wave motion (Figure 4-11.)

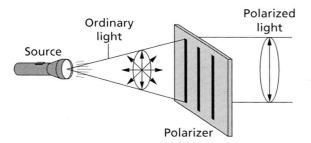

Figure 4-11. When ordinary light strikes a Polaroid filter, only waves vibrating in one plane pass through.

Experiments have determined that light can be polarized. Certain filters *(Polaroids)* allow only waves vibrating in one particular plane to pass through them; all others are blocked. This could occur only if light were a transverse wave. In addition, light can be polarized by reflection.

Diffraction Light passing through a single slit produces a pattern of bright and dark bands on a screen behind the slit. This pattern, however, is different from that produced by two slits. Directly opposite the slit there is a wide, bright band known as the *central maximum*. On either side of this dominant band of light appear much narrower and fainter bands separated by areas of darkness (Figure 4-12). This is known as a diffraction pattern.

Diffraction results from interference. Since the slit opening must have some size, the part of

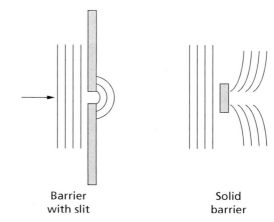

Figure 4-12. Diffraction occurs when parallel waves encounter a barrier.

the wavefront that passes through the slit is not a single point but a collection of points. Each point in this collection acts as a source of expanding waves, in accordance with Huygen's principle. Therefore, behind the slit, many overlapping waves interfere with each other, in some places constructively, in others destructively. The bright and dark bands on the screen reveal the arrangement of the areas of constructive and destructive interference.

With the unaided eye, the fainter and narrower bands on either side of the central maximum cannot be detected. Instead, we see the dominant central maximum with what appears to be fuzzy boundaries. Since the width of the central maximum grows with distance from the slit, it appears as if light spreads out upon passing through an opening or behind an obstruction. Thus, diffraction is defined as the spreading of light behind an obstacle.

The width of the central maximum band increases as the size of the slit shrinks. The narrower the slit, the wider is the central maximum. The width of the central maximum also increases as the wavelength increases.

QUESTIONS

72. Orange light has a frequency of 5.0×10^{14} hertz in a vacuum. What is the wavelength of this light? (1) 1.5×10^{23} m (2) 1.7×10^6 m (3) 6.0×10^{-7} m (4) 2.0×10^{-15} m

73. How much time does it take light from a flash camera to reach a subject 6.0 meters across a room? (1) 5.0×10^{-9} s (2) 2.0×10^{-8} s (3) 5.0×10^{-8} s (4) 2.0×10^{-7} s

74. Which wave phenomenon makes it possible for a player to hear the sound from a referee's whistle in an open field even when standing behind the referee? (1) diffraction (2) Doppler effect (3) reflection (4) refraction

75. The spreading of a wave into the region behind an obstruction is called (1) diffraction (2) absorption (3) reflection (4) refraction

76. In a vacuum, light with a frequency of 5.0×10^{14} hertz has a wavelength of (1) 6.0×10^{-21} m (2) 6.0×10^{-7} m (3) 1.7×10^6 m (4) 1.5×10^{23} m

77. Which diagram below best represents the phenomenon of diffraction?

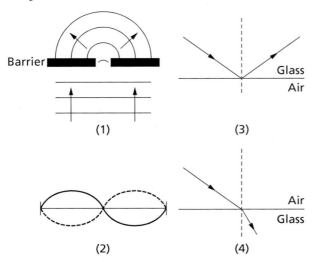

78. Radio waves diffract around buildings more than light waves do because, compared to light waves, radio waves (1) move faster (2) move slower (3) have a higher frequency (4) have a longer wavelength

79. Which wave characteristic is the same for all types of electromagnetic radiation traveling in a vacuum? (1) speed (2) wavelength (3) period (4) frequency

80. A wave is diffracted as it passes through an opening in a barrier. The amount of diffraction that the wave undergoes depends on both the (1) amplitude and frequency of the incident wave (2) wavelength and speed of the incident wave (3) wavelength of the incident wave and the size of the opening (4) amplitude of the incident wave and the size of the opening

81. At STP, sound with a wavelength of 1.66 meters has a frequency of (1) 2 Hz (2) 20 Hz (3) 200 Hz (4) 2000 Hz

82. Waves pass through a 10.-centimeter opening in a barrier without being diffracted. This ob-

servation provides evidence that the wavelength of the waves is (1) much shorter than 10. cm (2) equal to 10. cm (3) longer than 10. cm, but shorter than 20. cm (4) longer than 20. cm

83. The pattern of bright and dark bands observed when monochromatic light passes though two narrow slits is due to (1) polarization (2) reflection (3) refraction (4) interference

84. Which is *not* in the electromagnetic spectrum? (1) light waves (2) radio waves (3) sound waves (4) x-rays

85. Which characterizes a polarized wave? (1) transverse and vibrating in one plane (2) transverse and vibrating in all directions (3) circular and vibrating at random (4) longitudinal and vibrating at random

86. Whether a wave is longitudinal or transverse may be determined by its ability to be (1) diffracted (2) reflected (3) polarized (4) refracted

87. Electrons oscillating with a frequency of 2.0×10^{10} hertz produce electromagnetic waves. These waves would be classified as (1) infrared (2) visible (3) microwave (4) x-ray

88. Compared to the period of a wave of red light, the period of a wave of green light is (1) less (2) greater (3) the same

The Behavior of Light

In the previous section, you learned that light can be diffracted and polarized. Light also can be reflected; that is, it can bounce off a surface. If light was not reflected, we would be able to see only those objects that produce their own light. Light can also be refracted; it can be bent at the boundary between two media.

Reflection

The Law of Reflection Light (or any wave) that is reflected off a surface obeys the *Law of Reflection: the angle of incidence is equal to the angle of reflection.* The **angle of incidence**, θ_i, is defined as the angle between the incident ray and the **normal**, a line perpendicular to the surface (Figure 4-13). The **angle of reflection**, θ_r, is defined as the angle between the normal and the reflected ray. The Law of Reflection says that $\theta_i = \theta_r$. If a ray of light strikes a surface at an angle of 40° from the normal, it will be reflected at an angle of 40° from the normal on the opposite side. The incident ray,

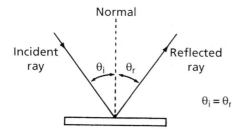

Figure 4-13. The Law of Reflection.

the reflected ray, and the normal all lie in the same plane.

Regular Reflection In real life, a light source typically emits a multitude of rays in different directions. The law of reflection applies to every individual pencil-thin ray of light that strikes a surface (Figure 4-14). If the surface is smooth, all the reflected rays can be extended to one point behind the surface as illustrated in Figure 4-14. To an eye intercepting these reflected rays, the rays all appear to be coming from that point. The point looks like the source of light. We say that the eye sees an image of the source behind the surface at that point.

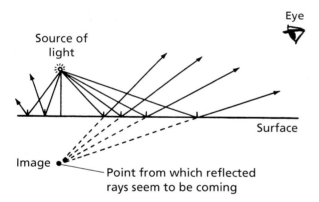

Figure 4-14. Regular reflection.

If the source is not just a point of light but an extended object, an image is formed of the entire object. The image is erect (upright), left-right reversed, equal in size to that of the object, and each part of the image appears as far behind the surface as the corresponding part of the object is in front of the surface. We see this every time we look at ourselves in a plane mirror.

Diffuse Reflection If the rays emitted by a light source encounter a rough, irregular surface, the reflected rays are scattered in different directions, as shown in Figure 4-15. The reflected rays do not seem to be emanating from any one point and no

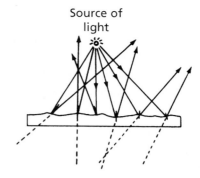

Figure 4-15. Diffuse reflection.

image appears. Diffuse reflection allows us to see objects that do not produce their own light.

Refraction

The Law of Refraction When light (or any wave) passes obliquely (at an angle other than 90°) from one medium into another, it is undergoes **refraction**. That is, the light bends at the boundary between the two media. The direction of this bending depends on the change in speed experienced by the light as it goes from the first medium into the second medium. If the speed of light is slower in the second medium, as is the case when light passes from air into glass, the light bends toward the normal. In this case, the angle between the refracted ray and the normal (the **angle of refraction**) is smaller than the angle of incidence (Figure 4-16a). On the other hand, if the speed of light is greater in the second medium, as is the case when light passes from water to air, the light bends away from the normal. In this case, the angle of refraction is greater than the angle of incidence (Figure 4-16b). If light strikes a boundary between two media perpendicularly, that is, along the normal, no bending takes place and the light passes straight through.

The Absolute Index of Refraction The amount of bending of light, whether toward or away from the normal, depends on the ratio of the

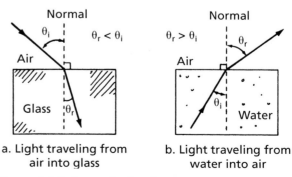

a. Light traveling from air into glass

b. Light traveling from water into air

Figure 4-16. Laws of refraction.

speed of light in the first medium to its speed in the second medium.

The **absolute index of refraction**, n, of a medium is defined as the ratio of the speed of light in vacuum (c) to the speed of light in the medium (v). This is expressed by the equation

$$n = \frac{c}{v}$$

Since the speed of light varies in different media, it follows that different substances have different absolute indices of refraction. The absolute index of refraction of vacuum is 1.00. Since light travels slower in all material media than it does through vacuum, all substances have absolute indices of refraction greater than 1.00. The absolute index of refraction of air is only slightly greater than 1.00, since the speed of light in air is only slightly less than in vacuum. A table of n values for several substances is given in Table 4-1 and in the *Reference Tables for Physical Setting/Physics*, which are an appendix to this book. Indices of refraction have no units.

Table 4-1.
Absolute indices of refraction.

Material	Absolute Index of Refraction ($f = 5.09 \times 10^{14}$ Hz)
Air	1.0
Corn oil	1.47
Diamond	2.42
Ethyl alcohol	1.36
Glass, crown	1.52
Glass, flint	1.66
Glycerol	1.47
Lucite	1.50
Quartz, fused	1.46
Sodium chloride	1.54
Water	1.33
Zircon	1.92

Sample Problem

3. The speed of yellow light in crown glass is 1.97×10^8 m/s. What is the absolute index of refraction of crown glass?

Solution:

$$n = \frac{c}{v}$$

$$n = \frac{3.0 \times 10^8 \text{ m/s}}{1.97 \times 10^8 \text{ m/s}} = 1.52$$

Snell's Law The angle of refraction for any given angle of incidence is found by using *Snell's Law*:

$$n_1 \sin \theta_1 = n_2 \sin \theta_2$$

where θ_1 is the angle of incidence, θ_2 is the angle of refraction, n_1 is the absolute index of refraction of the first medium, and n_2 is the absolute index of refraction of the second medium.

Snell's law can also be expressed in terms of the velocities of light in the two media, as follows:

$$\frac{\sin \theta_1}{\sin \theta_2} = \frac{v_1}{v_2}$$

where v_1 is the speed of light in the first medium and v_2 is the speed of light in the second medium.

Note that for any two substances, the ratio n_2/n_1 is equal to v_1/v_2. This ratio is referred to as the **relative index of refraction** of the particular pair and order of media. Also note that

$$\frac{n_2}{n_1} = \frac{v_1}{v_2} = \frac{\lambda_1}{\lambda_2}$$

Sample Problem

4. A ray of light strikes crown glass at an angle of 30° from the normal. The absolute index of refraction of crown glass is 1.52, and the speed of light in the glass is 1.97×10^8 m/s. What is the angle of refraction?

Solution:

Method I

$$n_1 \sin \theta_1 = n_2 \sin \theta_2$$

$$(1.00)(\sin 30°) = (1.52)(\sin \theta_2)$$

$$\sin \theta_2 = \frac{0.5}{1.52} = 0.33$$

$$\theta_2 = 19°$$

Method II

$$\frac{\sin \theta_1}{\sin \theta_2} = \frac{v_1}{v_2}$$

$$\frac{\sin 30°}{\sin \theta_2} = \frac{3.0 \times 10^8 \text{ m/s}}{1.97 \times 10^8 \text{ m/s}}$$

$$\frac{0.5}{\sin \theta_2} = 1.52$$

$$\sin \theta_2 = 0.33$$
$$\theta_2 = 19°$$

Critical Angle. When light crosses a boundary and its speed is greater in the second medium, the

Law of Refraction states that it bends away from the normal. The angle of refraction is then greater than the angle of incidence. Therefore, at such boundaries, there exists an angle of incidence for which the corresponding angle of refraction is 90°. At this angle of incidence, called the **critical angle**, θ_c, the ray emerges parallel to the boundary (Figure 4-17).

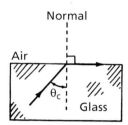

Figure 4-17. Critical angle.

Critical angles do not exist when the speed of light is slower in the second medium. The light then bends toward the normal and away from the boundary.

Total Internal Reflection The critical angle is the largest angle of incidence for which Snell's law works when light crosses a boundary where its speed is greater in the second medium. For all angles of incidence greater than the critical angle, no refraction occurs and the light does not enter the second medium. Instead, the rays are reflected back into the first medium, obeying the Law of Reflection (Figure 4-18). This phenomenon is known as *total internal reflection*.

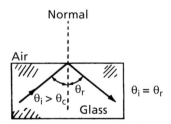

Figure 4-18. Total internal reflection.

For example, consider a ray of light that enters perpendicularly into a 45°-45°-90° glass prism (Figure 4-19). No refraction occurs at the first boundary (into the glass) since the light strikes that boundary perpendicularly. The light continues straight into the glass and approaches the opposite boundary, on the way out of the glass, with an angle of incidence of 45°. This is larger than the critical angle of glass-to-air, which is 42°. The light does not leave the prism but is internally reflected. The angle of reflection is equal to the angle of inci-

dence, and the light is reflected 45° from, and on the other side of, the normal.

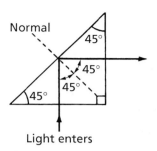

Figure 4-19. A ray of light strikes a 45°-45°-90° prism perpendicularly.

Dispersion White light is *polychromatic*. That is, it consists of waves of different colors and frequencies. In some media, called *dispersive media*, waves of different frequency travel at different speeds. Since the amount of refraction at a boundary depends on the ratio of the speed of light in the first medium to that in the second medium, it follows that at some boundaries different colors undergo different amounts of bending and have different absolute indices of refraction. When white light passes obliquely into a dispersive medium, each of the different colors and frequencies in the mixture undergoes a different amount of bending. The net result is that the colors are separated, or dispersed, by the dispersive medium.

Both glass and water are dispersive media for light. A ray of white light passing obliquely through glass emerges not as a single ray of white, but as separate rays of red, orange, yellow, green, blue, and violet. The red, which travels the fastest through glass, and therefore has the smallest index (since $n = c/v$), is refracted the least. Violet, which travels slowest, and has the greatest index of all the colors, is refracted the most (Figure 4-20).

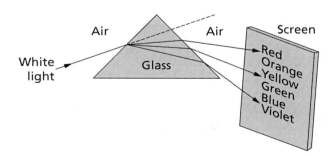

Figure 4-20. Dispersion.

A **nondispersive medium** is one in which the speed of the wave does not depend on the frequency. A vacuum is nondispersive for light.

QUESTIONS
PART A

89. As a sound wave passes from water, where the speed is 1.49×10^3 meters per second, into air, the wave's speed (1) decreases and its frequency remains the same (2) increases and its frequency remains the same (3) remains the same and its frequency decreases (4) remains the same and its frequency increases

90. In a certain material, a beam of monochromatic light ($f = 5.09 \times 10^{14}$ hertz) has a speed of 2.25×10^8 meters per second. The material could be (1) crown glass (2) flint glass (3) glycerol (4) water

91. Which ray diagram correctly represents the phenomenon of refraction?

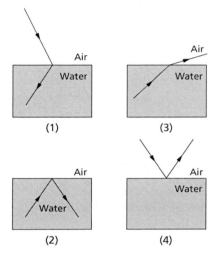

(1)   (3)

(2)   (4)

92. A beam of monochromatic light travels through flint glass, crown glass, Lucite, and water. The speed of the light beam is slowest in (1) flint glass (2) crown glass (3) Lucite (4) water

93. The diagram below shows a ray of light passing from air into glass at an angle of incidence of 0°.

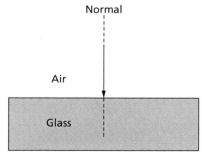

Which statement best describes the speed and direction of the light ray as it passes into the glass? (1) Only speed changes. (2) Only direction changes. (3) Both speed and direction change. (4) Neither speed nor direction changes.

94. A ray of monochromatic light is incident on an air–sodium chloride boundary as shown in the diagram below. At the boundary, part of the ray is reflected back into the air and part is refracted as it enters the sodium chloride.

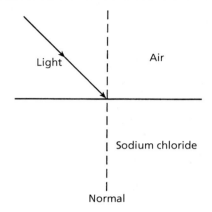

Compared to the ray's angle of refraction in the sodium chloride, the ray's angle of reflection in the air is (1) smaller (2) larger (3) the same

95. A ray of monochromatic light ($f = 5.09 \times 10^{14}$ hertz) in air is incident at an angle of 30.° on a boundary with corn oil. What is the angle of refraction, to the nearest degree, for this light ray in the corn oil? (1) 6° (2) 30° (3) 20° (4) 47°

96. A laser beam is directed at the surface of a smooth, calm pond as represented in the diagram below.

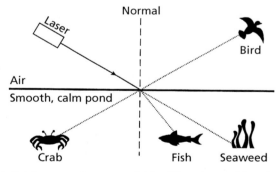

Which organisms could be illuminated by the laser light? (1) the bird and the fish (2) the bird and the seaweed (3) the crab and the seaweed (4) the crab and the fish

97. The speed of light ($f = 5.09 \times 10^{14}$ hertz) in a transparent material is 0.75 times its speed in air. The absolute index of refraction of the material is approximately (1) 0.75 (2) 1.3 (3) 2.3 (4) 4.0

98. The following diagram shows wavefronts spreading into the region behind a barrier.

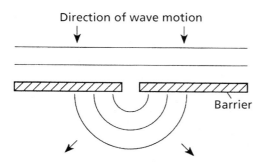

Direction of wave motion

Barrier

Which wave phenomenon is represented in the diagram? (1) reflection (2) refraction (3) diffraction (4) standing waves

99. The diagram below represents the wave pattern produced by two sources located at points A and B.

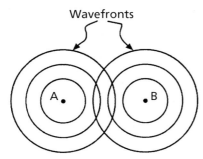

Wavefronts

Which phenomenon occurs at the intersections of the circular wavefronts? (1) diffraction (2) interference (3) refraction (4) reflection

Base your answers to questions 100 through 102 on the following diagram, which represents a ray of monochromatic light incident upon the surface of plate X. The values of n in the diagram represent absolute indices of refraction.

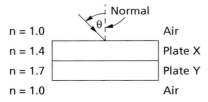

n = 1.0	Air
n = 1.4	Plate X
n = 1.7	Plate Y
n = 1.0	Air

100. The speed of the light ray in plate X is approximately (1) 1.8×10^8 m/s (2) 2.1×10^8 m/s (3) 2.5×10^8 m/s (4) 2.9×10^8 m/s

101. Compared with angle θ, the angle of refraction of the light ray in plate X is (1) smaller (2) greater (3) the same

102. Compared with angle θ, the angle of refraction of the ray emerging from plate Y into the air will be (1) smaller (2) greater (3) the same

103. Which wave phenomenon could *not* be demonstrated with a single wave pulse? (1) a

standing wave (2) diffraction (3) reflection (4) refraction

Base your answers to questions 104 and 105 on the diagram below, which represents a light ray traveling from air to Lucite to medium Y and back into air.

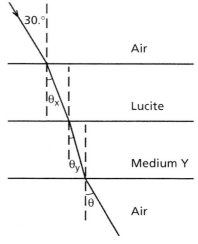

104. The sine of angle x is (1) 0.333 (2) 0.500 (3) 0.707 (4) 0.886

105. Light travels *slowest* in (1) air, only (2) Lucite, only (3) medium Y, only (4) air, Lucite, and medium Y

106. The diagram below represents a ray of monochromatic light ($f = 5.09 \times 10^{14}$ hertz) passing from medium X ($n = 1.46$) into fused quartz.

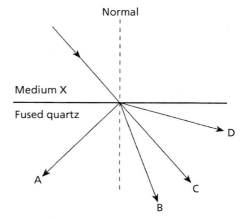

Which path will the ray follow in the quartz? (1) A (2) B (3) C (4) D

107. A monochromatic ray of light ($f = 5.09 \times 10^{14}$ hertz) traveling in air is incident upon medium A at an angle of 45°. If the angle of refraction is 29°, medium A could be (1) water (2) fused quartz (3) Lucite (4) flint glass

Physics in Your Life

Seismic Waves

When you think of waves you might think of waves on a rope or waves in a body of water. You may even imagine sound or light waves. But there are other types of waves that can travel through Earth and eventually cause changes on Earth's surface. These waves are called seismic waves, and they are produced by earthquakes.

An earthquake occurs when layers of rock beneath Earth's surface move as a result of built-up stress. The disturbance during this motion travels in the form of waves that spread out in every direction from the point where the earthquake occurs, known as the focus.

There are three general types of seismic waves: primary waves, secondary waves, and surface waves. Primary waves, or P waves, are longitudinal waves. P waves travel faster than secondary or surface waves and are therefore detected before the others. The medium for these waves is rock, which experiences compressions and rarefactions much like air does as a sound wave travels through it.

Secondary, or S waves, are transverse waves. S waves vibrate from side to side and push the ground back and forth or up and down. As a result, they shake structures on Earth's surface. Unlike P waves, S waves cannot travel through liquids. Therefore, S waves cannot travel through Earth's liquid core to the opposite side of the planet.

Surface waves are sometimes formed when P waves and S waves reach Earth's surface. These waves are a combination of longitudinal and transverse waves. They move in an almost circular pattern along the surface. Although surface waves travel more slowly than either P waves or S waves, they are responsible for the most severe ground movements during an earthquake.

Questions

1. Compare primary, secondary, and seismic waves.

2. A seismograph is a device that detects seismic waves. Explain what characteristic of the waves is related to the amount of energy carried by the wave.

3. Which is the only type of seismic wave that can travel through Earth's core? Explain.

4. In what order do the three different types of earthquake waves arrive at any point removed some distance from the center of the quake?

5. Which type of wave is the cause of the most severe shaking of buildings during an earthquake?

Chapter Review Questions

PART A

1. If the frequency of a periodic wave is doubled, the period of the wave will be (1) halved (2) doubled (3) quartered (4) quadrupled

2. How are electromagnetic waves that are produced by oscillating charges and sound waves that are produced by oscillating tuning forks similar? (1) Both have the same frequency as their respective sources. (2) Both require a matter medium for propagation. (3) Both are longitudinal waves. (4) Both are transverse waves.

3. Wave motion in a medium transfers (1) energy, only (2) mass, only (3) both mass and energy (4) neither mass nor energy

4. What happens to the frequency and the speed of an electromagnetic wave as it passes from air into glass? (1) The frequency decreases and the speed increases. (2) The frequency increases and the speed decreases. (3) The frequency remains the same and the speed increases. (4) The frequency remains the same and the speed decreases.

5. When observed from Earth, the wavelengths of light emitted by a star are shifted toward the red end of the electromagnetic spectrum. This redshift occurs because the star is (1) at rest relative to Earth (2) moving away from Earth (3) moving toward Earth at decreasing speed (4) moving toward Earth at increasing speed

6. Two pulses, A and B, travel toward each other along the same rope, as shown below.

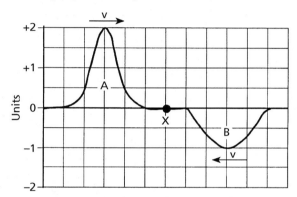

When the centers of the two pulses meet at point X, the amplitude at the center of the resultant pulse will be (1) +1 unit (2) +2 units (3) 0 (4) −1 unit

7. A single vibratory disturbance moving through a medium is called a (an) (1) node (2) antinode (3) standing wave (4) pulse

8. Radio waves and gamma rays traveling in space have the same (1) frequency (2) amplitude (3) period (4) speed

9. Which pair of terms best describes light waves traveling from the sun to Earth? (1) electromagnetic and transverse (2) electromagnetic and longitudinal (3) mechanical and transverse (4) mechanical and longitudinal

10. Which quantity is the product of the absolute index of refraction of water and the speed of light in water? (1) wavelength of light in a vacuum (2) frequency of light in water (3) sine of the angle of incidence (4) speed of light in a vacuum

11. A sonar wave is reflected from the ocean floor. For which angles of incidence do the wave's angle of reflection equal its angle of inci-

dence? (1) angles less than 45°, only (2) an angle of 45°, only (3) angles greater than 45°, only (4) all angles of incidence

12. A student in a band notices that a drum vibrates when another instrument emits a certain frequency note. This phenomenon illustrates (1) reflection (2) resonance (3) refraction (4) diffraction

13. The diagram below shows two pulses, A and B, approaching each other in a uniform medium.

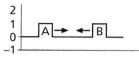

Which diagram best represents the superposition of the two pulses?

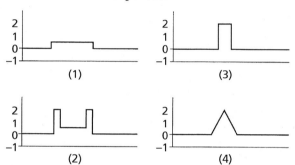

14. A beam of monochromatic light ($f = 5.09 \times 10^{14}$ hertz) passes through parallel sections of glycerol, medium X, and medium Y as shown in the diagram below.

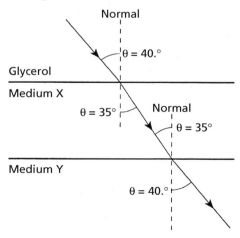

What could medium X and medium Y be? (1) X could be flint glass and Y could be corn oil. (2) X could be corn oil and Y could be flint glass. (3) X could be water and Y could be glycerol. (4) X could be glycerol and Y could be water.

15. A spaceship is moving away from Earth when a 2.00×10^6 hertz radio signal is received from Earth. Compared to the frequency of the signal sent from Earth, the frequency of the signal received by the spaceship is (1) lower (2) higher (3) the same

PART B-2

16. The diagram below shows a plane wave passing through a small opening in a barrier.

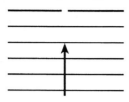

On a diagram *on a blank sheet of paper*, sketch four wavefronts after they have passed through the barrier.

Base your answers to questions 17 through 19 on the information and diagram below in which three waves, A, B, and C, travel 12 meters in 2.0 seconds through the same medium.

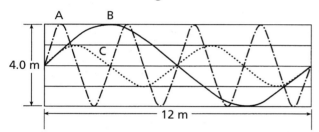

17. What is the amplitude of wave *C*?

18. What is the period of wave *A*?

19. What is the speed of wave *B*?

Base your answers to questions 20 and 21 on the information and diagram below.

In the diagram, a light ray, *R*, strikes the boundary of air and water.

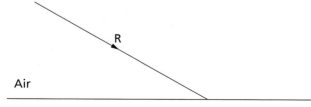

20. Using a protractor, determine the angle of incidence.

21. Using a protractor and straightedge, draw the reflected ray on a diagram on your own paper.

Base your answers to questions 22 through 24 on the information and diagram below.

A light ray with a frequency of 5.09×10^{14} hertz traveling in air is incident at an angle of $40.°$ on an air-water interface as shown. At the interface, part of the ray is refracted as it enters the water and part of the ray is reflected from the interface.

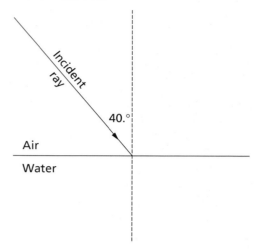

22. Calculate the angle of refraction of the light ray as it enters the water. Show all work, including the equation and substitution with units.

23. On a diagram *on a blank sheet of paper*, using a protractor and straightedge, draw the refracted ray. Label this ray "Refracted ray."

24. On a diagram *on a blank sheet of paper*, using a protractor and straightedge, draw the reflected ray. Label this ray "Reflected ray."

Base your answers to questions 25 through 28 on the diagram below, which represents a ray of monochromatic light ($f = 5.09 \times 10^{14}$ hertz) in air incident on flint glass.

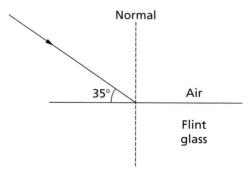

25. Determine the angle of incidence of the light ray in air.

26. Calculate the angle of refraction of the light ray in the flint glass. Show all work, including the equation and substitution with units.

27. Using a protractor and straightedge, draw the refracted ray on a diagram *on your own paper*.

28. What happens to the light from the incident ray that is *not* refracted or absorbed?

29. An FM radio station broadcasts its signal at a frequency of 9.15×10^7 hertz. Determine the wavelength of the signal in air.

30. A ray of light traveling in air is incident on an air-water boundary as shown below.

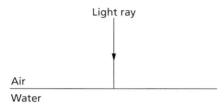

On a diagram on *a blank sheet of paper,* draw the path of the ray in the water.

31. A ray of monochromatic light with a frequency of 5.09×10^{14} hertz is transmitted through four different media, listed below.

A. corn oil
B. ethyl alcohol
C. flint glass
D. water

Rank the four media from the one through which the light travels at the slowest speed to the one through which the light travels at the fastest speed. Use the letters in front of each medium to indicate your answer.

32. The diagram below represents a transverse wave moving along a string.

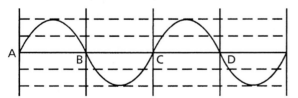

On a diagram *on a blank sheet of paper,* draw a transverse wave that would produce complete destructive interference when superimposed with the original wave.

33. A beam of light travels through medium X with a speed of 1.80×10^8 meters per second. Calculate the absolute index of refraction of medium X. Show all work, including the equation and substitution with units.

Base your answers to questions 34 and 35 on the information below.

A student plucks a guitar string and the vibrations produce a sound wave with a frequency of 650 hertz.

34. The sound wave produced can best be described as a (1) transverse wave of constant amplitude (2) longitudinal wave of constant frequency (3) mechanical wave of varying frequency (4) electromagnetic wave of varying wavelengths

35. Calculate the wavelength of the sound wave in air at STP. Show all work, including the equation and substitution with units.

36. Rubbing a moistened finger around the rim of a water glass transfers energy to the glass at the natural frequency of the glass. Which wave phenomenon is responsible for this effect?

37. Calculate the wavelength in a vacuum of a radio wave having a frequency of 2.2×10^6 hertz. Show all work, including the equation and substitution with units.

38. Two monochromatic, coherent light beams of the same wavelength converge on a screen. The point at which the beams converge appears dark. Which wave phenomenon best explains this effect?

39. Exposure to ultraviolet radiation can damage skin. Exposure to visible light does not damage skin. State *one* possible reason for this difference.

Base your answers to questions 40 through 42 on the diagram below which shows a ray of monochromatic light (f = 5.09 × 10¹⁴ hertz) passing through a flint glass prism.

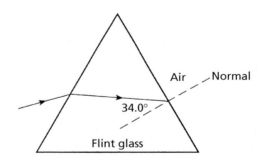

40. Calculate the angle of refraction of the light ray as it enters the air from the flint glass prism. Show all calculations, including the equation and substitution with units.

41. Using a protractor and a straightedge, construct the refracted light ray in the air on a diagram *on your own paper*.

42. What is the speed of the light ray in flint glass? (1) 5.53×10^{-9} m/s (2) 1.81×10^8 m/s (3) 3.00×10^8 m/s (4) 4.98×10^8 m/s

43. Determine the color of a ray of light with a wavelength of 6.21×10^{-7} meter.

Base your answers to questions 44 and 45 on the information below.

A periodic transverse wave has an amplitude of 0.20 meter and a wavelength of 3.0 meters.

44. On a grid, draw at least one cycle of this periodic wave.

45. If the frequency of this wave is 12 hertz, what is its speed? (1) 0.25 m/s (2) 12 m/s (3) 36 m/s (4) 4.0 m/s

PART C

Base your answers to questions 46 and 47 on the passage below.

Shattering Glass

An old television commercial for audio recording tape showed a singer breaking a wine glass with her voice. The question was then asked if this was actually her voice or a recording. The inference is that the tape is of such high quality that the excellent reproduction of the sound is able to break glass. This is a demonstration of resonance. It is certainly possibly to break a wine glass with an amplified singing voice. If the frequency of the voice is the same as the natural frequency of the glass, and the sound is loud enough, the glass can be set into a resonant vibration whose amplitude is large enough to surpass the elastic limit of the glass. But the inference that high-quality reproduction is necessary is not justified. All that is important is that the frequency be recorded and played back correctly. The waveform of the sound can be altered as long as the frequency remains the same. Suppose, for example, that the singer sings a perfect sine wave, but the tape records it as a square wave. If the tape player plays the sound back at the right speed, the glass will still receive energy at the resonance frequency and will be set into vibration leading to breakage, even though the tape reproduction was terrible. Thus, this phenomenon does not require high-quality reproduction and, thus, does not demonstrate the quality of the recording tape. What it does demonstrate is the quality of the tape player, in that it played back the tape at an accurate speed!

46. List *two* properties that a singer's voice must have in order to shatter a glass.

47. Explain why the glass would not break if the tape player did not play back at an accurate speed.

Base your answers to questions 48 and 49 on the information below.

A transverse wave with an amplitude of 0.20 meter and wavelength of 3.0 meters travels toward the right in a medium with a speed of 4.0 meters per second.

48. On a copy of the diagram *on a blank sheet of paper*, place an X at each of *two* points that are in phase with each other.

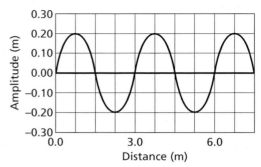

49. Calculate the period of the wave. Show all work, including the equation and substitution with units.

Base your answers to questions 50 through 52 on the information below.

A periodic wave traveling in a uniform medium has a wavelength of 0.080 meter, an amplitude of 0.040 meter, and a frequency of 5.0 hertz.

50. Determine the period of the wave.

51. On a grid *on a blank sheet of paper*, starting at point *A*, sketch a graph of *at least one* complete cycle of the wave showing its amplitude and period.

52. Calculate the speed of the wave. Show all work, including the equation and substitution with units.

Base your answers to questions 53 through 55 on the information and diagram below.

A monochromatic beam of yellow light, AB, is incident upon a Lucite block in air at an angle of 33°.

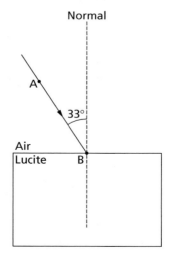

53. Calculate the angle of refraction for incident beam *AB*. Show all work, including the equation and substitution with units.

54. Using a straightedge, a protractor, and your answer from question 54, draw an arrow on a diagram *on your own paper* to represent the path of the refracted beam.

55. Compare the speed of the yellow light in air to the speed of the yellow light in Lucite.

Base your answers to questions 56 and 57 on the information and diagram below.

A ray of light of frequency 5.09×10^{14} hertz is incident on a water-air interface as shown in the diagram below.

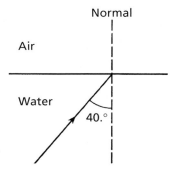

56. Calculate the angle of refraction of the light ray in air. Show all work, including the equation and substitution with units.

57. Calculate the speed of the light while in the water. Show all work, including the equation and substitution with units.

Base your answers to questions 58 and 59 on the information and graph below.

Sunlight is composed of various intensities of all frequencies of visible light. The graph represents the relationship between light intensity and frequency.

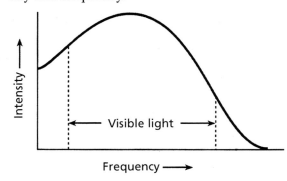

58. Based on the graph, which color of visible light has the lowest intensity?

59. It has been suggested that fire trucks be painted yellow-green instead of red. Using information from the graph, explain the advantage of using yellow-green paint.

Base your answers to questions 60 through 62 on the information and diagram below.

A ray of light passes from air into a block of transparent material *X* as shown in the diagram below.

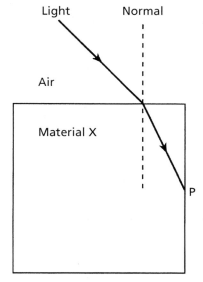

60. Measure the angles of incidence and refraction to the nearest degree for this light ray at the air–material *X* boundary

61. Calculate the absolute index of refraction of material *X*. Show all work, including the equation and substitution with units.

62. The refracted light ray is reflected from the material *X*–air boundary at point *P*. Using a protractor and straightedge, draw the reflected ray from point *P on your own paper*.

Enrichment
Wave Phenomena

DOUBLE-SLIT INTERFERENCE PATTERN

When coherent light waves pass through two slits in a barrier on their way to a screen, a pattern of bright and dark bands forms on the screen. The bands are equally spaced and of the same width. This pattern is the result of alternating lines of constructive and destructive interference.

The distance between two adjacent bands of light, x, and the distance between the barrier and the screen, L, are related to the distance between the slits, d, and the wavelength of the light, λ, by the formula

$$\frac{\lambda}{d} = \frac{x}{L}$$

When using this relationship, you must take care to express all distances in the same unit of length, for example, the meter. (See Figure 4E-1)

Sample Problem

E1. A beam of red light is made to pass through two slits that are $.4.0 \times 10^{-3}$ m apart. On a screen 1.0 meter distant from the slits an interference pattern appears with bands of light separated by 1.8×10^{-4} m. What is the wavelength of the light?

Solution:

$$\frac{\lambda}{d} = \frac{x}{L}$$

$$\frac{\lambda}{4.0 \times 10^{-3} \text{ m}} = \frac{1.8 \times 10^{-4} \text{ m}}{1.0 \text{ m}}$$

$$\lambda = 7.2 \times 10^{-7} \text{ m}$$

In doing this experiment with various colors of light, it has been found that the spacing between the bands, x, is different for different colors of light. This can only mean that the wavelength, λ, is different for different colors.

Calculating the Critical Angle

When light crosses a boundary and its speed is greater in the second medium, the Law of Refraction states that it bends away from the normal. The angle of refraction is then greater than the angle of incidence. Therefore, at such boundaries, there exists an angle of incidence for which the corresponding angle of refraction is 90°. At this angle of incidence, called the *critical angle*, θ_c, the ray emerges parallel to the boundary (Figure 4-17 on

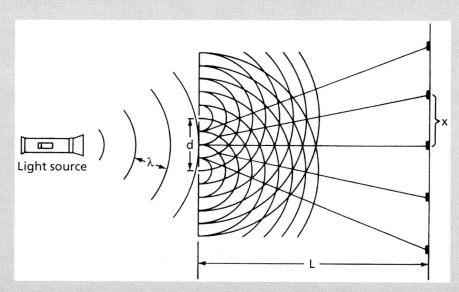

Figure 4E-1. Interference pattern generated by light passing through two narrow slits.

p. 153). The value of θ_c depends on the absolute indices of refraction of the two media and is therefore different for different media.

$$\sin \theta_C = \frac{n_2}{n_1}$$

If the first medium is some material and the second is vacuum (or air), the value of θ_c is given by the formula

$$\sin \theta_C = \frac{1}{n}$$

where n is the absolute index of refraction of the first medium.

Sample Problem

E2. What is the critical angle when light passes from diamond to air? (The absolute index of refraction of diamond is 2.42.)

Solution:

$$\sin \theta_C = \frac{1}{n} = \frac{1}{2.42} = 0.41$$

$$\theta_C = 24°$$

Questions

Base your answer to question E1 on the following diagram, which shows light from a monochromatic source incident on a screen after passing through a double slit.

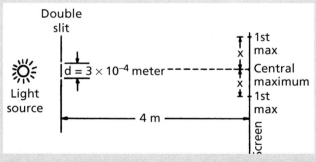

E1. What is the wavelength of the light source if the distance between the central light band and the next light band is 0.01 meter? (1) 6.7×10^5 m (2) 8.3 m (3) 3.3×10^{-1} m (4) 7.5×10^{-7} m

Base your answers to questions E2 through E6 on the following diagram.

The diagram represents two parallel slits 2.0×10^{-4} meter apart that are illuminated by parallel rays of monochromatic light of wavelength 6.0×10^{-7} meter. The interference pattern is formed on a screen 2.0 meters from the slits.

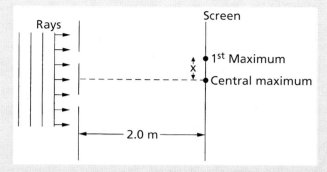

E2. Distance x is (1) 6.0×10^{-3} m (2) 6.0×10^{-7} m (3) 3.0×10^{-3} m (4) 3.0 m

E3. The difference in path length for the light from each of the two slits to the first maximum is (1) λ (2) 2λ (3) $\lambda/2$ (4) 0

E4. If the wavelength of the light passing through the slits is doubled, the distance from the central maximum to the first maximum will (1) decrease (2) increase (3) remain the same

E5. If the screen is moved closer to the slits, the distance between the central maximum and the first maximum will (1) decrease (2) increase (3) remain the same

E6. If the distance between the slits is decreased, the distance between the central maximum and the first maximum will (1) decrease (2) increase (3) remain the same

Base your answer to question E7 on the following diagram, which represents monochromatic light incident upon a double slit in barrier A, producing an interference pattern on screen B.

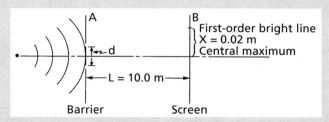

E7. If $x = 0.02$ meter, $L = 10.0$ meters, and the wavelength of the incident light is a 5.0×10^{-7} meter, the distance d between the slits is (1) 2.5×10^{-4} m (2) 2.0×10^{-2} m (3) 2.5×10^{-2} m (4) 4.0×10^{-5} m

E8. Light ($\lambda = 5.9 \times 10^{-7}$ meter) travels through a solution. If the absolute index of refraction of the solution is increased, the critical angle will (1) decrease (2) increase (3) remain the same

GEOMETRIC OPTICS

Plane Surfaces

We learned earlier (p. 151) that rays of light from a point-source that are reflected off a smooth surface appear to come from a single point behind the surface. As a result, the eye sees an image of the source behind the surface. Since the image is not produced by rays of light actually coming to a point, but merely seem to be coming from a point, we call it a **virtual image**.

Careful study of Figure 4E-2 reveals that the distance between the source of light and the surface (the object distance, d_o) is equal to the distance between the surface and the image (the image distance, d_i). These distances are measured along the line that connects the object (the source) and the image. That line is perpendicular to the surface.

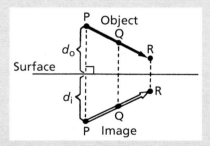

Figure 4E-2. Reflection from a plane surface creates a virtual image.

If the source is not just a point of light but an extended object, an image is formed of the entire object. The image is erect, reversed, equal in size to that of the object, and appears as far behind the surface as the corresponding part of the object is in front of the surface (Figure 4E2). We see this every time we look at ourselves in a plane mirror.

Converging Lenses

A lens that is thicker in the middle than at the edges is said to be **convex** in shape. A line drawn perpendicularly to the plane of such a lens and through its center is referred to as the **principal axis** of the lens (Figure 4E-3). A ray of light traveling along the principal axis is not refracted, because it enters and leaves the lens perpendicularly to its surfaces. Rays of light that pass through the center of the lens not along the principal axis are refracted, but emerge parallel to their original di-

rection. If the lens is thin, however, these rays emerge, for all practical purposes, lined up with their original direction.

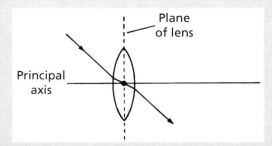

Figure 4E-3. A convex lens.

If we apply the law of refraction to each of a group of rays that strike the lens parallel to the principal axis, we find that the lens acts to bring these rays together, to a point, on the other side of the lens. Convex lenses are therefore referred to as **converging lenses**, and the point where the incoming parallel rays meet is known as the **focal point** or **principal focus** of the lens. The focal point is situated on the principal axis (Figure 4E-4). The distance between the center of the lens and the focal point, called the **focal length** of the lens, depends upon the material the lens is made of, its size and curvature. It is, therefore, different for different lenses.

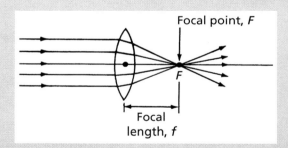

Figure 4E-4. Light rays striking a convex lens parallel to the principal axis converge at the focal point.

Real Images Rays of light that come from an object situated on one side of a convex lens and not too distant from it, do not, however, approach the lens parallel to each other. Ray 1 from the top of the object-source of light in Figure 4E-5 happens to be traveling parallel to the principal axis and is refracted in such a way that it passes through the focal point, F, on the opposite side of the lens. But rays 2 and 3 that emanate from the same point are not parallel to the principal axis. They, too, are refracted by the lens, on their way in and again on the way out, but these refractions do not direct them to the focal point.

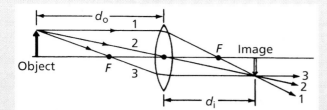

Figure 4E-5. A convex lens with the object distance greater than the focal length.

Instead, all the rays coming from the top of the object-source are made to come together, by the action of the lens, at some point other than the focal point. The same happens to the rays that come from every other part of the object-source.

The points of convergence of rays from different parts of the object-source are arranged, near each other, in the same way that the object parts are arranged in the object. The net result is that a screen placed at the appropriate distance from the lens reveals a clear image of the entire object.

This phenomenon occurs when the object distance (the distance between the object and the center of the lens) is greater than the focal length. Since this image is created by the actual convergence of rays of light (unlike the image formed by plane mirrors) we call it a **real image**. Real images are always inverted, and can be projected on a screen or film in a camera.

Virtual Images When the object distance is less than the focal length, the refracted rays of light do not come together and cannot form a real image. Instead, they diverge as they emerge from the other side of the lens. However, to the eye it seems as if those rays are coming from one point. As a result, the eye sees a **virtual image** of the object on the same side of the lens as the object (Figure 4E-6). Virtual images are always upright, and cannot be projected onto a screen.

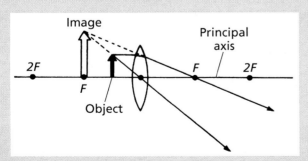

Figure 4E-6. A convex lens with the object distance less than the focal length.

In the case of an object distance equal to the focal length, no image of any type is formed. Just

as incoming parallel rays meet at the focal point, so rays coming from the focal point emerge from the lens parallel to each other. They neither converge to form a real image nor diverge to form a virtual image (Figure 4E-7).

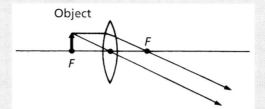

Figure 4E-7.

Size and Location of Image The location of an image produced by a convex lens can be determined from the formula

$$\frac{1}{d_o} + \frac{1}{d_i} = \frac{1}{f}$$

where d_o is the object distance, d_i is the distance between the lens and the image (known as the image distance) and f is the focal length. All of these distances are measured from the center of the lens, along the principal axis. (Figure 4E-5)

Once the position of the image is known, its size (length or width) can be determined from the formula

$$\frac{s_o}{s_i} = \frac{d_o}{d_i}$$

where s_o is the size of the object-source and s_i is the size of the image.

Alternatively, the size and location of the image can be determined by drawing a ray diagram. Start by placing the object-source on the principal axis. Draw one ray from the top of the object parallel to the principal axis, and another from the top of the object through the center of the lens. The first ray emerges from the lens to pass through the focal point; the second goes straight through the lens. If the rays converge, the image is real and the top of the inverted image coincides with the point where the drawn rays meet (on the side of the lens opposite that of the object). If the rays diverge, the image is virtual and the top of the upright image coincides with the point where the divergent rays seem to be coming from (on the same side of the lens as the object). The image distance and size can then be determined by measurement. If the rays neither converge nor diverge, no image of any type is formed.

Sample Problems

E3. An 8-cm-tall object is situated 30 cm from a convex lens whose focal length is 10 cm. How far from the lens does an image of the object appear? How tall is the image?

Solution:

a. Formula method

Image distance

$$\frac{1}{d_o} + \frac{1}{d_i} = \frac{1}{f}$$

$$\frac{1}{d_i} = \frac{1}{f} - \frac{1}{d_o}$$

$$\frac{1}{d_i} = \frac{1}{10 \text{ cm}} - \frac{1}{30 \text{ cm}} = \frac{2}{30 \text{ cm}}$$

$$d_i = 15 \text{ cm}$$

Image size

$$\frac{s_o}{s_i} = \frac{d_o}{d_i}$$

$$s_i = \frac{d_i \times s_o}{d_o}$$

$$s_i = \frac{15 \text{ cm} \times 8 \text{ cm}}{30 \text{ cm}} = 4 \text{ cm}$$

b. Ray diagram method (see Figure 4E-8)

E4. What is the **magnification** (ratio of image size to object size) of a magnifying glass (convex lens) whose focal length is 6 cm if an object is placed 2 cm behind it?

Solution:

$$\frac{1}{d_o} + \frac{1}{d_i} = \frac{1}{f}$$

$$\frac{1}{d_i} = \frac{1}{f} - \frac{1}{d_o}$$

$$\frac{1}{d_i} = \frac{1}{6 \text{ cm}} - \frac{1}{2 \text{ cm}} = \frac{-2}{6} \text{ cm}$$

$$\frac{1}{d_i} = \frac{-1}{3} \text{ cm}$$

$$d_i = -3 \text{ cm}$$

$$\frac{s_o}{s_i} = \frac{d_o}{|d_i|}$$

$$\frac{s_o}{s_i} = \frac{2}{|-3|} = \frac{2}{3}$$

$$\frac{s_i}{s_o} = \frac{3}{2} = 1.5$$

Magnification is 1.5.

Note: When the image distance is a negative number, a virtual image is formed. The absolute value of the image distance is then used in the formula

$$\frac{s_o}{s_i} = \frac{d_o}{d_i}$$

The eye, camera, telescope, microscope, projector, and magnifying glass all make use of convex lenses.

Diverging Lenses

A lens that is thinner in the middle than at the edges is **concave** in shape. Applying the law of refraction to each of a group of rays that strike such a lens parallel to its principal axis, we find that the lens acts to separate the rays (Figure 4E-9). Such lenses are therefore called **diverging lenses**.

The point from which incident rays parallel to the principal axis appear to be coming from after they are refracted by a diverging lens is called the **virtual focal point** of the lens. The distance between the virtual focal point and the center of the lens is the focal length. The formulas, definitions,

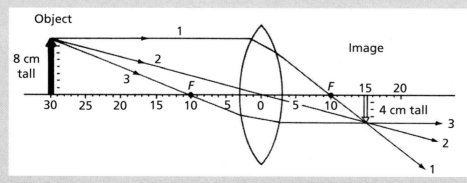

Object

8 cm tall

Image

4 cm tall

Rules

1. In parallel, out through F
2. In through center, out straight
3. In through F, out parallel

Figure 4E-8.

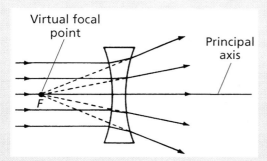

Figure 4E-9. A concave lens.

and symbols introduced above for converging lenses also apply to diverging lenses. The only exception is that the focal length, f, of a concave lens is assigned a negative value. The procedures for drawing ray diagrams are similar to those used with converging lenses.

Since diverging lenses do not bring rays of light together, they cannot form real images. However, an eye that intercepts the diverging rays sees them as coming from one point behind the lens. In this way, diverging lenses produce virtual images on the same side of the lens as the object. These images are erect and smaller than the object (Figure 4E-10). Table 4E-1 summarizes the information about lenses.

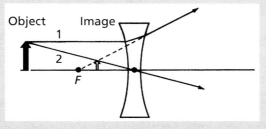

Figure 4E-10. Diverging lenses produce virtual images on the same side as the object.

Concave Mirrors

A small segment of a sphere with a reflecting surface on the inside of the sphere is called a **concave mirror**. The point that corresponds to the center of the sphere is called the **center of curvature** of the mirror. The distance between the center of curvature, C, and the mirror is known as the **radius of curvature**, R. The line that connects the center of curvature to the geometric center of the mirror is known as the *principal axis* of the mirror (Figure 4E-11). Rays of light that pass through the center of curvature from any direction, including that of the principal axis, strike the mirror perpendicularly and are reflected perpendicularly back onto themselves (as dictated by the Law of Reflection).

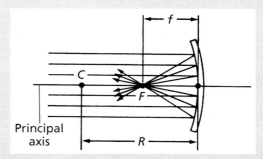

Figure 4E-11. A concave mirror.

If we apply the Law of Reflection to each of a group of rays that strike the mirror parallel to the principal axis, we find that the mirror acts to converge those rays to a point. Concave mirrors are therefore called **converging mirrors**, and the point where the incident parallel rays meet is called the *focal point* or *principal focus* of the mirror. The focal point and the center of curvature are situated on the principal axis.

Table 4E-1.

Summary Table of Lenses

Type of Lens	Object Distance	Real (inverted) Virtual (erect)	Image Size	Position of Image	Image Distance
Converging	Infinity	Real	Much smaller	Opposite	$d_i = f$
	$d_o > 2f$	Real	Smaller	Opposite	$d_i > f$
					$d_i < 2f$
	$d_o = 2f$	Real	Same	Opposite	$d_i = 2f$
	d_o between f & $2f$	Real	Larger	Opposite	$d_i > 2f$
	$d_o = f$	No image	—	—	—
	$d_o < f$	Virtual	Larger	Same	d_i negative & $|d_i| > d_o$
Diverging	Any value	Virtual	Smaller	Same	d_i negative
					$|d_i| > d_o$
					$|d_i| < |f|$

Note: When d_i is positive, the image is on opposite side of lens and real; when d_i is negative, the image is on the same side of the lens as the object and virtual.

The distance between the focal point and the mirror (measured along the principal axis) is the *focal length, f*, of the mirror. This distance is different for different mirrors but is always equal to half the radius of curvature of the mirror ($f = \frac{1}{2}R$). The focal point is always found midway between the mirror and the center of curvature.

Images by Concave Mirrors

Rays of light that come from an object situated on the reflecting side of a concave mirror at distances greater than the focal length are made to converge at a point on the same side of the mirror as the object. That point is *not* the focal point, however, since these rays do not strike the mirror parallel to the principal axis. A real, inverted image of the object appears on a screen placed at the point of convergence (Figure 4E-12).

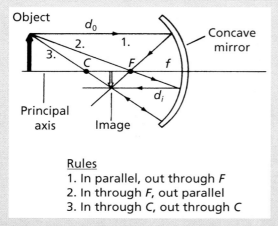

Rules
1. In parallel, out through *F*
2. In through *F*, out parallel
3. In through *C*, out through *C*

Figure 4E-12. A real image produced by a concave mirror.

As in the case of lenses, the distance between the mirror and the image, d_i, is related to the distance between the mirror and the object, d_o, and the focal length, *f*, by the formula

$$\frac{1}{d_o} + \frac{1}{d_i} = \frac{1}{f}$$

The focal length is positive and can be replaced by $f = \frac{1}{2}R$ Thus we obtain

$$\frac{1}{d_o} + \frac{1}{d_i} = \frac{2}{R}$$

The size of the image is related to the size of the object by the formula

$$\frac{s_o}{s_i} = \frac{d_o}{d_i}$$

All these distances are measured along the principal axis to the center of the reflecting surface. As in the case for lenses, the size and location of the image created by a mirror can also be found by drawing an appropriate ray diagram (Figure 4E-12).

If the object-source of light is placed at the focal point ($d_o = f$), the reflected rays emerge parallel to each other, and no image is created; the formula yields no solution for d_i (Figure 4E-13). This provides a mechanism for focusing all the rays of light emitted by a source in one direction, a technique that is widely used in searchlights and car headlights.

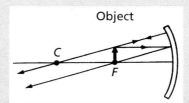

Figure 4E-13.

If the object is placed closer to the mirror than the focal point ($d_o < f$), the reflected rays diverge, and no real image is created. Instead, the eye sees the diverging rays as coming from one point on the opposite side of the mirror. A virtual, upright image appears on the side of the mirror opposite that of the object. This is indicated by the negative value for d_i obtained by applying the formula

$$\frac{1}{d_o} + \frac{1}{d_i} = \frac{1}{f}$$

in such situations (Figure 4E-14).

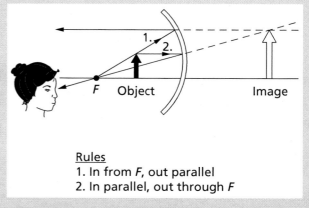

Rules
1. In from *F*, out parallel
2. In parallel, out through *F*

Figure 4E-14. A virtual image formed by a concave mirror.

Convex Mirrors

A small segment of a sphere with a reflecting surface on the outside is called a **convex mirror**. If we apply the Law of Reflection to each of a group of rays that strike the mirror parallel to the principal axis, we find that the mirror acts to separate the rays. Convex mirrors are therefore called **diverging mirrors** (Figure 4E-15).

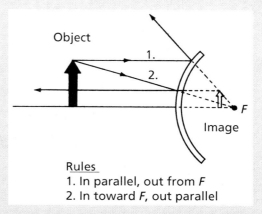

Rules
1. In parallel, out from F
2. In toward F, out parallel

Figure 4E-16. Diverging mirrors create virtual images.

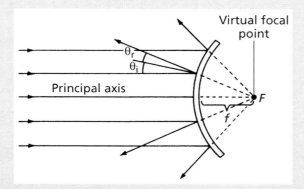

Figure 4E-15. A convex mirror.

The point where the incident rays parallel to the principal axis seem to be coming from after being reflected by the mirror is the *virtual focal point* of the mirror. The distance between this point and the center of the diverging mirror is the *focal length* of the mirror. The formulas, definitions, and symbols described earlier for converging mirrors are also applicable to diverging mirrors, as are the procedures for drawing ray diagrams. The only exception is that the focal length, f, is negative for diverging mirrors.

Since diverging mirrors do not bring rays of light together, they cannot form real images. Instead, diverging mirrors create virtual images on the side of the mirror opposite that of the object. These images are erect and smaller than the object (Figure 4E-16).

Diverging mirrors are widely used as outside rear-view mirrors for cars and as security mirrors in stores, where a wide field of view is desirable.

Table 4E-2 summarizes the information about mirrors.

Defects in Lenses and Mirrors

Since different colors of light undergo different amounts of refraction when entering and leaving a dispersive medium such as a lens (p. 153), it follows that rays of light of different color do not converge at the same point after passing through a lens. Violet light, whose wavelength is the shortest, converges closest to the lens; red light, whose wavelength is the largest, converges farthest from the lens. Different colors, as a result, have different focal lengths even when passing through the same lens.

This fact must be taken into account when images are made with an instrument, such as a camera, whose operation is based on refraction. If the light is monochromatic, the image distance (distance between lens and film) can be adjusted from

Table 4E-2.

Summary Table of Mirrors

Type of Mirror	Object Distance	Real (inverted) Virtual (erect)	Image Size	Position of Image	Image Distance
Converging	Infinity	Real	Much smaller	Same	$d_i = f$
	$d_o > 2f$	Real	Smaller	Same	$d_i > f$ $d_i < 2f$
	$d_o = 2f$	Real	Same	Same	$d_i = 2f$
	d_o between f & $2f$	Real	Larger	Same	$d_i > 2f$
	$d_o = f$	No image	—	—	
	$d_o < f$	Virtual	Larger	Opposite	d_i negative & $|d_i| > d_o$
Diverging	Any value	Virtual	Smaller	Opposite	d_i negative $|d_i| < d_o$ $|d_i| < f$

Note: When d_i is positive, the image is real and on the same side of the mirror as the object; when d_i is negative, the image is virtual and on the opposite side of the mirror.

color to color. In the case of infrared cameras, however, this would necessitate a rather large adjustment. Instead, the lens is changed to one with a different focal length.

If the light consists of a mixture of colors, as in the case of sunlight and incandescent bulbs, the various colors form separate images near one another. This results in a blurred image—a defect known as chromatic aberration. The problem can be rectified by joining the converging lens to a diverging one made of a different material.

Another defect, one that affects both lenses and mirrors, is known as spherical aberration. Rays of light that pass through the edges of a lens, or that are reflected near the edges of a mirror, do not meet all the other rays at the focal point or in the image. This problem is corrected by covering the edges of the lens with a diaphragm, or by making the mirror parabolic instead of spherical in shape.

Questions

E9. As an object is moved closer to a plane mirror, the distance between the image and the mirror will (1) decrease (2) increase (3) remain the same

E10. Which diagram best represents the reflection of an object *O* by plane mirror *M*?

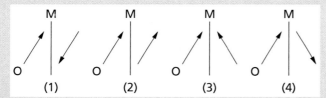

E11. Which graph best represents the relationship between the image size and the object size for an object reflected in a plane mirror?

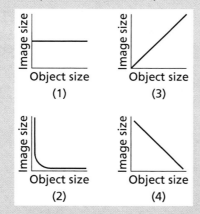

Base your answers to questions E12 through E16 on the following diagram, which represents an object placed 0.20 meter from a converging lens with a focal length of 0.15 meter.

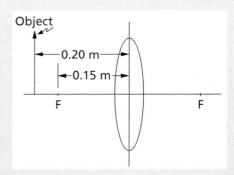

E12. Which phenomenon best describes the image formation by the lens? (1) diffraction (2) dispersion (3) polarization (4) refraction

E13. The image produced by the lens is (1) enlarged and real (2) enlarged and erect (3) diminished and virtual (4) diminished and inverted

E14. If the object distance were increased, the image would become (1) larger and erect (2) smaller and virtual (3) smaller, only (4) larger, only

E15. If the object were placed 0.10 meter from the lens, the image would be (1) enlarged and inverted (2) real and inverted (3) reduced and real (4) virtual and erect

E16. Which monochromatic light, when used to illuminate the object, would produce the *smallest* image distance? (1) red (2) yellow (3) green (4) blue

E17. In which direction does most of the light in ray *R* pass? (1) *A* (2) *B* (3) *C* (4) *D*

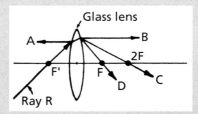

Base your answers to questions E18 through E21 on the following diagram, which represents an object that is 0.2 meter high. The object is located 0.5 meter from a converging lens with a focal length of 1.0 meter.

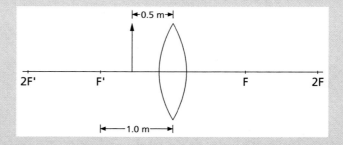

E18. The image would be described as (1) real and inverted (2) real and erect (3) virtual and inverted (4) virtual and erect

E19. The smallest image of the object would be produced by the lens when the object is located at (1) 0.5F (2) 2F (3) 3F (4) 4F

E20. If the object were moved toward the lens from the position shown in the diagram, the distance from the lens to the image would (1) decrease (2) increase (3) remain the same

E21. If the object were moved toward the lens from the position shown in the diagram, the size of the image would (1) decrease (2) increase (3) remain the same

Base your answers to questions E22 through E26 on the following information and diagram.

The diagram represents a converging lens made of Lucite, which is used to focus the parallel monochromatic yellow light rays shown. F and F' are the principal foci.

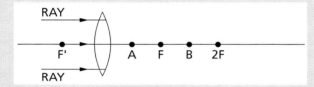

E22. The rays will pass through point (1) A (2) B (3) F (4) 2F

E23. If an object is placed between F' and the lens, the image formed would be (1) real and smaller (2) real and larger (3) virtual and smaller (4) virtual and larger

E24. If an object that is placed 0.04 meter to the left of the lens will produce a real image at a distance of 0.08 meter to the right of the lens, the focal length of the lens is approximately (1) 0.015 m (2) 0.027 m (3) 0.040 m (4) 0.080 m

E25. As the light emerges from the lens, its speed will (1) decrease (2) increase (3) remain the same

E26. The Lucite lens is replaced by a flint glass lens of identical shape. Compared with the focal length of the Lucite lens, the focal length of the flint glass lens will be (1) smaller (2) larger (3) the same

E27. The diagram below represents light rays approaching a diverging lens parallel to the principal axis.

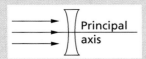

Which of the following diagrams best represents the light rays after they have passed through the diverging lens?

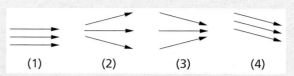

E28. A light ray is incident upon a diverging lens as shown in the diagram below.

Which of the following diagrams best represents the path of the ray after it enters the lens?

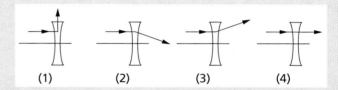

E29. In the following diagram of a concave mirror, which point represents the center of curvature?

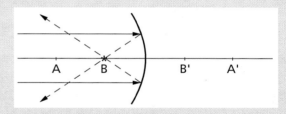

(1) A (2) B (3) B' (4) A'

E30. The following diagram represents a spherical mirror with three parallel light rays approaching. Which light ray will be reflected normal to the surface of the mirror?

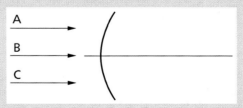

(1) A, only (2) B, only (3) C, only (4) all of the rays

E31. Which of the following are diverging instruments? (1) concave lenses and concave mirrors (2) concave lenses and convex mirrors (3) convex mirrors and convex lenses (4) convex lenses and concave mirrors

E32. The image created by a concave lens (1) can appear on either side of the lens (2) is always on the same side of the lens as the object (3) is always on the opposite side of the lens as the object

Base your answers to questions E33 through E36 on the following information.

A 10-centimeter-tall arrow is situated 12 centimeters from a concave lens whose focal length is 6 centimeters.

E33. The image distance is closest to (1) 6 cm (2) 18 cm (3) 4 cm (4) 2 cm

E34. The image length is closest to (1) 30 cm (2) 1.2 cm (3) 2.5 cm (4) 3.33 cm

E35. The image is (1) real and inverted (2) virtual and upright (3) real and upright (4) virtual and inverted

E36. If the object is moved closer to the lens, the image (1) moves away from the lens (2) remains in place (3) moves closer to the lens

Base your answers to questions E37 through E42 on the following information.

A 10-centimeter-tall arrow is situated in front of a concave mirror at the center of curvature of the mirror.

E37. The image distance will be (1) equal to the focal length (2) equal to the object distance (3) twice the object distance (4) equal to one-half the focal length

E38. The height of the image will be closest to (1) 10 cm (2) 5 cm (3) 20 cm (4) 15 cm

E39. The image will be (1) real and upright (2) virtual and upright (3) virtual and inverted (4) real and inverted

E40. If the arrow is moved farther away from the mirror, the image (1) moves away from the mirror (2) remains in place (3) moves closer to the mirror (4) disappears

E41. A source of light is placed at the focal point in front of a concave mirror. The reflected rays (1) are parallel to one another (2) meet at the center of curvature (3) diverge (4) meet at the focal point

E42. An object is placed closer to a concave mirror than is the focal point. The image (1) appears on the same side of the mirror as the object (2) appears on the opposite side of the mirror as the object (3) does not appear anywhere

CHAPTER 5

Modern Physics

The Dual Nature of Light

Two Models of Light

In the mid 1800s, scientists were convinced that the age-old question, "What is light?" had been answered conclusively. Light, as you learned in Chapter 4, is an electromagnetic wave. Polarization and diffraction offered proof of the wave nature of light. In addition, the speed of light was the same as that of other electromagnetic waves. By the late 1800s, however, certain experiments showed that light behaved as though it consisted of particles. These apparently contradictory models of light (wave and particle) took decades to resolve.

The Photoelectric Effect

When electromagnetic radiation strikes certain materials, particularly metals, electrons are ejected from them and escape into the space around the materials. This phenomenon is known as the **photoelectric effect**. Materials that behave in this manner are said to be **photoemissive**, and the emitted electrons are referred to as **photoelectrons**. This effect is the basis of the photocell that powers your solar-powered calculator.

The more intense the electromagnetic radiation that strikes a photoemissive material, the more photoelectrons ejected per second. A brighter beam of light, for example, causes more photoelectrons to be emitted per second than a dimmer one. Increasing the intensity of the radiation, however, does not result in more energetic photoelectrons. Instead, the kinetic energy of the emitted electrons depends on the frequency of the incident radiation and on the type of photoemissive material. The higher the frequency, the greater the energy of the photoelectrons.

The Quantum Theory

In 1905, Albert Einstein explained the photoelectric effect. He proposed that light and all forms of electromagnetic radiation consisted of particles called **photons**. This proposal was based on Max Planck's **quantum theory**, which states that electromagnetic radiation is emitted in discrete amounts, or **quanta**, of energy. Einstein extended this idea: not only was electromagnetic radiation *emitted* in discrete amounts of energy, it was also *absorbed* in discrete amounts, because electromagnetic radiation consists of particles, each carrying a discrete amount of energy.

The energy of each particle, or photon, of light or any type of electromagnetic radiation is directly proportional to the frequency of the radiation. It is found, according to Einstein, by using the formula

$$E = hf$$

where E is the energy, in joules, h is Planck's constant, 6.63×10^{-34} J·s, and f is the frequency of the radiation, in hertz. This formula indicates that photons of higher frequency have more energy than those of lower frequency.

Since all forms of electromagnetic radiation travel at the speed of light and satisfy the relationship $c = \lambda f$, the energy of a photon can also be expressed as

$$E = \frac{hc}{\lambda} = hf$$

where λ is the wavelength and c is the speed of light. Thus, the energy of a photon is inversely proportional to its wavelength.

In the photoelectric effect, each photon acts individually on one electron. A photon gives either all of its energy, equal to hf, to the electron it interacts with, or none. An electron that gains no energy remains in the material. One that absorbs all of the energy of the photon may or may not escape from the particular material, depending on how much energy it absorbed and how much it needs to escape.

Sample Problems

1. The frequency of orange light is 4.82×10^{14} Hz. What is the energy of the photons?

Solution:

$$E = hf$$

$$E = 6.63 \times 10^{-34} \text{ J} \cdot \text{s} \times 4.82 \times 10^{14} \text{ Hz}$$

$$E = 3.20 \times 10^{-19} \text{ J}$$

2. The wavelength of a certain color of light is 6.1×10^{-7} m. What is the energy of the photons? What is the color of the light?

Solution:

$$E_{photon} = \frac{hc}{\lambda}$$

$$E_{photon} = \frac{6.63 \times 10^{-34} \text{ J} \cdot \text{s} \times 3.00 \times 10^8 \text{ m/s}}{6.1 \times 10^{-7} \text{ m}}$$

$$E_{photon} = 3.26 \times 10^{-19} \text{ J} = 3.3 \times 10^{19} \text{ J}$$

$$E_{photon} = hf$$

$$f = \frac{E}{h}$$

$$f = \frac{3.3 \times 10^{-19} \text{ J}}{6.63 \times 10^{-34} \text{ J} \cdot \text{s}}$$

$$f = 4.98 \times 10^{14} \text{ Hz}$$

Referring to the *Physics Reference Tables*, you find that the light is orange.

Photon Momentum

The idea that light and other electromagnetic radiation consist of particles was enhanced by experiments conducted by Arthur Compton in 1922. Compton aimed beams of x-rays at electrons and showed that both energy and momentum are conserved in photon-particle collisions, just as they are in collisions between ordinary particles. (See Figure 5-1.) The energy and momentum lost by the photons are gained by the electrons they collide with. This implies that photons are like particles in every way—they have momentum

and exert force when they collide with other particles. The difference between photons and ordinary particles is that photons have no rest mass—they cannot exist in a state of rest.

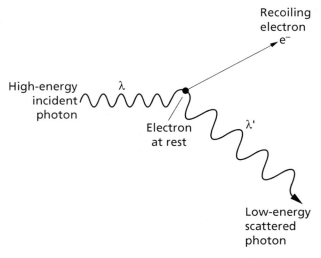

Figure 5-1. When a high-energy, short-wavelength photon collides with an electron, energy and momentum are conserved. The photon loses energy and momentum, as it becomes a low-energy, longer wavelength photon, while the electron gains kinetic energy and momentum.

Matter Waves

In 1924, Louis de Broglie proposed that if electromagnetic waves have particle properties, then moving particles should have wave properties. In other words, there ought to be wave-particle duality. He based this idea on his belief that nature is symmetrical. Experimental evidence soon confirmed his ideas. Two American physicists, C. J. Davisson and L. H. Germer, found that beams of electrons produce diffraction patterns, just as waves do. This electron diffraction supported the theory that matter has a wave nature. This was later found to be true for protons and neutrons as well. The waves associated with moving particles are known as **matter waves (de Broglie waves)**.

For subatomic particles with extremely small mass, the wavelength of the matter wave is observable. For objects of greater mass, however, the de Broglie wavelength is negligibly small, cannot be detected, and for all practical purposes, can be ignored.

QUESTIONS

PART A

1. A photon of light carries (1) energy, but not momentum (2) momentum, but not energy

(3) both energy and momentum (4) neither energy nor momentum

2. Which characteristic of electromagnetic radiation is directly proportional to the energy of a photon? (1) wavelength (2) period (3) frequency (4) path

3. Light of wavelength 5.0×10^{-7} meter consists of photons having an energy of (1) 1.1×10^{-48} J (2) 1.3×10^{-27} J (3) 4.0×10^{-19} J (4) 1.7×10^{-5} J

4. Wave-particle duality is most apparent in analyzing the motion of (1) a baseball (2) a space shuttle (3) a galaxy (4) an electron

5. The energy of a photon is inversely proportional to its (1) wavelength (2) speed (3) frequency (4) phase

6. Which phenomenon best supports the theory that matter has a wave nature? (1) electron momentum (2) electron diffraction (3) photon momentum (4) photon diffrection

7. Which two characteristics of light can best be explained by the wave theory of light? (1) reflection and refraction (2) reflection and interference (3) refraction and diffraction (4) interference and diffraction

8. A photon of which electromagnetic radiation has the most energy? (1) ultraviolet (2) x-ray (3) infrared (4) microwave

9. The graph below represents the relationship between the energy and the frequency of photons.

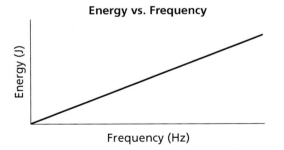

Energy vs. Frequency

The slope of the graph would be (1) 6.63×10^{-34} J·s (2) 6.67×10^{-11} N·m²/kg² (3) 1.60×10^{-19} J (4) 1.60×10^{-19} C

Base your answers to questions 10 and 11 on the following diagram, which represents monochromatic light hitting a photoemissive surface A. Each photon has 8.0×10^{-19} J of energy. B represents the particle emitted when a photon strikes surface A.

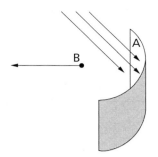

10. What is particle *B*? (1) an alpha particle (2) an electron (3) a neutron (4) a proton

11. What is the frequency of the incident light? (1) 1.2×10^{15} Hz (2) 3.7×10^{15} Hz (3) 5.3×10^{15} Hz (4) 8.3×10^{15} Hz

The Atom

With the discovery of electrons and protons, scientists turned their attention to the arrangement of these particles within the fundamental unit of matter—the atom. In the simplest model, which came to be known as the "plum pudding" model, the protons and electrons were thought to be mixed together and uniformly distributed within the entire atom. The mutual attraction between the unlike charges overcomes the repulsion between the like charges, forming a tightly bound and cohesive unit. This occurs because the distances between the unlike charges is smaller (they are arranged one beside the other) than the distances between the like charges (they are arranged diagonally across from each other). In this model, no location within the atom is favored to occupy more of one type of charge than the other. Further research proved that this model was not correct.

The Rutherford Experiment

To further probe into the structure of the atom, Ernest Rutherford conducted a series of experiments in which he aimed a beam of alpha (α) particles at thin sheets of metal foil. Alpha particles, emitted by radioactive substances, consist of two protons and two neutrons and are therefore positively charged. They are identical to the nuclei of helium atoms. Rutherford found that most of the alpha particles passed through the metal foil without being deflected. A small percentage, however, were deflected, or scattered, through angles ranging from 0° to 180°. Some even reversed course and bounced back.

Planetary Model of the Atom

To explain his results, Rutherford proposed that an atom is a sphere made up of mostly empty space with a tiny heavy core, or *nucleus*, at the center, where most of the atom's mass and all of its positive charge are located. The negative charges of the atom make up the outer shell. To explain why this model of the atom does not collapse onto itself under the influence of the electric attraction between the protons in the nucleus and the electrons in the shell, Rutherford proposed that the electrons must be in motion around the nucleus. The electric attraction provides the centripetal force necessary to keep the electrons orbiting.

Since alpha (α) particles are positively charged, they must be fragments of the nuclei of the atoms that emit them. The reason most of the alpha particles go straight through the foil is that most of the space occupied by each atom in the foil is empty. The electrons in the shells are unable to deflect the alpha particles, because alpha particles are much heavier than electrons. To the extent that a nucleus in the foil exerts a repulsive electric force on an alpha particle, another nucleus in a neighboring atom exerts a force in the opposite direction. These forces cancel each other.

However, a small percentage of the alpha particles in the beam get close enough to the heavier nuclei in the foil to experience a repulsive force that deflects the lighter α particles. An alpha particle headed straight into a nucleus is stopped by the repulsive force before it collides head-on and is turned back. An alpha particle that, while coming close to a nucleus, is not quite aiming for a head-on collision, is forced into a hyperbolic path away from the nucleus (Figure 5-2). Such alpha particles emerge deflected from their original path through some number of degrees between zero and 180.

In assuming the atom has a dense central nuclear core, Rutherford obtained excellent agreement between theory and experiment. The number of alpha particles deflected through different scattering angles matched the theoretically predicted results in every case.

Rutherford's experiments yielded a good "order of magnitude" estimate for the size of the nucleus. The radius of a typical nucleus is of the order of 10^{-14} meter. This is one ten-thousandth the radius of a typical atom, which is about 10^{-10} meter. Clearly, the atom is mostly empty space.

Rutherford's model of the atom, while successful in explaining scattering phenomena, did not resolve all the mysteries of atomic behavior. One such unresolved mystery is that of *emission spectra*. Gases under low pressure, when heated to incandescence, emit electromagnetic radiation of only certain wavelengths and frequencies and no others. The radiation in the visible light range, for example, can be separated and identified by passing the light through a prism (Figure 5-3). Lines of specific colors (wavelengths) appear, separated by gaps of darkness that correspond to missing colors. The same is true of the radiation outside the visible light range, such as radio, infrared, ultraviolet, and x-ray emissions. Specific wavelengths are emitted; wavelengths in between are not.

The radiation emitted by hydrogen consists of wavelengths that have been grouped into "series" named after their discoverers. The Balmer series is one such group; it includes the visible light wavelengths in the hydrogen spectrum (lines of red, green, and violet).

Even more curious is the fact that no two elements emit the same set of wavelengths. Each emits a unique combination, called the **emission spectrum** of that element. Unknown gases can be identified by their emission spectra, much as people can be identified by their fingerprints.

If electromagnetic radiation of mixed wavelength (such as white light) is passed through an

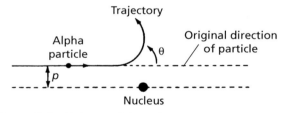

Figure 5-2. Alpha-particle trajectory.

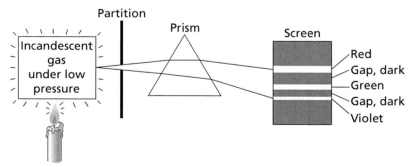

Figure 5-3. Emission spectrum.

unheated gas under low pressure, we find that certain specific wavelengths are missing from the mix of light that makes it through the gas. The list of absent colors (wavelengths) is identical to the emission spectrum of the gas and appears as a set of dark lines on a bright background. The dark lines represent the specific wavelengths that have been absorbed by the gas and constitute the **absorption spectrum** of the gas. (Figure 5-4.)

Rutherford's model does not explain why only certain wavelengths are emitted and not others, and why every element emits a different set of wavelengths.

The Bohr Atom

To explain such mysteries as the emission spectra of the elements, Niels Bohr proposed a planetary model with the electrons restricted as to where in the atom they can be. According to Bohr, the electrons in the shell of an atom are not free to be at any distance from the nucleus, nor are they allowed to have any amount of energy. Instead, they are restricted to certain distances and energy values. Each type of atom has a unique set of allowed orbits, each at a certain distance from the nucleus, and the electrons are confined to those orbits. The total energy of an electron, kinetic plus potential, is determined by the orbit it is in and no other amounts of energy are permitted to the electrons.

As an electron circles around the nucleus in one of its allowed orbits, it neither loses nor gains energy. It is said to be in a **stationary state**. If that state is also the lowest possible orbit, with the least amount of energy, the electron is said to be in the **ground state**.

These ideas contradict the classical laws of electromagnetism (p. 123). An orbiting electron must experience a centripetal acceleration and accelerating charges are supposed to emit electromagnetic waves and, in so doing, lose energy. The electrons in their stationary state, therefore, should continue to lose energy and gradually spiral inward toward the nucleus, until the atom collapses. However, atoms do not collapse and, un-

less excited, do not radiate electromagnetic waves. Clearly, the classical laws of physics required modification, and Bohr provided it.

Under ordinary conditions, the electrons in an atom are in the lowest available orbits, those with the lowest energy levels. However, the orbit and energy of the electrons can be raised through the process of **excitation**. This occurs when atoms absorb energy through heating, collision with particles, or irradiation. According to Bohr, an atom will absorb only an amount of energy equal to the difference between allowed energy levels. Other amounts of energy cannot be absorbed since doing so would place the electrons between allowed orbits. The energy values accepted by an atom are referred to as the excitation energies of the atom. Since different atoms have different sets of allowed orbits, they also have different excitation energies.

The only exception is the amount of energy equal to or greater than that needed to remove an electron from an atom. This amount of energy, known as the **ionization potential**, is equal to the difference between the energy of the electron at infinity and its energy in the ground state. Once the electron is removed, it no longer belongs to the atom, and restrictions on its energy no longer apply.

An electron that jumps to a higher orbit is said to be in an **excited state**. Soon after becoming excited, such an electron falls to a lower orbit (and lower energy level) by emitting and losing some or all of the energy it gained. As in the case of absorbed energy, the amount of energy emitted must be equal to the difference in energy between allowed orbits.

In summary: *atoms can emit and absorb energy only in quantized amounts*. Since absorption and emission are accomplished by electrons rising and falling between the same fixed energy levels, the quanta of energy that can be absorbed are the same as those that can be emitted. The energy lost by a falling electron is emitted in the form of a photon of electromagnetic radiation.

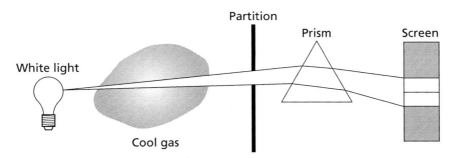

Figure 5-4. Production of an absorption spectrum.

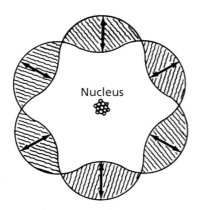

Figure 5-5. The matter wave of an electron in an allowed orbit is a standing wave.

Hydrogen Spectrum Explained

The next step was to develop a procedure to determine the allowed energy levels of atoms, so that their emission spectra could be explained and predicted. This Bohr succeeded in doing only for the hydrogen atom.

Bohr's successful idea was this: The matter wave associated with an electron in an allowed orbit is a standing wave (Figure 5-5). In other words, the circumference of an allowed orbit is equal to a whole number of wavelengths of the matter wave associated with the electron in that orbit. A helpful way to represent the allowed energy levels of an atom is by drawing an energy-level diagram. Figure 5-6 shows the allowed energy levels for hydrogen and mercury atoms.

According to Bohr, luminous gases emit electromagnetic radiation because electrons in their excited atoms descend from higher energy levels to lower energy levels. The energy lost by a descending electron becomes the energy of an emitted photon. Since a photon has energy $E = hf$ or $E = hc/\lambda$, the emitted radiation is of a specific wavelength, frequency, and color.

The frequency of an emitted photon is provided by the formula

$$E_{photon} = hf = E_i - E_f$$

where E_i is the energy of the electron in the higher (initial) orbit and E_f is the energy of the electron in the lower (final) orbit.

Bohr demonstrated that each of the wavelengths (colors) in the hydrogen spectrum corresponds to photons emitted by electrons that descend from one of his prescribed energy levels to another. The Balmer series, for example, consists of wavelengths that are emitted by electrons that

Energy Level Diagrams

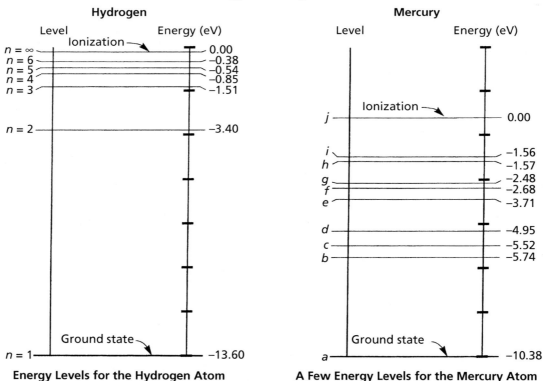

Figure 5-6. Energy levels for the hydrogen atom and some of the energy levels of the mercury atom.

descend from the third, fourth, fifth, and higher energy levels to the second energy level. Bohr even predicted the existence of series that had not yet been observed, and those series were later found. Every one of their wavelengths matched Bohr's prediction!

The fact that different atoms have different emission spectra implies that they have different sets of allowed energy levels for their electrons. However, Bohr was unable to develop a basis for determining the energy levels of atoms other than hydrogen, nor was he able to explain all aspects of the hydrogen atom, such as why some colors in its spectrum are brighter than others. Much progress has been made in this area since Bohr's time.

Sample Problem

3. Find the frequency, wavelength, and color of the radiation emitted by a hydrogen atom as its excited electron falls from third to the second energy level (from energy-level diagram $E_3 = -1.51$ eV, $E_2 = -3.40$ eV).

Solution:

$$E_{photon} = hf = E_i - E_f$$
$$hf = E_3 - E_2$$

$$6.63 \text{ J} \cdot \text{s} \times f = -1.51 \text{ eV} \left(1.60 \times 10^{-19} \frac{\text{J}}{\text{eV}} \right)$$
$$- (-3.40 \text{ eV}) \left(1.60 \times 10^{-19} \frac{\text{J}}{\text{eV}} \right)$$

(changing electron volts to joules)

$$f = \frac{3.02 \times 10^{-19}}{6.63 \times 10^{-34}} = 4.56 \times 10^{14} \text{ Hz}$$

$$\lambda f = c$$

$$\lambda = \frac{c}{f}$$

$$\lambda = \frac{3.00 \times 10^8 \text{ m/s}}{4.56 \times 10^{14} \text{ Hz}} = 6.58 \times 10^{-7} \text{ m}$$

The emitted radiation is red light (from the color-frequency table in the *Physics Reference Tables*).

The Electron Cloud Model

The most recent model of the atom is based on the principles of quantum mechanics and is referred to as the *electron cloud model*. The electron cloud model proposes that electrons in atoms do not have precisely described positions and mo-

menta. Instead, only the probability of finding an electron at a specific position with a specific momentum is provided by the laws of nature. The shape of the probability distribution of an electron depends on the number of electrons in the atom and the atom's energy. The region of most probable electron location is known as a *state*, and each electron in an atom occupies a state. No more than two electrons can be in the same state at the same time.

The electron cloud model does not contradict the Bohr model; it casts it in a different light. For example, the electron cloud model's most probable position for the single electron in the ground state of a hydrogen atom coincides with Bohr's lowest allowed orbit. According to the cloud model, the electron is not *prohibited* from being outside that orbit, but the probability of its being inside the orbit is much greater than its being found outside the orbit.

QUESTIONS

12. White light is passed through a cloud of cool hydrogen gas and then examined with a spectroscope. The dark lines observed on a bright background are caused by (1) the hydrogen emitting all frequencies in white light (2) the hydrogen absorbing certain frequencies of the white light (3) diffraction of the white light (4) constructive interference

13. After electrons in hydrogen atoms are excited to the $n = 3$ energy state, how many different frequencies of radiation can be emitted as the electrons return to the ground state? (1) 1 (2) 2 (3) 3 (4) 4

14. Which phenomenon provides evidence that the hydrogen atom has discrete energy levels? (1) emission spectra (2) photoelectric effect (3) alpha particle scattering (4) natural radioactive decay

15. Compared with the amount of energy required to excite an atom, the amount of energy released by the atom when it returns to the ground state is (1) less (2) greater (3) the same

16. If an orbiting electron falls to a lower orbit, the total energy of that atom will (1) decrease (2) increase (3) remain the same

17. An electron in a mercury atom drops from energy level i to the ground state by emitting a

single photon. This photon has an energy of (1) 1.56 eV (2) 8.82 eV (3) 10.38 eV (4) 11.94 eV

18. The bright-line emission spectrum of an element can best be explained by (1) electrons transitioning between discrete energy levels in the atoms of that element (2) protons acting as both particles and waves (3) electrons being located in the nucleus (4) protons being dispersed uniformly throughout the atoms of that element

19. How much energy is required to move an electron in a mercury atom from the ground state to energy level h? (1) 1.57 eV (2) 8.81 eV (3) 10.38 eV (4) 11.95 eV

20. What is the minimum energy needed to ionize a hydrogen atom in the $n = 2$ energy state? (1) 13.6 eV (2) 10.2 eV (3) 3.40 eV (4) 1.89 eV

21. An atom changing from an energy state of -0.54 eV to an energy state of -0.85 eV will emit a photon whose energy is (1) 0.31 eV (2) 0.54 eV (3) 0.85 eV (4) 1.39 eV

22. The lowest energy state of an atom is called its (1) ground state (2) ionized state (3) initial energy state (4) final energy state

23. A hydrogen atom emits a photon with an energy of 1.63×10^{-18} J as it changes to the ground state. The radiation emitted by the atom would be classified as (1) infrared (2) ultraviolet (3) blue light (4) red light

24. How much energy is needed to raise a hydrogen atom from the $n = 2$ energy level to the $n = 4$ energy level? (1) 10.2 eV (2) 2.55 eV (3) 1.90 eV (4) 0.65 eV

25. A hydrogen atom is excited to the $n = 3$ state. In returning to the ground state, the atom could *not* emit a photon with an energy of (1) 1.9 eV (2) 10.2 eV (3) 12.1 eV (4) 12.75 eV

26. If a hydrogen atom absorbs 1.9 eV of energy, it could be excited from energy level (1) $n = 1$ to $n = 2$ (2) $n = 1$ to $n = 3$ (3) $n = 2$ to $n = 3$ (4) $n = 2$ to $n = 4$

The Nucleus: Protons and Neutrons

The nucleus of an atom contains most of the mass and all of the positive charge of the atom. The particles in the nucleus, protons and neutrons, are called **nucleons**. A proton carries one elementary unit of positive charge (1.60×10^{-19} C), equal in amount to the negative charge on an electron. In addition, a proton has approximately 1800 times

the mass of an electron. Neutrons carry no net charge and have approximately the same mass as protons.

Extensive experimentation led to the discovery that, in addition to the proton, neutron, and electron, many other subatomic particles exist. These include the **neutrino** (a particle with no charge and much less mass than the electron), the various types of **mesons** (with masses between that of the electron and the proton), and the **hyperons** (with masses greater than that of a neutron). Furthermore, each subatomic particle has an **antiparticle**—a particle with the same mass but opposite charge. For example, the **positron** (or antielectron) has the mass of an electron but is positively charged. The antiproton has a negative charge.

The Standard Model

The Standard Model classifies all particles into photons, hadrons, or leptons (Figure 5-7). This division is based on how the particles respond to various forces. Hadrons are further subdivided into baryons and mesons on the basis of their composition. Recent theories and experiments indicate that the heavier baryons (protons, neutrons, and hyperons), as well as the intermediate mass mesons, are themselves composed of various combinations of constituent particles called **quarks**. **Baryons** are composed of three quarks, and *mesons* are composed of two quarks. Quarks carry an amount of charge equal to one-third or two-thirds of the elementary unit of charge, positive or negative. For each of the six quark types experimentally confirmed at this time, there exists an *antiquark* particle. Mesons are composed of a quark and an antiquark. Physicists refer to the different quark types as "flavors" and have named them the *down-quark, up-quark, strange-quark, charm-quark, bottom-quark,* and *top-quark*.

An up-quark (u) has a charge of $+2/3$ (that is, two-thirds of the elementary unit of charge on a proton), the down-quark (d) has a charge of $-1/3$ and the strange-quark (s) also has a charge of $-1/3$. A proton consists of a combination of two up-quarks and one down-quark (symbolized as *uud*). Its total charge therefore is $(+2/3) + (+2/3) + (-1/3)$ which adds to $+1$ unit of elementary charge. A neutron consists of the combination *udd*. Its total charge is $(+2/3) + (-1/3) + (-1/3)$ which adds to zero, again the expected amount.

The Nuclear Forces

In addition to the forces of gravity and electromagnetism, there exist in nature other forces,

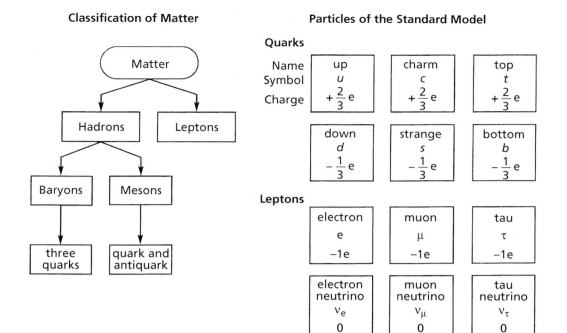

Classification of Matter

Particles of the Standard Model

Quarks

Name	up	charm	top
Symbol	u	c	t
Charge	$+\frac{2}{3}e$	$+\frac{2}{3}e$	$+\frac{2}{3}e$

down	strange	bottom
d	s	b
$-\frac{1}{3}e$	$-\frac{1}{3}e$	$-\frac{1}{3}e$

Leptons

electron	muon	tau
e	μ	τ
$-1e$	$-1e$	$-1e$

electron neutrino	muon neutrino	tau neutrino
ν_e	ν_μ	ν_τ
0	0	0

Note: For each particle, there is a corresponding antiparticle with a change opposite that of its associated particle.

Figure 5-7. Particles of the Standard Model.

called nuclear forces. The **strong nuclear force**, like gravity, is an attractive force that acts between masses. However, unlike gravity, the nuclear force is effective across only extremely small distances. Particles separated by distances of 10^{-15} meter or less exert very strong, mutually attractive nuclear forces on each other. At distances of about 10^{-14} meter, the nuclear force is much weakened, and at 10^{-13} meter or more, it is practically nonexistent.

This strong nuclear attractive force overcomes the electric repulsion between the protons in a nucleus and prevents them from flying apart. The strong nuclear force provides the "glue" that binds the nucleons together into a tightly packed and cohesive unit.

The **weak nuclear force** is the interaction between electrons and protons or neutrons, needed to explain certain forms of nuclear decay.

The nuclear forces do work on particles they act on. The presence of a nuclear force therefore implies that nuclear potential energy exists. We refer to this form of potential energy simply as **nuclear energy**.

When a nucleus is formed, the strong nuclear force acts to bring the nucleons together. This means that the process of forming a nucleus involves a loss of nuclear energy, just as gravitational potential energy is lost when Earth acts to attract an object to itself. If this lost energy were to remain with the nucleons in some other form,

such as kinetic energy, a stable nucleus could not form. The nucleons, after coming together, would bounce off each other and the energy would be reconverted to nuclear potential energy, just as a falling object's kinetic energy becomes gravitational potential energy again when the object bounces elastically off the ground.

A stable nucleus, therefore, is formed only when the lost nuclear energy is removed from the nucleons as they come together. The energy lost by the coalescing nucleons is emitted as photons.

The Mass Defect

Experiments have revealed that the mass of every nucleus is less than the sum of its parts. The mass of the C-12 nucleus, for example, is less than the sum of the masses of 6 protons and 6 neutrons. This discrepancy exists for every one of the hundreds of different types of nuclei that exist. The difference between the mass of a nucleus and the sum of the masses of its nucleons is called the **mass defect**. What happened to the missing mass?

This puzzle was solved when Albert Einstein postulated the equivalence of mass and energy in his theory of special relativity. Mass is a form of energy, and energy is associated with mass. As an object's energy increases, its mass increases; as its energy decreases, its mass decreases. Since energy is removed from nucleons when they come together to form a nucleus, mass is also removed. Although the mass lost is small, it is significant and detectable when compared with the extremely small masses of nuclei.

The Mass-Energy Relationship

The mass associated with a specific amount of energy is provided by the formula

$$E = mc^2$$

where E is energy, in joules; m is mass, in kg; and c is the speed of light, 3.00×10^8 meters per second. Thus, the mass defect of a nucleus, in kg, times c^2 (the speed of light squared), is equal to the energy in joules lost by the formation of the nucleus.

Universal Mass Unit

It is customary to express the masses of nuclei and subatomic particles in terms of **universal mass units**, symbolized by the letter u. One universal mass unit is equal to 9.31×10^2 MeV (megaelectronvolts), or 1.66×10^{-27} kg. The mass of a single proton or neutron is slightly greater than 1 u.

Sample Problem

4A. How much energy is released when a helium nucleus (mass 6.64×10^{-27} kg) is formed from 2 protons and 2 neutrons?

Solution:

The *Reference Tables* give the mass of 1 proton (or 1 neutron) as 1.67×10^{-27} kg.

2 protons + 2 neutrons = $2(1.67 \times 10^{-27}$ kg$)$ + $2(1.67 \times 10^{-27}$ kg$)$ = 6.68×10^{-27} kg

The mass lost: 6.68×10^{-27} kg $- 6.64 \times 10^{-27}$ kg $= 0.04 \times 10^{-27}$ kg $= 4 \times 10^{-29}$ kg

Converting mass to energy:

$$E = mc^2$$

$$E = 4 \times 10^{-29} \text{ kg} \times (3.00 \times 10^8 \text{ m/s})^2$$

$$E = 3.6 \times 10^{-12} \text{ J} = 4 \times 10^{-12} \text{ J}$$

4B. The mass of a helium nucleus is 2.12×10^{-2} u less than the sum of the mass of the particles that formed it. How much energy was released in the formation of the helium nucleus?

Solution:

From the *Physics Reference Tables*, you know that u = 9.31×10^2 MeV; thus

$$(2.12 \times 10^{-2} \text{ u}) \times (9.31 \times 10^2 \text{ MeV}/u)$$
$$= 1.97 \times 10^1 \text{ MeV}$$

QUESTIONS

PART A

27. The force that holds protons and neutrons together is known as the (1) gravitational force (2) strong force (3) magnetic force (4) electrostatic force

28. A meson may *not* have a charge of (1) $+1e$ (2) $+2e$ (3) $0e$ (4) $-1e$

29. The tau neutrino, the muon neutrino, and the electron neutrino are all (1) leptons (2) hadrons (3) baryons (4) mesons

30. The energy equivalent of the rest mass of an electron is approximately (1) 5.1×10^5 J (2) 8.2×10^{-14} J (3) 2.7×10^{-22} J (4) 8.5×10^{-28} J

31. The strong force is the force of (1) repulsion between protons (2) attraction between protons and electrons (3) repulsion between nucleons (4) attraction between nucleons

32. If a deuterium nucleus has a mass of 1.53×10^{-3} universal mass units less than its components, this mass represents an energy of (1) 1.38 MeV (2) 1.42 MeV (3) 1.53 MeV (4) 3.16 MeV

33. What type of nuclear force holds the protons and neutrons in an atom together? (1) a strong force that acts over a short range (2) a strong force that acts over a long range (3) a weak force that acts over a short range (4) a weak force that acts over a long range

34. What type of particle has a charge of 1.6×10^{-19} C and a rest mass of 1.67×10^{-27} kg? (1) proton (2) electron (3) neutron (4) alpha particle

35. Which describes the nuclear forces that hold nucleons together? (1) weak and long-range (2) weak and short-range (3) strong and long-range (4) strong and short-range

36. What type of force holds the nucleons of an atom together? (1) coulomb force (2) magnetic force (3) atomic force (4) strong nuclear force

37. When compared with the total mass of its nucleons, the mass of the nucleus is (1) less (2) greater (3) the same

38. How much energy would be produced if 1.0×10^{-3} kilogram of matter was entirely converted to energy? (1) 9.0×10^{13} J (2) 3.0×10^{16} J (3) 9.0×10^{16} J (4) 3.0×10^{19} J

39. The positron can best be described as a (1) positively charged electron (2) proton (3) positively charged hyperon (4) antiproton

40. The mass of the neutrino is (1) equal to that of an electron (2) less than that of an electron (3) equal to that of a proton (4) between that of an electron and a proton

41. Particles with mass between that of an electron and a proton are called (1) neutrinos (2) hyperons (3) mesons (4) positrons

42. Particles with mass greater than that of a neutron are called (1) neutrinos (2) antiprotons (3) mesons (4) hyperons

43. Which of the following are classified as baryons? (1) mesons and neutrinos (2) positrons and negative protons (3) positrons and electrons (4) protons, neutrons, and hyperons

44. According to the Standard Model, a proton is constructed of two up quarks and one down quark (*uud*) and a neutron is constructed of one up quark and two down quarks (*udd*). During beta decay, a neutron decays into a proton, an electron, and an electron antineutrino. During this process there is a conversion of a (1) *u* quark to a *d* quark (2) *d* quark to a meson (3) baryon to another baryon (4) lepton to another lepton

45. Which combination of quarks could produce a neutral baryon? (1) *cdt* (2) *cts* (3) *cdb* (4) *cdu*

Physics in Your Life

The Production of Laser Light

Many industries commonly employ lasers for everything from surgical tools to bar code readers. Laser is an acronym for *l*ight *a*mplification by *s*timulated *e*mission of *r*adiation. A laser is a device that can produce a very narrow, intense beam of monochromatic, coherent light. Monochromatic means that the light has one color, and therefore one wavelength. Coherent means that any cross section of the beam has the same phase. In contrast, an ordinary light source emits light in all directions and the emitted light is incoherent.

Our understanding of the action of a laser is based on quantum theory. A photon, a particle of light, can be absorbed by an atom if and only if its energy (*hf*) corresponds to the energy difference between an occupied energy level of the atom and an available excited state. If an atom is already in the excited state, it may jump spontaneously to the lower state with the emission of a photon. If, however, a photon with this same energy strikes the excited atom, it can stimulate the atom to make the transition to the lower state sooner. This phenomenon is called stimulated emission. When this occurs, the original photon is expelled from the atom along with a second photon of the same frequency. These two photons are exactly in phase and they are moving in the same direction.

Two photons are certainly not enough to produce useful light. However, a laser is designed to produce many identical photons. To accomplish this goal, the lasing material is placed in a narrow tube. A mirror is placed at one end of the tube and a partially transparent mirror is placed at the other end. When the photons strike the mirrors, most are reflected back. As they move in the opposite direction, they strike other excited atoms, causing the release of additional photons. As they bounce back and forth between the mirrors, they continue to stimulate the emission of more photons. A small percentage of photons pass through the partially transparent mirror. These photons make up the laser beam.

Questions

1. How does laser light differ from ordinary light? How is it the same?
2. Compare spontaneous emission to stimulated emission.
3. Why do lasers contain mirrors on each end? How are the mirrors different from one another?
4. Suggest reasons why laser light is useful for a variety of applications where precision is essential.
5. Some common lasers use a ruby rod consisting of Al_2O_3 with a small percentage of Al atoms replaced by chromium atoms. Other lasers use a mixture of helium and neon gases. How might the lasing material affect the properties of the resulting laser beam?

Chapter Review Questions

PART A

1. Which phenomenon best supports the theory that matter has a wave nature? (1) electron momentum (2) electron diffraction (3) photon momentum (4) photon diffraction

2. Compared to a photon of red light, a photon of blue light has a (1) greater energy (2) longer wavelength (3) smaller momentum (4) lower frequency

3. Compared with the total energy of the hydrogen atom in the ground state, the total energy of the atom in an excited state is (1) less (2) greater (3) the same

4. Which statement is true of the strong nuclear force? (1) It acts over very great distances. (2) It holds protons and neutrons together. (3) It is much weaker than gravitational forces. (4) It repels neutral charges.

5. Protons and neutrons are examples of (1) positrons (2) baryons (3) mesons (4) quarks

PART B-1

6. A photon of which electromagnetic radiation has the most energy? (1) ultraviolet (2) x-ray (3) infrared (4) microwave

7. The following diagram represents the bright-line spectra of four elements, A, B, C, and D, and the spectrum of an unknown gaseous sample.

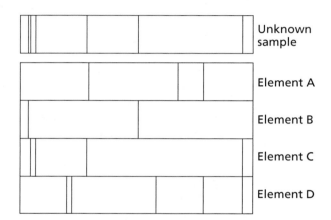

Based on comparisons of these spectra, which two elements are found in the unknown sample? (1) A and B (2) A and D (3) B and C (4) C and D

8. A hydrogen atom with an electron initially in the $n = 2$ level is excited further until the electron is in the $n = 4$ level. This energy level change occurs because the atom has (1) absorbed a 0.85-eV photon (2) emitted a 0.85-eV photon (3) absorbed a 2.55-eV photon (4) emitted a 2.55-eV photon

9. According to the Standard Model of Particle Physics, a meson is composed of (1) a quark and a muon neutrino (2) a quark and an antiquark (3) three quarks (4) a lepton and an antilepton

PART B-2

10. What prevents the nucleus of a helium atom from flying apart?

Base your answers to questions 11 and 12 on the following graph, which shows the maximum kinetic energy of photoelectrons as a function of the frequency the incident electromagnetic waves for two photoemissive metals, A and B.

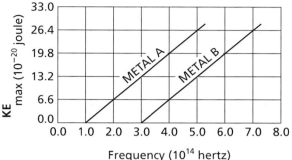

Frequency (10^{14} hertz)
Note: 1 hertz = 1 cycle / second

11. Calculate the slope of the line for metal B.

12. What is the meaning of the slope?

Base your answers to questions 13 and 14 on the information and equation below.

During the process of beta (β) emission, a neutron in the nucleus of an atom is converted into a proton, an electron, an electron antineutrino, and energy.

Neutron → proton + electron
            + electron antineutrino + energy

13. Based on conservation laws, how does the mass of the neutron compare to the mass of the proton?

14. Since charge must be conserved in the reaction shown, what charge must an electron antineutrino carry?

15. A photon has a wavelength of 9.00×10^{-10} meter. Calculate the energy of this photon in joules. Show all work, including the equation and substitution with units.

16. After a uranium nucleus emits an alpha particle, the total mass of the new nucleus and the alpha particle is less than the mass of the original uranium nucleus. Explain what happens to the missing mass.

Base your answers to questions 17 and 18 on the information below.

When an electron and its antiparticle (positron) combine, they annihilate each other and become energy in the form of gamma rays.

17. The positron has the same mass as the electron. Calculate how many joules of energy are released when they annihilate. Show all work, including the equation and substitution with units.

18. What conservation law prevents this from happening with two electrons?

19. What are the sign and charge, in coulombs, of an antiproton?

Base your answers to questions 20 and 21 on the information below.

A lambda particle consists of an up, a down, and a strange quark.

20. A lambda particle can be classified as a (1) baryon (2) lepton (3) meson (4) photon

21. What is the charge of a lambda particle in elementary charges?

22. How much energy, in megaelectronvolts, is produced when 0.250 universal mass unit of matter is completely converted into energy?

23. Explain why a hydrogen atom in the ground state can absorb a 10.2-electronvolt photon, but can *not* absorb an 11.0-electronvolt photon.

24. Exposure to ultraviolet radiation can damage skin. Exposure to visible light does not damage skin. State *one* possible reason for this difference.

Base your answers to questions 25 and 26 on the information and diagram below.

The diagram shows the collision of an incident photon having a frequency of 2.00×10^{19} hertz with an electron initially at rest.

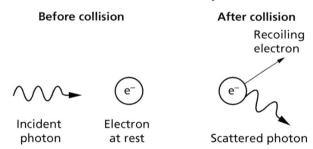

25. Calculate the initial energy of the photon. Show all calculations, including the equation and substitution with units.

26. What is the total energy of the two-particle system after the collision?

PART C

Base your answers to questions 27 through 30 on the information below.

An electron in a hydrogen atom drops from the $n = 3$ energy level to the $n = 2$ energy level.

27. What is the energy, in electronvolts, of the emitted photon?

28. What is the energy, in joules, of the emitted photon?

29. Calculate the frequency of the emitted radiation. Show all work, including the equation and substitution with units.

30. Calculate the wavelength of the emitted radiation. Show all work, including the equation and substitution with units.

Base your answers to questions 31 through 33 on the information below.

The light of the "alpha line" in the Balmer series of the hydrogen spectrum has a wavelength of 6.58×10^{-7} meter.

31. Calculate the energy of an "alpha line" photon in joules. Show all work, including the equation and substitution with units.

32. What is the energy of an "alpha line" photon in electronvolts?

33. Using your answer to question 32, explain whether or not this result verifies that the "alpha line" corresponds to a transition from energy level $n = 3$ to energy level $n = 2$ in a hydrogen atom.

Base your answers to questions 34 through 37 on the Energy Level Diagram for Hydrogen given in the Reference Tables.

34. Determine the energy, in electronvolts, of a photon emitted by an electron as it moves from the $n = 6$ to the $n = 2$ energy level in a hydrogen atom.

35. Convert the energy of the photon to joules.

36. Calculate the frequency of the emitted photon. Show all work, including the equation and substitution with units.

37. Is this the only energy and/or frequency that an electron in the $n = 6$ energy level of a hydrogen atom could emit? Explain your answer.

Base your answers to questions 38 and 39 on the following diagram, which shows some energy levels for an atom of an unknown substance.

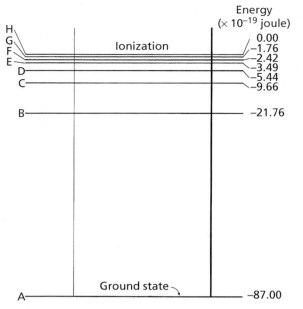

38. Determine the minimum energy necessary for an electron to change from the B energy level to the F energy level.

39. Calculate the frequency of the photon emitted when an electron in this atom changes from the F energy level to the B energy level. Show all work, including the equation and substitution with units.

Base your answers to questions 40 and 41 on the information below.

Louis de Broglie extended the idea of wave-particle duality to all of nature with his matter-wave equation, $\lambda = h/mv$, where λ is the particle's wavelength, m is its mass, v is its velocity, and h is Planck's constant.

40. Using this equation, calculate the de Broglie wavelength of a helium nucleus (mass = 6.7×10^{-27} kg) moving with a speed of 2.0×10^6 meters per second. Show all work, including the equation and substitution with units.

41. The wavelength of this particle is of the same order of magnitude as which type of electromagnetic radiation?

Enrichment
Modern Physics

THE PHOTOELECTRIC EFFECT

Each photoemissive material has a **threshold frequency**—a frequency below which no photoelectrons will be emitted, no matter how intense the radiation. For example, a material whose threshold frequency is that of yellow light emits no electrons when bombarded by red light or radio waves, no matter how intense the radiation. However, this material will emit many electrons when bombarded by even the faintest green light or by x-rays.

Figure 5-E1 shows the relationship between the maximum kinetic energy of photoelectrons, KE_{max} and the frequency of the incident radiation, f, for two different photoemissive materials, A and B. For every photoemissive material, the maximum kinetic energy of photoelectrons varies linearly with the frequency of the incident radiation. The slope of the graph is the same for all photoemissive materials and is equal to *Planck's constant, h* (6.63×10^{-34} J·s). The point at which the graph intercepts the *x*-axis is different for different photoemissive materials and represents the threshold frequency, f_0, of the material.

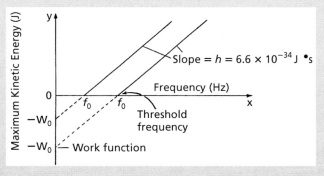

Figure 5-E1.

The relationship between KE_{max} and f can also be expressed mathematically as

$$KE_{max} = hf - W_0$$

where h is Planck's constant and W_0 is the absolute value of the y-intercept of the graph. This y-intercept is different for different photoemissive materials and is known as the *work function* of the material.

If light and the other electromagnetic radiations were just electromagnetic waves, the kinetic energy of the ejected electrons should depend on the intensity of the radiation, rather than the frequency, and no threshold frequency should exist. More intense radiation (brighter light) then would mean waves with greater amplitude that exert stronger electric and magnetic forces. As a result, the electrons should emerge with more kinetic energy when the radiation is more intense. But since frequency, not intensity, is the sole determining factor of the kinetic energy of emitted photoelectrons, to the point that a "threshold" for frequency exists, a new and improved model of light had to be devised.

The minimum amount of energy an electron needs to be able to escape from a photoemissive material is the **work function**, W_0, of that material. This is also the minimum amount of energy an electron loses as it escapes. Thus, if hf (photon energy) is greater than W_0, the electron picks up more than enough energy to escape. After absorbing hf joules of energy, the escaping electron loses a minimum of W_0 joules on the way out and emerges with, at most, $(hf - W_0)$ joules of kinetic energy. This is why the maximum kinetic energy of photoelectrons satisfies the relationship

$$KE_{max} = hf - W_0$$

Below the threshold frequency, hf is less than W_0. An electron hit by such a photon cannot escape from the material, even though it absorbs all the energy of the photon. Increasing the intensity of radiation does not help, because that only increases the number of photons hitting the material per second, not the amount of energy per photon. Each photon must still act individually on one electron and can impart only hf joules of energy—not enough to eject the electron. Very rarely do two photons strike the same electron in quick succession so that the electron absorbs a second dosage of energy before the first dosage is dissipated through the material.

The threshold frequency f_0 of a photoemissive material is thus the frequency at which the quantity hf is equal to the work function W_0. This leads to the equation

$$W_0 = hf_0$$

a formula that is confirmed by the KE_{max} versus f graph for any photoemissive material (such as the graph in Figure 5E-1). If we use the dotted portion of the graph between the x- and y-intercepts to calculate the slope, which we know is equal to h, Planck's constant, we get

$$\text{Slope} = \frac{\Delta y}{\Delta x} = \frac{W_0}{f_0} = h \quad \text{or} \quad W_0 = hf_0$$

Increasing the intensity of radiation increases the rate of emission of photoelectrons because more photons then strike the photoemissive material per second. If the photons are energetic enough to eject electrons (above the threshold), the more of them that hit the material per second, the greater the number of electrons emitted per second.

Sample Problems

E1. The threshold frequency of a photoemissive material is 2.00×10^{15} Hz. What is the work function of the material?

Solution:

$$W_0 = hf_0$$

$$W_0 = (6.63 \times 10^{-34} \text{ J} \cdot \text{s})(2.00 \times 10^{15} \text{ Hz})$$

$$W_0 = 1.33 \times 10^{-18} \text{ J}$$

E2. If radiation with frequency of 3.00×10^{15} Hz strikes the material, what is the maximum kinetic energy of the emitted photoelectrons?

Solution:

$$KE_{max} = hf - W_0$$

$$KE_{max} = (6.63 \times 10^{-34} \text{ J} \cdot \text{s})(3.00 \times 10^{15} \text{ Hz}) - (1.33 \times 10^{-18} \text{ J})$$

$$KE_{max} = 1.99 \times 10^{-18} \text{ J} - (1.33 \times 10^{-18} \text{ J})$$

$$KE_{max} = 6.59 \times 10^{-19} \text{ J}$$

E3. What is the energy of each incident photon?

Solution:

$$E = hf$$

$$E = (6.63 \times 10^{-34} \text{ J} \cdot \text{s})(3.00 \times 10^{15} \text{ Hz})$$

$$E = 1.90 \times 10^{-18} \text{ J}$$

Photon Momentum

The momentum, p, of a photon was shown by Compton to satisfy the relationship

$$p = \frac{h}{\lambda}$$

where h is Planck's constant and λ is the wavelength in meters. Combining this formula with $c = \lambda f$, yields

$$p = \frac{hf}{c} \quad \text{and} \quad p = \frac{E}{c}$$

Sample Problem

E4. What is the momentum of a photon of orange light (wavelength of 6.00×10^{-7} meter)?

Solution:

$$p = \frac{h}{\lambda}$$

$$p = \frac{6.63 \times 10^{-34} \text{ J} \cdot \text{s}}{6.00 \times 10^{-7} \text{m}}$$

$$p = 1.11 \times 10^{-27} \frac{\frac{\text{kg} \cdot \text{m}}{\text{s}^2} \cdot \text{m} \cdot \text{s}}{\text{m}}$$

$$p = 1.11 \times 10^{-27} \frac{\text{kg} \cdot \text{m}}{\text{s}}$$

The de Broglie Wavelength

The wavelength of a matter wave, known as the **de Broglie wavelength**, is provided by the formula

$$\lambda = \frac{h}{p}$$

where h is Planck's constant and p is the momentum of the particle in kg·m/s. Since the momentum of a particle is equal to mv, this formula can also be written as

$$\lambda = \frac{h}{mv}$$

For subatomic particles with extremely small mass, the wavelength of the matter wave is large enough to be observed and measured. For objects of greater mass, however, the de Broglie wavelength is negligibly small, cannot be detected, and, for all practical purposes, can be ignored.

Sample Problem

E5. What is the de Broglie wavelength of an electron moving at 3.00×10^6 m/s? (The mass of an electron is 9.11×10^{-31} kg.)

Solution:

$$\lambda = \frac{h}{p} = \frac{h}{mv}$$

$$\lambda = \frac{6.63 \times 10^{-34} \text{ J} \cdot \text{s}}{(9.11 \times 10^{-31} \text{ kg})(3.00 \times 10^6 \text{ m/s})}$$

$$\lambda = 2.43 \times 10^{-10} \text{ m}$$

Questions

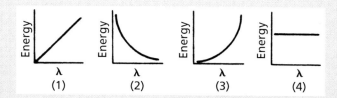

E1. A monochromatic light incident upon a photoemissive surface emits electrons. If the intensity of the incident light is increased, the rate of electron emission will (1) decrease (2) increase (3) remain the same

E2. When incident on a given photoemissive surface, which color of light will produce photoelectrons with the greatest energy? (1) red (2) orange (3) violet (4) green

E3. The threshold frequency of a metal surface is in the violet light region. What type of radiation will cause photoelectrons to be emitted from the metal's surface? (1) infrared light (2) red light (3) ultraviolet light (4) radio waves

E4. The work function of a photoelectric material can be found by determining the minimum frequency of light that will cause electron emission and then (1) adding it to the velocity of light (2) multiplying it by the velocity of light (3) adding it to Planck's constant (4) multiplying it by Planck's constant

E5. The work function of a metal is 4.2 eV. If photons with an energy of 5.0 eV strike the metal, the maximum kinetic energy of the emitted photoelectrons will be (1) 0 eV (2) 0.80 eV (3) 3.8 eV (4) 9.2 eV

E6. The threshold frequency for a photoemissive surface is 6.4×10^{14} Hz. Which color light, if incident upon the surface, may produce photoelectrons? (1) blue (2) green (3) yellow (4) red

E7. Compared with the energy of the photons of blue light, the energy of the photons of red light is (1) less (2) greater (3) the same

E8. The energy of a photon varies (1) directly as the wavelength (2) directly as the frequency (3) inversely as the frequency (4) inversely as the square of the frequency

E9. All of the following particles are traveling at the same speed. Which has the greatest wavelength? (1) proton (2) alpha particle (3) neutron (4) electron

E10. If the wave properties of a particle are difficult to observe, it is probably due to the particle's (1) small size (2) large mass (3) low momentum (4) high charge

E11. Which graph best represents the relationship between the energy of a photon and its wavelength?

E12. As the wavelength of a ray of light increases, the momentum of the photons (1) decreases (2) increases (3) remains the same

E13. Compared with the photon momentum of blue light, the photon momentum of red light is (1) less (2) greater (3) the same

E14. Which is conserved when a photon collides with an electron? (1) velocity (2) momentum, only (3) energy, only (4) momentum and energy

Base your answers to questions E15 through E19 on the following information.

Photons with an energy of 3.0 eV strike a metal surface and eject electrons with a maximum kinetic energy of 2.0 eV.

E15. The work function of the metal is (1) 1.0 eV (2) 2.0 eV (3) 3.0 eV (4) 5.0 eV

E16. If the photons had a higher frequency, what would remain constant? (1) the energy of the photons (2) the speed of the photons (3) the kinetic energy of the electrons (4) the speed of the electrons

E17. If the photon intensity were decreased, there would be (1) an increase in the energy of the photons (2) a decrease in the energy of the photons (3) an increase in the rate of electron emission (4) a decrease in the rate of electron emission

E18. Compared with the frequency of the 3.0-eV photons, the threshold frequency for the metal is (1) lower (2) higher (3) the same

E19. If a metal with a greater work function were used and the photon energy remained constant, the maximum energy of the ejected electrons would (1) decrease (2) increase (3) remain the same

Base your answers to questions E20 through E24 on the following information.

Photons of wavelength 2×10^{-7} meter are incident upon a photoemissive surface whose work function is 6.6×10^{-19} joule.

E20. The speed of the incident photons is approximately (1) 2.0×10^{-7} m/s (2) 6.6×10^{-19} m/s (3) 1.3×10^{-25} m/s (4) 3.0×10^8 m/s

E21. The maximum kinetic energy of the photoelectrons is approximately (1) 0 J (2) 3.3 $\times 10^{-19}$ J (3) 6.6 $\times 10^{-19}$ J (4) 9.9 $\times 10^{-19}$ J

E22. If the frequency of the incident photons is increased, the kinetic energy of the emitted photoelectrons will (1) decrease (2) increase (3) remain the same

E23. If the intensity of the incident photons is decreased, the rate of emission of photoelectrons will (1) decrease (2) increase (3) remain the same

E24. Photons of the same wavelength are incident upon a photoemissive surface with a lower work function. Compared with the original situation, the maximum kinetic energy of the photoelectrons emitted from the new surface would be (1) less (2) greater (3) the same

Base your answers to questions E25 through E29 on the following graph, which represents the maximum kinetic energy of photoelectrons as a function of incident electromagnetic frequencies for two different photoemissive metals, A and B.

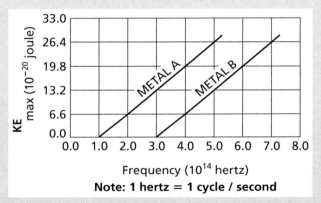

E25. The slope of each line is equal to (1) Bohr's constant (2) the photoelectric constant (3) Compton's constant (4) Planck's constant

E26. The threshold frequency for metal *A* is (1) 1.0 $\times 10^{14}$ Hz (2) 2.0 $\times 10^{14}$ Hz (3) 3.0 $\times 10^{14}$ Hz (4) 0.0 Hz

E27. The work function for metal *B* is closest to (1) 0.0 J (2) 2.0 $\times 10^{-19}$ J (3) 3.0 $\times 10^{-19}$ J (4) 1.5 $\times 10^{-14}$ J

E28. Compared with the work function for metal *B*, the work function for metal *A* is (1) less (2) greater (3) the same

E29. Monochromatic light with a period of 2.0 $\times 10^{-15}$ second is incident on both of the metals. Compared with the energy of the photoelectrons emitted by metal *A*, the energy of the photoelectrons emitted by metal *B* is (1) less (2) greater (3) the same

Base your answers to questions E30 through E32 on the following diagram, which represents monochromatic light incident upon photoemissive surface A. *Each photon has 8.0 $\times 10^{-19}$ joule of energy.* B *represents the particle emitted when a photon strikes surface* A.

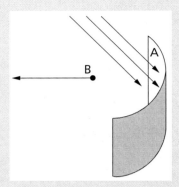

E30. What is particle *B*? (1) an alpha particle (2) an electron (3) a neutron (4) a proton

E31. If the work function of metal *A* is 3.2 $\times 10^{-19}$ J, the energy of particle *B* is (1) 3.0 $\times 10^{-19}$ J (2) 4.8 $\times 10^{-19}$ J (3) 8.0 $\times 10^{-19}$ J (4) 11 $\times 10^{-19}$ J

E32. The frequency of the incident light is approximately (1) 1.2 $\times 10^{15}$ Hz (2) 5.3 $\times 10^{-15}$ Hz (3) 3.7 $\times 10^{-15}$ Hz (4) 8.3 $\times 10^{-16}$ Hz

NUCLEAR REACTIONS

The number of protons in the nucleus is referred to as the **atomic number** of the atom and is symbolized by the letter *Z*. The atomic number is also the number of electrons when the atom is electrically neutral. Every element of the *Periodic Table of the Elements* has a unique atomic number. The number of nucleons, or the total number of protons and neutrons, is known as the **mass number** of the atom, and is symbolized by the letter *A*. The number of neutrons is thus equal to $A - Z$. Atoms with identical atomic numbers but different mass numbers are called **isotopes** of the same element, and their chemical behavior is identical to each other.

The nuclei of all carbon atoms, for example, contain six protons. The atomic number, *Z*, of carbon, therefore, is six. While most carbon atoms also contain six neutrons, some contain as many as 7, 8, 9, or 10 neutrons. A carbon atom, therefore, can have a mass number, *A*, of 12, 13, 14, 15, or 16.

We represent elements and their isotopes using the form $^A_Z X$, where *X* is the chemical symbol of the element, *Z* the atomic number, and *A* the mass number. The isotopes of carbon thus appear as follows:

$$^{12}_{6}C, ^{13}_{6}C, ^{14}_{6}C, ^{15}_{6}C, ^{16}_{6}C,$$

where C is the symbol for carbon.

Sometimes, the atomic number is omitted, and a nucleus is symbolized in the form of *X-A*. The isotopes of carbon then appear as follows:

C-12, C-13, C-14, C-15, C-16

Nuclei that possess the same number of protons and neutrons are said to belong to the same nuclear species, called a **nuclide**. Different nuclides with the same atomic number are, therefore, isotopes of the same element.

Observational Tools

Since nuclei are buried deep inside atoms, hidden by clouds of electrons, special tools are needed to investigate their structure and behavior. Much has been learned about nuclei by observing the particles they emit naturally or when bombarded by other particles.

Charged particles are detected by **Geiger counters** that convert the particles' presence into currents of charge and by **scintillation counters** that convert the particles' energy into photons of light. *Photographic plates* are used to record the shapes of the tracks made by charged particles in electric or magnetic fields. The **cloud chamber** is a particularly useful device that reveals the path of a charged particle by creating a trail of condensed vapor in its wake. Knowledge of the deflection experienced by a particle in an electric or magnetic field leads to knowledge of its charge to mass ratio.

Particle accelerators project charged particles at high velocity so that they have enough energy to penetrate, and perhaps smash, nuclei. When these charged particles strike the nucleus of an atom, they upset its stability and new particles may be produced. Positively charged particles, such as protons, are repelled and turned back by the positive nuclei, unless the particles are accelerated, giving them enough kinetic energy to overcome the electric repulsive force. Neutral particles, such as neutrons, need not be accelerated since they are not subject to this repulsion.

Different types of accelerators have been designed, but all use electric and magnetic fields to accelerate the charged particles that are to be aimed at nuclei. The **Van de Graaff generator** accelerates particles by driving them parallel to an intense electric field. In the **cyclotron** and *synchrotron*, charged particles are made to circle in a magnetic field and are accelerated once during every round trip when they pass through an electric field. In the **linear accelerator** the particles are continuously accelerated in an electric field as they travel in a straight line.

Natural Radioactivity

All nuclei with atomic numbers greater than 83 are unstable and disintegrate on their own. Many isotopes with atomic numbers less than 84 are also unstable. These nuclei eject parts of themselves and in so doing are changed, that is, they experience a **transmutation**, into new elements with different atomic and mass numbers. This naturally occurring transmutation is called **radioactivity**, and the transmutation of one element into another is called **radioactive decay**. The ejection of particles from radioactive nuclei is always associated with the emission of high-energy, short-wavelength photons, called **gamma rays**.

Alpha Decay

Different types of particles are ejected by different radioactive nuclei. In a process called **alpha decay**, an alpha particle, consisting of two protons and two neutrons, is emitted from a nucleus. The alpha particle is identical to the nucleus of a helium atom $^{4}_{2}He$. Thus, the emission of an alpha particle decreases the mass number by four and the atomic number by two. For example, radium-226 nuclei decay into radon-222 nuclei through alpha decay. This is represented by the following nuclear reaction equation:

$$^{226}_{88}Ra \rightarrow ^{222}_{88}Rn + ^{4}_{2}He$$

Note that the subscripts and superscripts on both sides of the equation are independently balanced ($86 + 2 = 88$, and $222 + 4 = 226$). The subscripts represent elementary units of positive charge (usually protons), and the superscripts represent mass numbers (protons plus neutrons). In all nuclear reactions, the sum of the positive charges and the sum of the mass numbers on both sides of the equation must be so balanced.

Beta Decay (Negative)

In **beta decay (negative)**, an electron is ejected from a nucleus. A neutron in the nucleus disintegrates into a proton and an electron; the proton remains in the nucleus, and the electron is ejected. This happens, for example, to thorium-234 nuclei as they decay into protactinium-234 nuclei, as follows:

$$^{234}_{90}Th \rightarrow ^{234}_{91}Pa + ^{0}_{-1}e$$

Since the electron carries one elementary unit of negative charge, we assign it a subscript of −1, and since its mass is insignificant compared with that of protons and neutrons, we assign it a superscript of 0. The process of beta (negative) decay increases the atomic number of the nucleus by one, but leaves the mass number intact.

Gamma Radiation

When a nucleus in an excited state returns to a state of lower energy, the energy difference is emitted in the form of gamma ray photons. Since photons have no charge and much less mass than a nucleon, these emissions change neither the atomic number nor the mass number of nuclei and produce no transmutation or decay.

If all the particles and photons involved in a nuclear reaction (those ejected and those remaining) are taken into account, the total mass-energy after the reaction is always equal to the total mass-energy before the reaction. This is referred to as the *law of conservation of mass-energy*.

Half-Life

Each radioactive nuclide decays at a unique rate. The time it takes for half the number of nuclei present in a sample to decay is known as the **half-life** of the radioactive substance. Half-lives range from as little as 10^{-22} seconds for some radioactive substances to as much as 10^{17} years for others. The rate of decay is independent of all environmental factors such as pressure, temperature, and chemical combination with other substances.

At the end of every half-life period, the amount of original material remaining is half the amount that was present at the beginning of that period. This relationship is expressed by the formula

$$m_f = \frac{m_i}{2^n}$$

where m_i is the initial mass of the radioactive material, m_f is the mass remaining, and n is the number of half-lives elapsed.

Sample Problem

E6. If 32 grams of a radioactive substance with a half-life of one hour are present at noon, how much of the substance will remain at 4 P.M.?

Solution:

$$m_f = \frac{m_i}{2^n} = \frac{32 \text{ g}}{2^4}$$

$$= \frac{32 \text{ g}}{16} = 2.0 \text{ g}$$

The number of nuclei that decay, and consequently the number of particles ejected, per time decreases as time goes on. In the sample problem, for example, 16 grams decayed in the first hour, 8 grams decayed during the second hour, 4 grams during the third, and so on. But the time it takes for half of the amount present at any time to decay is constant. It is the half-life of the material (1 hour).

Decay Series

Natural radioactivity usually proceeds in a succession of steps, called a *decay series*. After an atom of $^{238}_{92}U$ decays into $^{234}_{90}Th$, for example, the thorium atom decays into $^{234}_{91}Pa$, which then becomes $^{234}_{92}U$, which then decays into $^{230}_{90}Th$, and so on. A series comes to an end when the product of a decay step is stable (nonradioactive). The series that starts with $^{238}_{92}U$ ends, after 14 steps, with $^{206}_{82}Pb$.

It is customary to illustrate the steps of a decay series with arrows and dots on a mass number versus atomic number graph, as illustrated in Figure 5-E2. Every dot represents a nucleus with a particular mass number and atomic number. An arrow that points from one dot to another that is two units to the left and four units down (such as arrow A in the figure) indicates that the decay step decreases the atomic number by two and the mass number by four. Such an arrow, therefore, represents the emission of an alpha particle.

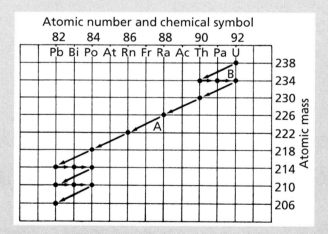

Figure 5-E2. The uranium-238 decay series.

An arrow that points from one dot to another that is one unit to the right (such as arrow B in the figure), indicates that the decay step increased the atomic number by one but left the mass number unchanged. Such an arrow, therefore, represents the emission of a beta particle (electron).

The 14-step series consists of 8 alpha particle emissions and 6 beta particle emissions.

Artificial Transmutation

The first artificial transmutation of one element to another was performed by Rutherford in 1919. Rutherford bombarded nitrogen with energetic alpha particles that were moving fast enough to overcome the electric repulsion between themselves and the target nuclei. The alpha particles collided with, and were absorbed by, the nitrogen nuclei, and protons were ejected. In the process oxygen and hydrogen nuclei were created. This reaction is summarized by the equation

$$\,^4_2\text{He} + \,^{14}_7\text{N} \rightarrow \,^{17}_8\text{O} + \,^1_1\text{H}$$

Positron Emission

Since 1919 many other artificially induced transmutations have been performed by the bombardment of nuclei with protons, neutrons, and other particles. Frequently the artificially created elements are themselves radioactive. For example, when energetic alpha particles are aimed at aluminum-27 nuclei, the aluminum nuclei turn into radioactive phosphorus-30 nuclei and emit neutrons. Over time the phosphorus nuclei decay into stable silicon-30 nuclei by emitting positrons. Positrons carry the same amount of mass and charge as electrons, but their charge is positive. They are referred to as *beta positive* (β^+) particles. [When the expression "beta particle" is used without specifying + or −, the reference is to electrons, or (β^-) particles.]

The reactions described above are represented by the following equations:

$$\,^4_2\text{He} + \,^{27}_{13}\text{Al} \rightarrow \,^{30}_{15}\text{P} + \,^1_0 n$$

$$\,^{30}_{15}\text{P} \rightarrow \,^{30}_{14}\text{Si} + \,^0_{+1} e$$

The $\,^0_{+1}e$ particle, or positron, comes from the breakup of a proton into a neutron and positron. The neutron remains in the nucleus and the positron is ejected. By replacing a proton in the nucleus with a neutron, the process of beta-positive decay decreases the atomic number by one and leaves the mass number unchanged.

Another example of an artificially created nucleus that is radioactive and emits positrons is copper-64.

$$\,^{64}_{29}\text{Cu} \rightarrow \,^{64}_{28}\text{Mg} + \,^0_{+1} e$$

An artificially created nucleus that emits beta-negative particles (electrons) is sodium-24.

$$\,^{24}_{11}\text{Na} \rightarrow \,^{24}_{12}\text{Mg} + \,^0_{-1} e$$

Electron Capture

A radioactive process that results in the emission of gamma rays is known as **electron capture**. It occurs when a nucleus "captures" and absorbs one of the atom's innermost electrons. The electron then unites with a proton to form a neutron. This decreases the atom's atomic number by one but leaves the mass number unchanged.

An example of such a reaction is the change of potassium-40 into argon-40:

$$\,^{40}_{19}\text{K} + \,^0_{-1} e \rightarrow \,^{40}_{18}\text{Ar}$$

Sample Problem

E7. In the following nuclear reaction equation, what type of particle is X?

$$\,^{214}_{82}\text{Pb} \rightarrow \,^{214}_{83}\text{Bi} + X$$

Solution:
Since the subscripts and superscripts must be independently balanced, X must have a subscript of -1 ($83 - 1 = 82$) and a superscript of 0 ($214 + 0 = 214$). With a mass number of zero and one elementary unit of negative charge, X must be a negative beta particle (electron).

The Neutron

Neutrons were discovered in 1932 by James Chadwick when he bombarded beryllium with energetic alpha particles. The beryllium nuclei were transmuted to carbon-12 nuclei and neutrons were emitted. This is represented by the equation

$$\,^4_2\text{He} + \,^9_4\text{Be} \rightarrow \,^{12}_6\text{C} + \,^1_0 n$$

Neutron emitters are often used to initiate nuclear reactions, because neutrons are very effective as nucleus smashers. Since neutrons carry no charge, they are not repelled electrically by nuclei and, therefore, need not be accelerated in order that they collide with the nuclei. If a neutron merely comes close to a nucleus, the nuclear attractive force acts to bend its path toward the nucleus and a direct hit results. This increases the odds enormously that a collision will occur, since even slow-moving neutrons that are not headed straight into a nucleus can collide with it.

Indeed, slow neutrons are more effective nucleus smashers than fast neutrons. The nuclear attractive force exerted by a nucleus is unable to capture a neutron that moves too fast, unless the neutron is headed straight into the nucleus. But slow neutrons are pulled into nuclei even if their trajectories would not otherwise lead them there.

Moderators

To increase the probability that neutrons will collide with nuclei, power plants that derive their energy from nuclear reactors employ moderators to slow down the neutrons. **Moderators** are materials whose atoms have light nuclei that do not combine with neutrons that collide with them. Typical moderators are hydrogen, deuterium, carbon, water, and paraffin. When fast-moving neutrons pass through these substances, they repeatedly collide with the nuclei and give up part of their energy. Soon the average kinetic energy of the neutrons is approximately equal to that of the atoms of the moderator (about $\frac{1}{25}$ eV). Such slow-moving neutrons are called *thermal neutrons*.

Nuclear Fission

When certain massive nuclei, such as uranium-235 or plutonium-239, are struck by slow thermal neutrons, the nuclei split into two smaller fragments, a number of neutrons are ejected, and energy is released. This process is called **fission**. The binding energy per nucleon of the two new lighter nuclei is greater than that of the original heavier nucleus. Thus, the reaction leads to a loss of more nuclear energy per nucleon—energy that is removed from the nucleons and converted to heat, photons, and other forms. The loss of nuclear energy corresponds to a loss of mass on the part of the nucleons. This mass is converted to the energy released in amounts provided by the relationship $E = mc^2$. A typical equation for this type of fission reaction is

$$_0^1 n + {}^{235}_{92}U \rightarrow {}^{141}_{56}Ba + {}^{92}_{36}Kr + 3{}_0^1n + Q \text{ (energy)}$$

The ejected neutrons are available for collisions with other nuclei in the sample of U-235. Those collisions, in turn, liberate more neutrons that collide with still more nuclei. If enough fissionable material is present, a self-sustaining **chain reaction** of collisions leading to more collisions can occur with each collision splitting a nucleus and releasing energy. In this way a large amount of energy can be released in an uncontrolled manner, as happens when an atomic bomb explodes. The rate at which fission reactions take place, however, can be controlled so that a uniform, continuous, and usable output of energy occurs. Today this process is the basis of the operation of hundreds of nuclear reactors around the world. The minimum amount of fissionable material necessary to initiate a chain reaction is called the **critical mass**.

Reactors

The most common method of preventing a chain reaction from getting out of control is the insertion of *control rods* between small pellets of the fissionable material referred to as the **nuclear fuel**. These rods consist of substances such as boron and cadmium that absorb neutrons. Control rods effectively prevent many of the neutrons that escape from one pellet of fissioning material from reaching a neighboring pellet, thereby slowing down the chain reaction. This controlling mechanism must be carefully adjusted to ensure that the chain reaction is neither shut down nor allowed to accelerate.

Reactors must also have circulating *coolants* to carry the heat generated by the nuclear reaction away from the reactor to the turbines or heat exchangers. Without such coolants, the temperature in the reactor would rise to dangerous levels. Common coolants include water, air, helium, carbon dioxide, molten sodium, and molten lithium. In some reactors, the coolant also serves as the moderator. Heavy water (in which the two hydrogen atoms are H-2, not H-1) is frequently used for both functions.

Reactors must also be *shielded*. An internal shield of steel is used to prevent escaping radiation from damaging the walls of the reactor. An external shield of high density concrete is used to prevent injury to the personnel who operate the plant and to the population in the vicinity of the plant.

A major problem that plagues all reactors is waste disposal. The products of fission are highly radioactive and many remain so for a long time. Therefore, they cannot be discarded without precautions. Underground storage in special containers in isolated areas provides only a temporary solution. More permanent solutions have yet to be found.

The uranium used in nuclear reactors contains more U-238 than U-235, because uranium ore found underground is approximately 97% U-238, and the process of increasing the U-235 content of the uranium pellets is a difficult and expensive one. When a U-238 atom absorbs a neutron, its nucleus does not fission. Instead, it is converted to plutonium-239, a highly radioactive substance that does fission when struck by a neutron. Some reactors treat this plutonium as waste, while others use it as fuel to generate yet more energy. The latter type of reactor is known as a **breeder reactor**.

Nuclear Fusion

Nuclear energy is also released when certain light nuclei are united to become one heavier nucleus. During this process of **nuclear fusion**, the binding energy per nucleon is increased and nuclear energy and mass are lost by the nuclei and released in other forms. It is believed that the sun's enormous energy output derives from the fusion of various hydrogen isotopes into helium nuclei. A typical fusion reaction of this type is

$$^3_1H + {}^1_1H \rightarrow {}^4_2He + Q$$

Similar fusion reactions are responsible for the uncontrolled release of energy that occurs when a hydrogen bomb explodes.

Fusion reactions release much more energy per nucleon than do fission reactions, but they are more difficult to initiate. Unlike neutrons, nuclei are charged and repel each other. Bringing them together to the point that the short-range nuclear attractive force overcomes the repulsive electric force and fuses the nuclei together requires very high temperatures and pressures.

Since the difficulty of bringing nuclei together increases with the amount of charge on the nuclei, the best materials to use for fusion reactions are nuclei with the smallest possible atomic number. As a result, the isotopes of hydrogen, H-1, H-2 (deuterium), and H-3 (tritium) are used for this purpose.

Controlled Fusion

Scientists continue to be challenged by the task of designing a reactor that would use fusion reactions to liberate energy in a controlled, useable, and efficient manner. Since the oceans of the earth provide an almost limitless supply of hydrogen isotopes (every water molecule contains two hydrogen atoms), and more energy is liberated per nucleon by fusion than by any other process, such a reactor would be a monumental achievement. Thus far, however, the problem of containing the extremely high temperatures needed to initiate a fusion reaction poses an unresolved obstacle to the construction of such a reactor.

Questions

E33. The nucleus of isotope *A* of an element has a larger mass than the nucleus of isotope *B* of the same element. Compared with the number of protons in the nucleus of isotope *A*, the number of protons in the nucleus of isotope *B* is (1) less (2) greater (3) the same

E34. Which is an isotope of $^{44}_{21}Sc$? (1) $^{44}_{20}Ca$ (2) $^{46}_{20}Ca$ (3) $^{46}_{21}Sc$ (4) $^{44}_{22}Ti$

E35. An atom consists of 9 protons, 9 electrons, and 10 neutrons. The number of nucleons in this atom is (1) 0 (2) 9 (3) 19 (4) 28

E36. The total number of neutrons in the nucleus of any atom is equal to the (1) mass number of the atom (2) atomic number of the atom (3) atomic number minus the mass number (4) mass number minus the atomic number

E37. A pair of isotopes is (1) $^{238}_{92}U$ and $^{239}_{92}U$ (2) $^{239}_{93}Np$ and $^{239}_{92}U$ (3) $^{235}_{92}U$ and $^{239}_{94}Pu$ (4) $^{239}_{93}Np$ and $^{239}_{94}Pu$

E38. Which device makes visual observation of the path of a charged particle possible? (1) Geiger counter (2) Van de Graaff generator (3) cyclotron (4) cloud chamber

E39. Which device could be used to give a positively charged particle sufficient kinetic energy to penetrate the nucleus of an atom? (1) electroscope (2) Geiger counter (3) cloud chamber (4) Van de Graaff generator

E40. Which of the following is used to accelerate a charged particle? (1) a photographic plate (2) an electroscope (3) a cyclotron (4) a cloud chamber

E41. Of the following particles, the one that cannot be accelerated in an atom-smashing machine is the (1) proton (2) neutron (3) alpha particle

E42. Which group of particles can *all* be accelerated by a cyclotron? (1) alpha particles, electrons, and neutrons (2) electrons, neutrons, and protons (3) protons, alpha particles, and electrons (4) neutrons, protons, and alpha particles

Base your answers to questions E43 and E44 on the following nuclear equation.

$$^{226}_{88}Ra \rightarrow {}^{222}_{86}Rn + {}^4_2He$$

E43. What is represented by ^4_2He? (1) an alpha particle (2) a beta particle (3) a gamma ray (4) a positron

E44. This equation represents the process of (1) alpha decay (2) beta decay (3) fission (4) fusion

E45. In which reaction does *X* represent a beta particle?

(1) $^{234}_{92}U \rightarrow {}^{230}_{90}Th + X$

(2) $^{214}_{84}Pa \rightarrow {}^{210}_{82}Pb + X$

(3) $^{226}_{88}Ra \rightarrow {}^{222}_{86}Rn + X$

(4) $^{214}_{82}Ra \rightarrow {}^{214}_{83}Bi + X$

E46. In the decay series of U-238, the change from Th-234 to U-234 involves the emission of (1) neutrons (2) positrons (3) alpha particles (4) beta particles

E47. How many beta particles are given off when one atom of U-238 completely disintegrates to Pb-206? (1) 6 (2) 8 (3) 10 (4) 14

Base your answers to questions E48 through E50 on the following information.

$^{131}_{53}I$ initially decays by emission of beta particles.

E48. Beta particles are (1) protons (2) electrons (3) neutrons (4) electromagnetic waves

E49. When $^{131}_{53}I$ decays by beta emission, it becomes (1) $^{130}_{53}I$ (2) $^{129}_{51}Sb$ (3) $^{131}_{54}Xe$ (4) $^{135}_{54}Xe$

E50. The half-life of $^{131}_{53}I$ is 8 days. After 24 days, how much of a 100.-gram sample of $^{131}_{53}I$ would remain? (1) 0 g (2) 12.5 g (3) 25.0 g (4) 50.0 g

Base your answers to questions E51 through E55 on Figure 5-E2.

E51. Which change is the result of the loss of two negative beta particles? (1) U to Th (2) Th to U (3) Pb to Bi (4) Bi to Pb

E52. Which of the following pairs of isotopes is found in the Uranium Disintegration Series? (1) $^{226}_{84}Po$ and $^{218}_{84}Po$ (2) $^{226}_{88}Ra$ and $^{222}_{88}Ra$ (3) $^{214}_{83}Bi$ and $^{210}_{83}Bi$ (4) $^{239}_{92}U$ and $^{238}_{92}U$

E53. Which particle is emitted as $^{234}_{90}Th$ changes to $^{234}_{91}Pa$ (1) a neutron (2) an alpha particle (3) a proton (4) a negative beta particle

E54. When a nucleus emits an alpha particle, the mass number of the nucleus (1) decreases (2) increases (3) remains the same

E55. As a sample of uranium disintegrates, the half-life of the remaining uranium (1) decreases (2) increases (3) remains the same

Base your answers to questions E56 through E60 on the following graph, which represents the disintegration of a sample of a radioactive element. At time t = 0 the sample has a mass of 4.0 kilograms.

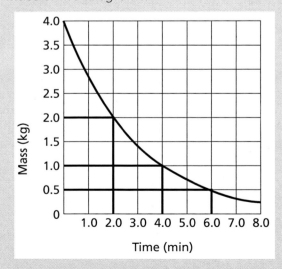

E56. What mass of the material remains at 4.0 minutes? (1) 1 kg (2) 2 kg (3) 0 kg (4) 4 kg

E57. What is the half-life of the isotope? (1) 1.0 min (2) 2.0 min (3) 3.0 min (4) 4.0 min

E58. How many half-lives of the isotope occurred during 8.0 minutes? (1) 1 (2) 2 (3) 8 (4) 4

E59. How long did it take for the mass of the sample to reach 0.25 kilogram? (1) 1 min (2) 5 min (3) 3 min (4) 8 min

E60. If the mass of this material had been 8.0 kilograms at time $t = 0$ its half-life would have been (1) less (2) greater (3) the same

Base your answers to questions E61 through E65 on the following information.

In the equation $^{221}_{87}Fr \rightarrow X + \gamma + Q$, the letter X represents the nucleus produced by the reaction, γ represents a gamma photon, and Q represents additional energy released in the reaction.

E61. Which nucleus is represented by X? (1) $^{217}_{85}X$ (2) $^{221}_{88}X$ (3) $^{220}_{87}X$ (4) $^{221}_{87}X$

E62. The rest mass of the gamma ray photon is approximately (1) one atomic mass unit (2) the mass of a proton (3) the mass of a neutron (4) zero

E63. If the energy Q equals 9.9×10^{-13} joule, the mass equivalent of this energy is (1) 0 kg (2) 9.1×10^{-31} kg (3) 1.1×10^{-29} kg (4) 3.3×10^{-21} kg

E64. The sample of $^{221}_{87}Fr$ (half-life = 4.8 minutes) will decay to one-fourth of its original amount in (1) 4.8 min (2) 9.6 min (3) 14.4 min (4) 19.2 min

E65. The gamma photon makes a collision with an electron at rest. During the interaction, the momentum of the photon will (1) decrease (2) increase (3) remain the same

E66. A certain radioactive isotope has a half-life of 2 days. If 8 kilograms of the isotope is placed in a sealed container, how much of the isotope will be left after 6 days? (1) 1 kg (2) 2 kg (3) 0.5 kg (4) 4 kg

Base your answers to questions E67 through E71 on the following information and nuclear equations.

When nitrogen is bombarded with protons, the first reaction that occurs is $^{14}_{7}N + ^{1}_{1}H \rightarrow ^{15}_{8}O + X$. The oxygen produced is radioactive with a half-life of 0.10 second, and decays in the following manner: $^{15}_{8}O \rightarrow ^{15}_{7}N + Y$.

E67. The first reaction is an example of (1) alpha decay (2) beta decay (3) induced transmutation (4) natural radioactivity

E68. In the first reaction, X represents (1) an alpha particle (2) a beta particle (3) a neutron (4) a gamma photon

E69. In the second reaction, Y represents (1) an electron (2) a neutron (3) a positron (4) a proton

E70. If a 4.0-kilogram sample of $^{15}_{8}O$ decays for 0.40 second, the mass of $^{15}_{8}O$ remaining will be (1) 1.0 kg (2) 2.0 kg (3) 0.50 kg (4) 0.25 kg

E71. As the amount of $^{15}_{8}O$ decreases, the half-life (1) decreases (2) increases (3) remains the same

Base your answers to questions E72 through E75 on the following nuclear equations.

$$^{27}_{13}Al + ^{4}_{2}He \rightarrow ^{30}_{15}P + X + energy$$

$$^{30}_{15}P \rightarrow ^{30}_{14}Si + Y + energy$$

E72. The first equation indicates that the radioactive phosphorus is produced by bombarding $^{27}_{13}A$ with (1) neutrons (2) positrons (3) alpha particles (4) protons

E73. In the first equation, particle X is (1) a neutron (2) an electron (3) a positron (4) a neutrino

E74. In the second equation, particle Y is (1) an alpha particle (2) a neutron (3) an electron (4) a positron

E75. The number of neutrons in the nucleus of $^{27}_{13}Al$ is (1) 13 (2) 14 (3) 27 (4) 40

Base your answers to questions E76 through E79 on the following nuclear equation.

$$^{30}_{15}P \rightarrow ^{A}_{Z}Si + ^{0}_{+1}X$$

E76. In the equation, X represents (1) a positron (2) an electron (3) a proton (4) a gamma photon

E77. What is the value of A in the equation? (1) 28 (2) 29 (3) 30 (4) 31

E78. What is the value of Z in the equation? (1) 14 (2) 15 (3) 16 (4) 17

E79. The nucleus of $^{30}_{15}P$ has (1) 30 protons (2) 30 neutrons (3) 15 nucleons (4) 15 neutrons

E80. The function of the moderator in a nuclear reactor is to (1) absorb neutrons (2) slow down neutrons (3) speed up neutrons (4) produce extra neutrons

E81. An atom of U-235 splits into two nearly equal parts. This is an example of (1) alpha decay (2) beta decay (3) fusion (4) fission

E82. When a nucleus captures an electron, the atomic number of the nucleus (1) decreases (2) increases (3) remains the same

E83. In a nuclear reactor, control rods are used to (1) slow down neutrons (2) speed up neutrons (3) absorb neutrons (4) produce neutrons

E84. During nuclear fusion, energy is released as a result of the (1) splitting of heavy nuclei (2) combining of heavy nuclei (3) combining of light nuclei (4) splitting of light nuclei

E85. When a neutron is emitted from a nucleus (1) the atomic number decreases (2) the atomic mass decreases (3) the atomic number does not change (4) both (2) and (3)

E86. When a beta particle is emitted from a nucleus (1) the atomic number decreases (2) the atomic number increases (3) the atomic mass decreases (4) the atomic mass increases

E87. In the nuclear reaction $^{9}_{4}Be + ^{4}_{2}He \rightarrow ^{12}_{6}C + W$, the symbol W represents (1) an electron (2) a proton (3) a neutron (4) an alpha particle

E88. In the nuclear reaction $^{239}_{92}U \rightarrow ^{239}_{93}Np + X$, the symbol X represents (1) a deuteron (2) an electron (3) a gamma ray (4) a proton

E89. In the nuclear reaction $^{4}_{2}He + ^{14}_{7}N \rightarrow ^{17}_{8}O + Z$, the symbol Z represents (1) a proton (2) a neutron (3) an alpha particle (4) an electron

E90. When fission occurs (1) energy is liberated (2) the products weigh less than the original material (3) mass is converted into energy (4) all of these

E91. Fusion is produced by (1) a chain reaction (2) neutron bombardment (3) intense heat (4) none of these

E92. The process of fusion is accompanied by (1) no change in mass (2) a loss in mass (3) a gain in mass (4) the transmutation of helium to hydrogen

E93. The nuclear reaction believed to be taking place in the sun and to be responsible for its release of energy is known as (1) condensation (2) fission (3) fusion (4) radiation

E94. An element suitable for fusion is (1) uranium (2) plutonium (3) hydrogen (4) all of these

E95. The nuclear raw materials for a fusion reaction have a total mass of 3.0067 grams. The products of this reaction may have a mass of approximately (1) 3.0065 g (2) 3.0067 g (3) 3.0069 g (4) 6.0134 g

E96. An atom that undergoes electron capture emits (1) a positron (2) an electron (3) a gamma photon (4) a positron and a neutrino

E97. Air can be used in a nuclear reactor as a (1) coolant (2) control on the chain reaction (3) moderator (4) shield

E98. In some reactors the coolant also serves as the (1) moderator (2) shield (3) control (4) all of the above

E99. Steel and concrete can be used in nuclear reactors as (1) moderators (2) control rods (3) shields (4) coolants

E100. Which of the following does not undergo fission? (1) U-235 (2) U-238 (3) plutonium-239 (4) all of the above

E101. Some reactors breed plutonium by bombarding which of the following with neutrons? (1) U-235 (2) U-238 (3) plutonium-239 (4) barium-141

E102. Fusion requires a high temperature to initiate because (1) atomic nuclei repel each other (2) neutrons are not charged (3) gravity must be overcome (4) mass must be converted to energy

E103. Which of the following nuclei would be most difficult to get to "fuse" together? (1) hydrogen-2 ($_1^2H$) (2) carbon-12 ($_6^{12}C$) (3) helium-4 ($_2^4He$) (4) lithium-7 ($_3^7Li$)

Laboratory Skills

Students are expected to master a number of skills as part of the Physical Setting: Physics course. Some of these are general skills, applicable to any scientific investigation, while others are specific laboratory procedures.

General Skills

1. Apply basic mathematics to data to arrive at an appropriate solution to a problem.
2. Formulate a question or define a problem for investigation and develop a hypothesis to be tested.
3. Collect, organize, and graph data.
4. Make predictions based on experimental data.
5. Formulate generalizations or conclusions based on an investigation.
6. Demonstrate safety skills in handling equipment, using chemicals, heating materials, setting up electric circuits, and working with radioactive substances.

Examples of Specific Laboratory Procedures

1. Determine the change in length of a spring as a function of force. Graph the experimental data.
2. Determine the period of a pendulum for a given mass and a given length.
3. Set up series and parallel circuits, each consisting of a power supply and two resistors. Determine the current through and the potential difference across each resistor and the circuit as a whole.
4. Map a magnetic field using a compass and a permanent magnet or electromagnet.
5. Determine the path of a light ray passing from air through another medium and back into air. Draw the ray diagram.
6. Formulate inferences about the contents of a black box (a sealed system into which one cannot see) by making external observations.

General Skills

1. Apply basic mathematics to data to arrive at an appropriate solution to a problem. You should be able to determine the precision of instruments used in an investigation. You should also be able to record data to the correct number of significant digits based on the precision of the instrument used. (See the Introduction section, pp. 1–2.)

You should be able to organize collected data into appropriate categories, graph the data when applicable, and arrive at valid relationships based on the data. Based on prior knowledge, you should be able to estimate answers. In general, the four basic operations—addition, subtraction, multiplication, and division—are all that are necessary to solve a problem. Occasionally, squaring and finding the square root are part of the solution. Examples of these skills are presented later in this section (see the spring experiment).

Frequently, graphed data produces a straight line with a y-intercept of zero. You should be able to calculate the slope of such a line by using the formula slope $= \Delta y/\Delta x$. The value of the slope is often an important physical constant.

2. Formulate a question or define a problem for investigation and develop a hypothesis to be tested. Before proceeding with an experiment, you should define the problem under investigation and formulate a hypothesis. This helps to structure the experiment and to clarify its purpose. For example, one hypothesis that might be formulated when investigating a swinging pendulum might be as follows: *If the length of a pendulum is increased while its angle of swing and mass are kept constant, then the period will also increase.*

Note some key characteristics of the above hypothesis:

    a) It has an If . . . , then . . . format.
    b) It describes the experiment to be done.
    c) It specifies what measurements will be made.
    d) It proposes an outcome, which may or may not be true.

These are features of an ideal hypothesis.

Note also that the above hypothesis requires that the length of the pendulum will be changed while its angle of swing and mass will be kept constant. During an experiment, you should change only one variable at a time when possible and hold all others constant. This helps to regulate the experiment. This is also called the definition of a **controlled experiment**.

3. Collect, organize, and graph data. During an experiment, raw data are collected. These data may be recorded in a log, chart, or data table. It is important to organize the chart or data table before beginning the experiment. For example, suppose you are measuring the voltage across a circuit every 10 seconds. You will only have time to enter your voltage reading before the next reading must be taken. If you had to enter the elapsed time, too, you might not be able to record the data quickly enough. Therefore, prepare as much of the table in advance as possible.

Graphed data often displays a relationship between variables that cannot be ascertained easily from raw data alone. Choose the size of the intervals on both axes appropriately so that your graph is neither too small nor too large. Once chosen, keep the intervals uniform. Axes should be labelled with their variable name and unit. By convention, the dependent variable should be placed on the y-axis (the vertical axis). The independent variable should be placed on the x-axis (the horizontal axis). The graph should have a title.

In many cases in physics, it is unnecessary for the line to run through every data point on the graph. A reasonable best-fit line suffices to establish the relationship between the variables being plotted. A best-fit line is one that follows the trend of the data but might leave some data points off the line on either side. When asked to draw a best-fit line, do not just connect the dots.

4. Make predictions based on experimental data. Graphing is an excellent method of presenting data clearly. Graphs also help to predict values that are not data points through *interpolation* and *extrapolation*.

For example, consider the graph of experimental data in Figure LS-1 that shows change of length of a spring plotted as a function of the force on the spring. From the graph you can predict that a force of 27 newtons will stretch the spring approximately 0.13 meter (prediction A in Figure LS-1). If the spring were to be stretched to 0.225 m, you can predict that a force of 45 N will be needed (prediction B in the figure). These are examples of **interpolation**. These predicted values

are *within* the range of the original data, namely 0–50 N and 0–0.25 m.

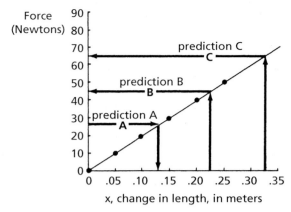

Figure LS-1. Force vs. change in length of a spring.

You could also predict how much force will be needed to stretch the spring to 0.33—approximately 65 N (prediction C in the figure). This is an example of **extrapolation**. The predicted values are *outside* the range of the original data. Extrapolation must be done cautiously. It is reasonable to predict values that lie slightly beyond the ends of the range of original data, but not too far beyond them. It would not be reasonable to extend the graph to predict a force needed to stretch the spring 0.70 m. That amount of stretch might cause the spring to be deformed or snap.

5. Formulate generalizations or conclusions based on an investigation. The results of an experiment are collected and analyzed. For a conclusion to be meaningful, the experiment must be repeated many times, and all the data obtained must be included in the analysis. The scope of the conclusion must be limited by the experimental data.

It is important to have confidence in data obtained. The best way to do this is to make several trials of each measurement and, if the results are close, determine an average value. Below are four measurements of the period of a pendulum whose length is 10.0 cm:

Trial 1—0.636 s/swing
Trial 2—0.632 s/swing
Trial 3—0.855 s/swing
Trial 4—0.635 s/swing

Trial 3 seems out of line with the other measurements and may indicate a procedural error. All the data should be rechecked and the measurements redone.

6. Safety in the Laboratory. You should practice the following safety precautions in the laboratory.

- Do not handle equipment or chemicals until you have been given specific instructions by your teacher.

- Report at once any equipment in the laboratory that appears to be unusual (broken, cracked, frayed), or any activity that appears to be proceeding in an abnormal fashion.

- Lab workspace must be kept clean and uncluttered. Clothing and hair must not restrict your movement, interfere with equipment, or present potential hazards. When using chemicals, wear safety glasses and aprons. Do not play or run in the laboratory.

- Familiarize yourself with the location and use of the fire extinguisher, fire blanket, and eye baths.

- Some physics experiments may involve the firing of projectiles. Make sure the area where the projectile will be "shot" is clear. Take care in loading, cocking, and firing the spring mechanism. Give a warning signal before firing.

- When using chemicals, never taste, touch, or inhale them. Never work with a combustible chemical if there is an open flame nearby.

- When heating material, use proper equipment such as gloves or tongs to handle hot objects. If you are heating material in a test tube, be sure the open end is not pointed toward anyone.

- Electrical wiring should be checked for fraying, bare metal exposure, and loose connections. Always have a circuit checked out by your instructor. Whenever possible use low voltage and low current arrangements. Never work with electricity if you or the working area are wet. Do not touch "live" wires. Turn the electricity off first.

- Always use tongs when handling radioactive material. Keep radioactive sources at arm's distance.

- Be aware of the symbols often used to indicate hazards (Figure LS-2).

Specific Laboratory Procedures

1. Determine the change in length of a spring as a function of force. Graph the experimental data. The procedure for performing this laboratory investigation is as follows:

   a) Mark the position of the bottom of the unstretched spring as shown by position A in Figure LS-3.

Use safety goggles and aprons.

Noxious or poisonous vapors.

Danger of breakage, as with glass.

Open flame.

Poisonous.

Material is caustic or corrosive.

Electric shock caution.

Potentially explosive.

Radioactive.

Figure LS-2. Hazard symbols.

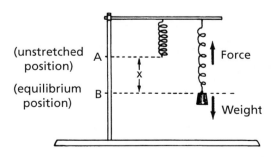

Figure LS-3.

b) Suspend a known weight (in newtons) from the spring. (If only the mass of the suspended object is provided, convert mass to weight by using the formula $W = mg$.) Wait for equilibrium to be established. At equilibrium the weight (acting downward) is equal to the force exerted by the spring (acting upward).

c) Mark the new position of the bottom of the stretched spring, as shown by position B in Figure LS-3.

d) Measure the distance (in meters) between points A and B with a ruler. This distance is the change in the length of the spring.

e) Repeat steps (a) through (d) with at least five different weights. In each case record the force, F (which is equivalent to the weight) in one column and the corresponding change in length, x, in another column. Make sure the columns are properly labeled, as in the sample in Table LS-1.

Table LS-1.

Force (F) (N)	Change in length (x) (m)
0.30	0.0201
0.60	0.0402
1.15	0.0701
1.50	0.1003
2.20	0.1500

f) The relationship between force and change in length is best visualized by plotting the data points on a graph. Axes should be labeled properly and an appropriate scale chosen for each. Choose scales so that all data points appear on the graph but are as far apart as possible (see Figure LS-4).

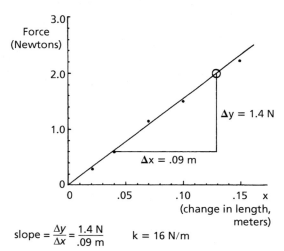

$$\text{slope} = \frac{\Delta y}{\Delta x} = \frac{1.4 \text{ N}}{.09 \text{ m}} \qquad k = 16 \text{ N/m}$$

Figure LS-4.

g) A straight-line graph with a y-intercept of zero indicates that $F = kx$, where k is a constant equal to the slope of the line. The force, F, exerted by the spring is directly proportional to the change in its length, x (Hooke's Law). The value of k as determined from the graph in Figure LS-4 is 16. N/m. (Note that the line does not run precisely through every data point. It is the best-fit line for the data obtained.)

2. Determine the period of a pendulum for a given mass and a given length.

The procedure for performing this laboratory investigation is as follows:

a) Record the length of the pendulum, measured from the top of the string to the center of the bob.

b) Pull the bob to one side and, while holding the string taut, measure the angle formed between the vertical and the string with a protractor (angle A in Figure LS-5).

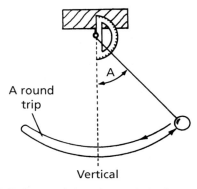

Figure LS-5. Determining the period of a pendulum.

c) Release the bob of the pendulum and start the stopwatch at the same time.

d) Allow the pendulum to swing through twenty complete round trips (see Figure LS-5).

e) At the instant the twentieth round trip is completed (when the bob returns to its starting point) stop the watch. Record the time to the correct precision.

f) The period, T, is the time it takes for the pendulum to complete one round trip. Divide the total time measured in step (e) by 20.0 to find the period of the pendulum for the given mass and length.

g) To verify that each of the 20 swings took the same amount of time, repeat the experiment by holding the bob at different angles. The period, T, turns out to be independent of the amplitude, angle A.

3. Set up series and parallel circuits, each consisting of a power supply and two resistors. Determine the current through and the potential difference across each resistor and the circuit as a whole.

a) Set up a series circuit as shown in Figure LS-6.

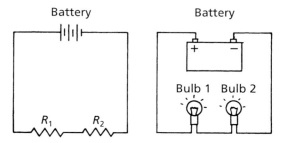

Figure LS-6. A series circuit.

b) Current, in amperes, is measured by inserting an ammeter in series with the branch of the circuit whose current is to be measured. In the case of a series circuit there is only one branch. To find the current through either resistor or through the circuit as a whole, insert the ammeter in series anywhere in the circuit. (See Figure LS-7.) Care must be taken to connect the terminals of the ammeter correctly. If those terminals are labeled plus and minus, the plus terminal of the ammeter should be connected to the wire coming from the plus terminal of the power supply. Likewise, the minus terminal of the ammeter should be connected to the wire coming from the minus terminal

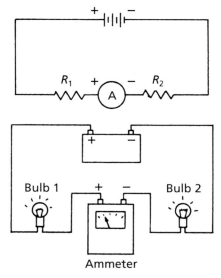

Figure LS-7. An ammeter in a series circuit.

of the power supply. If the ammeter is hooked up incorrectly the needle on the scale will move in the wrong direction, below the zero mark.

c) Read the current on the scale and record the value to the correct number of significant digits. The reading on the ammeter scale in Figure LS-8, for example, should be recorded as 0.66 ampere. (The smallest marked intervals on the scale represent 0.1 ampere each, so the precision of the ammeter is 0.01 ampere. See Mathematical Skills section.)

Figure LS-8. Reading an ammeter scale.

d) Potential difference, in volts, is measured by connecting a voltmeter in parallel with the circuit element whose potential difference is to be measured. To find, for example, the potential difference across bulb 1 in Figure LS-6, connect the voltmeter in parallel with that bulb, as in Figure LS-9. To measure the potential difference of the circuit as a whole, connect the voltmeter in parallel with the power supply (see Figure

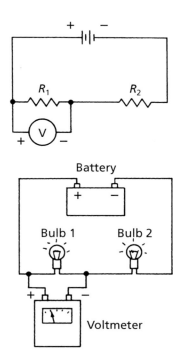

Figure LS-9. Finding the potential difference across bulb 1.

LS-10). Again, care must be taken to connect the terminals of the voltmeter properly (follow the same procedure as in step b). Record the potential difference to the correct number of significant figures.

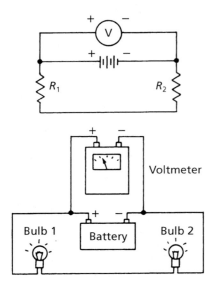

Figure LS-10. Finding the potential difference of the series circuit as a whole.

e) Set up a parallel circuit as shown in Figure LS-11.

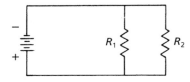

Figure LS-11. A parallel circuit.

f) To measure the current through each resistor (bulb) connect the ammeter in series with each branch. (See Figure LS-12.) To

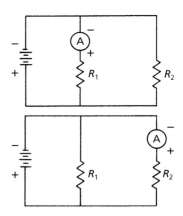

Figure LS-12. (top) Ammeter measures current through resistor 1. (bottom) Ammeter measures current through resistor 2.

measure the current in the circuit as a whole, connect the ammeter in series with the main line of the circuit. (See Figure LS-13.)

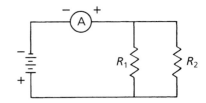

Figure LS-13. Ammeter measures current in the circuit as a whole.

g) To measure the potential difference across either resistor or the circuit as a whole, connect the voltmeter in parallel with all three, as shown in Figure LS-14.

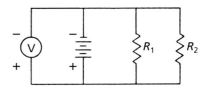

Figure LS-14. Measuring potential difference in a parallel circuit.

4. Map a magnetic field using a compass and a permanent magnet or electromagnet.

a) Use the compass as a "test magnet." Place it in various locations in the vicinity of the permanent magnet or electromagnet and observe the direction in which the N-pole of the compass points. At each location draw a small arrow near the compass indicating the direction assumed by the N-pole of the compass. (See Figure LS-15.) This is the direction of the magnetic field at each location.

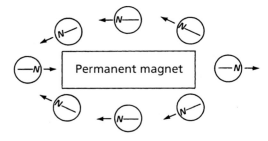

Figure LS-15.

b) After a sufficient number of arrows have been drawn, the magnetic field can be "mapped." Draw field lines so that they pass through or near the arrows in a parallel manner. Deduce the north and south end of the magnet from the direction of the field lines. (See Figure LS-16.)

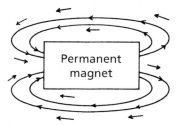

Figure LS-16.

5. Determine the path of a light ray passing from air through another medium and back into air. Draw the ray diagram.

a) Place a rectangular block of glass or Lucite on a piece of paper. The paper should rest on top of a piece of cardboard or Styrofoam.
b) Trace the shape of the block on the paper with a sharp-pointed pencil. Remove the block; then draw and label point P_1, as shown in Figure LS-17.

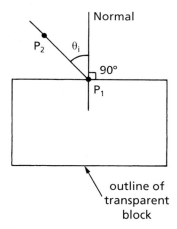

Figure LS-17.

c) Using a ruler and protractor, draw a line through point P_1, perpendicular to the top line of the traced block outline. (See Figure LS-17.) Label this line the "normal."
d) Using a ruler, draw a second line through point P_1 at some angle, θ, from the normal. Label this angle θ_i, the angle of incidence.
e) Draw and label point P_2 on the second line, as shown in Figure LS-17, at a distance of approximately 5 cm from point P_1.
f) Insert a pin through each of the points P_1 and P_2. Make sure each pin stands vertically. Place the block back on the paper so that its edges match the original tracing.

g) Lower your head until the plane of the paper is at eye level. With only one eye open, look through the transparent block at the two pins. Move your head until the pins appear to be lined up one behind the other. Place a ruler on the paper so that it is aimed from your eye to the image of the lined-up pins. Trace a line in this direction on the paper all the way to the block. (See Figure LS-18.)

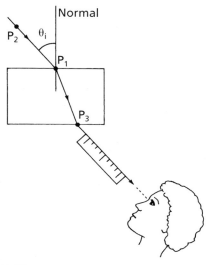

Figure LS-18.

h) Remove the block; then draw and label point P_3 (see Figure LS-18) where the traced line meets the block.
i) Draw a straight line connecting points P_1 and P_3. The light ray coming from P_2 in the direction of P_1 passed through the transparent block in the direction of line $\overline{P_1P_3}$ and emerged at point P_3 to enter your eye. The path from P_2 to P_1 to P_3 to your eye constitutes the ray diagram in this case. It should appear bent, or refracted, twice, when entering and exiting the block.

6. Formulate inferences about the contents of a black box (a sealed system into which one cannot see) by making external observations.
Use any available tool to determine the nature or characteristics of the contents of the box (other than opening it and looking inside). A compass should reveal whether the contents are magnetic. A charged pith ball can be helpful in determining whether the contents are charged or electrically neutral. Shaking the box and listening to the sounds made, weighing it, smelling it, measuring its temperature, and a host of other techniques all provide clues to help solve the mystery of what is in the box.

REVIEW QUESTIONS

PART B-1

1. The mass of a high school football player is approximately (1) 10^0 kg (2) 10^1 kg (3) 10^2 kg (4) 10^3 kg

2. A high school physics student is sitting in a seat reading this question. The magnitude of the force with which the seat is pushing up on the student to support the student is closest to (1) 0 N (2) 60 N (3) 600 N (4) 6000 N

3. An egg is dropped from a third-story window. The distance the egg falls from the window to the ground is closest to (1) 10^0 m (2) 10^1 m (3) 10^2 m (4) 10^3 m

4. What is the approximate mass of an automobile? (1) 10^1 kg (2) 10^2 kg (3) 10^3 kg (4) 10^6 kg

5. What is the approximate mass of a pencil? (1) 5.0×10^{-3} kg (2) 5.0×10^{-1} kg (3) 5.0×10^0 kg (4) 5.0×10^1 kg

6. The approximate height of a 12-ounce can of root beer is (1) 1.3×10^{-3} m (2) 1.3×10^{-1} m (3) 1.3×10^0 m (4) 1.3×10^1 m

7. The diameter of a United States penny is closest to (1) 10^0 m (2) 10^{-1} m (3) 10^{-2} m (4) 10^{-3} m

8. What is the approximate width of a person's little finger? (1) 1 m (2) 0.1 m (3) 0.01 m (4) 0.001 m

PART B-2

Base your answers to questions 9 through 12 on the information and table below.

In a laboratory exercise, a student kept the mass and amplitude of swing of a simple pendulum constant. The length of the pendulum was increased and the period of the pendulum was measured. The student recorded the data in the table below.

Length (m)	Period (s)
0.05	0.30
0.20	0.90
0.40	1.30
0.60	1.60
0.80	1.80
1.00	2.00

Directions: Using the information in the table, construct a graph on a grid, following the directions given.

9. Label each axis with the appropriate physical quantity and unit. Mark an appropriate scale on each axis.

10. Plot the data points for period versus pendulum length.

11. Draw the best-fit line or curve for the data graphed.

12. Using your graph, determine the period of a pendulum whose length is 0.25 meter.

Base your answers to questions 13 through 16 on the information and data table below.

In an experiment, a student measured the length and period of a simple pendulum. The data table lists the length (l) of the pendulum in meters and the square of the period (T^2) of the pendulum in seconds[2].

Length (l) (m)	Square of Period (T^2) (s^2)
0.100	0.410
0.300	1.18
0.500	1.91
0.700	2.87
0.900	3.60

Directions: Using the information in the data table, construct a graph on a grid, following the directions below.

13. Plot the data points for the square of period versus length.

14. Draw the best-fit straight line.

15. Using your graph, determine the time in seconds it would take this pendulum to make one complete swing if it were 0.200 meter long.

16. The period of a pendulum is related to its length by the formula $T^2 = 4\pi^2 l/g$, where g represents the acceleration due to gravity. Explain how the graph you have drawn could be used to calculate the value of g. You do *not* need to perform any actual calculations.

PART C

17. Explain how you would experimentally determine the spring constant of a particular spring. Your explanation must include: (a) measurements required (b) equipment needed (c) any equation(s) needed

Science, Engineering, Technology, and Society

The Physics of Baseball

What comes to mind when you hear the word *physics*? Perhaps you picture a bright green baseball field. Maybe you envision a pitcher throwing a baseball to home plate. Possibly you hear the crack of a bat hitting a ball and the roar of the crowd. That wasn't what you were thinking? Maybe it should be. After all, physics and baseball go hand in hand.

Pitched Balls

For years, many people thought that a curveball was just an optical illusion. Physicists, however, have long known that a spinning ball curves in flight. In fact, Isaac Newton wrote a paper on this subject back in 1671. Then, in 1852, the German physicist Gustav Magnus investigated the topic further.

According to Magnus, a ball moving through air interacts with a thin layer of air known as the boundary layer. The air in the boundary layer peels away from the surface of the ball. This creates a region of low pressure, known as a wake, behind the ball. The difference in pressure between the front and back of the ball creates a backward force on the ball. This slows its forward motion. This is the normal air resistance, or drag, on the ball.

If the baseball is spinning as it moves, the boundary layer separates at different points on opposite sides of the ball. It separates further upstream on the side of the ball that is turning into the airflow. It separates further downstream on the side of the ball turning backward. As a result, the air flowing around the ball is deflected slightly sideways. This results in an uneven, or asymmetrical, wake behind the ball. The effect is a pressure difference across the ball. The resulting lateral force pushes the ball sideways. The lateral force, which is at right angles to the forward motion of the ball, is known as the Magnus force.

The strength of the Magnus force depends on how fast the ball spins and the forward speed of the ball. The faster the spin and speed, the greater the strength of the Magnus force.

The Magnus force is also proportional to the density of the air. This explains why pitchers face a greater challenge in Denver. There, at more than a mile above sea level, the air is thinner. The drag on a baseball is therefore less than in a city closer to sea level. Pitches curve less, which makes it easier for batters to hit them. Scientists calculate that a ball curves about 25 percent less in Denver than in Boston or New York.

In addition, baseballs travel farther once they are hit in air that is less dense. A baseball hit in Denver will travel farther than the same ball had it been hit in New York.

The stitches on a baseball play a part in its motion. They gather up air as the ball spins. In this way, they increase the thickness of the boundary layer. They also allow the pitcher to grasp the ball in such a way that it can spin. This combination increases the Magnus force.

The direction of the Magnus force depends on the direction of spin. By controlling the direction of the spin, a pitcher can make the Magnus force point in any direction. If, for example, the pitcher gives the ball a clockwise rotation, the ball will experience a leftward force from the pitcher's perspective. This will cause the ball to curve away from a right-hand batter. A ball with a counterclockwise rotation will curve in the opposite direction.

A baseball can be made to drop downward as well. A thrown ball has a natural downward curve due to the force of gravity. If the pitcher gives the ball topspin, the Magnus force will act toward the ground. This causes the ball to curve more sharply toward the ground. This is known as a breaking ball.

If, instead, the ball is given a backspin, the Magnus force points away from the ground. For the ball to rise upward, the Magnus force would have to be greater than the weight of the ball. The spin required to produce a force of this magnitude is greater than can be achieved by pitching a ball. Therefore, a ball with backspin does not actually rise, but it falls less than the batter expects it to. Technically, there is no such thing as a rising fastball. This is a bit of an illusion.

Ball Trajectory

When a ball is hit with a bat, the momentum of the bat is transferred to the ball. As the ball flies away, it is affected by drag and gravity. In an ideal situation, the combination of these forces causes the ball to follow a parabolic path.

The horizontal distance the ball travels depends on the initial velocity and the angle at which it is hit. A ball hit at 100 feet per second at an angle of 45 degrees will travel about 312 feet. If the angle is changed to 35 degrees, the ball travels only 294 feet.

Who hit the farthest ball in baseball history? Mickey Mantle still holds that record for a ball he hit in 1953. Mantle's ball traveled 460 feet before hitting a sign. The ball would have landed 560 feet from home plate had the sign not been there.

Sweet Spot

When a batter hits a ball, he or she feels a vibration as the bat makes contact with the ball. There is one spot on the bat, however, where the vibration is reduced to the point that the batter does not feel it. If the ball hits this spot, known as the sweet spot, the batter is almost unaware of the collision. If the ball hits far from the sweet spot, the jarring of the hands can be almost painful. The sweet spot is the point of contact that will make the ball travel the farthest.

The sweet spot results from the vibrations of a bat. To understand this, think about what happens when a spring is pushed down and let go. The spring vibrates, or oscillates, back and forth. The same thing happens to a baseball bat. If enough force is exerted on the bat, it will oscillate as well.

The oscillations of a bat can be described by a wave. Like all natural objects, a bat has a resonant frequency. This is the frequency that will produce the greatest amplitude in the wave. A bat is not symmetric. Hitting a bat at different places will result in different frequencies and amplitudes.

When a bat is hit, two waves are produced. One wave is produced when the ball strikes the bat. The other is produced when the ball leaves the bat. At antinodes, the waves meet constructively. This means they add together. If the ball hits the bat at the antinode, it will cause the greatest vibration on the bat. The bat will vibrate so much that it will sting or even break. Antinodes are at the head and mid-point of the bat.

At nodes, the waves meet destructively and cancel out. If the ball hits at a node, the oscillation will stop. This node forms a sweet spot on the bat. In actuality, tests show that a baseball bat has several of these nodes close together. As a result, the sweet spot might be better described as a sweet zone.

The sweet spot is located about 17 cm from the end of the barrel of the bat. It is sometimes known as the "fat part" of the bat because it is the thickest part of the bat. The sweet spot of a wooden bat is slightly smaller than that on an aluminum bat.

How does this explain why the ball travels farthest when it collides with the sweet spot? The more the bat oscillates, the more energy that is wasted. The maximum output of the bat results when there are no oscillations. This occurs at the sweet spot.

Questions

1. Research whether a ball thrown on the moon would travel faster than a ball thrown with the same force on Earth?

2. What factors determine the magnitude of the Magnus force on a baseball?

3. Explain why baseballs don't curve as much in Denver as they do in New York.

4. What is the shape of the path through which a baseball travels after being hit? Why?

5. What type of interference occurs at the sweet spot of a baseball bat? Cite evidence from the text.

Solar Explosions

It is afternoon and you are quietly reading a book. Suddenly, the lights go out and so does the radio you had playing in the background. The sun is shining and there is no wind. What could be the cause of the blackout? Surprising, the answer might be found millions of kilometers away on the surface of the sun.

Like other stars, the sun is basically a glowing ball of hot gases. It is powered by thermonuclear reactions in its center, or core. Energy released by fusing hydrogen atoms into helium atoms radiates through the middle layer, known as the radiative zone. The energy then bubbles to the surface of the third layer, or convective zone. Although extremely hot, the surface of the sun is considerably cooler than the core.

Solar Flares

Occasionally, great bursts of energy are released by the sun. Although the exact cause is not completely understood, scientists do know that these bursts are related to the sun's magnetic field. The sun's magnetic field is stronger in some places than in others. In addition, the magnetic field lines of force can become twisted, storing energy, much like a twisted rubber band. When the stored energy is released, huge amounts of hot gas spray into space, this is a solar flare.

A solar flare produces a rapid change in the brightness of the sun. Solar flares extend out of the layer of the sun known as the corona. The corona is the outermost layer of the sun's atmosphere. It is hotter than the surface, or photosphere, of the sun.

The corona is made up of rarefied gas normally at a temperature of several million kelvins. However, during a flare, the temperature reaches 10 or 20 million kelvins. Data indicates that it can be as high as 100 million kelvins. The area of the sun covered by a solar flare is about as large as Earth.

A solar flare is one of the greatest explosions in the solar system. It releases an amount of energy equivalent to an explosion of 100 megaton hydrogen bombs. The radiation that is emitted spans the entire electromagnetic spectrum from long-wavelength radio waves to short-wavelength gamma rays. It also emits particles that include protons, neutrons, electrons, and alpha particles, which are the nuclei of helium atoms.

Some of the radiation will be deflected or captured by Earth's magnetic field. Some of the radiation will simply pass Earth by. An additional amount may be absorbed by the atmosphere.

Measuring Solar Flares

Scientists use special telescopes to detect solar flares according to their optical and radio emissions. Some emissions, such as x-rays and gamma rays, require telescopes in space because these emissions do not penetrate Earth's atmosphere.

Sunspots

Predicting when a solar flare will occur is extremely difficult. They depend on changes in the sun's magnetic field in ways that scientists do not completely understand. The frequency of solar flares can vary greatly. There can be as many as several in the same day or as few as one each week.

What scientists do know is that solar flares coincide with the sun's 11-year cycle of sunspots. Sunspots usually occur in pairs where magnetic fields break through the sun's surface. Field lines leave through one sunspot and return through the other. Because the magnetic field slows the radiation of heat, sunspots are cooler than areas around them. This is why they appear as dark patches.

By studying sunspot activity, scientists can gather clues about solar flares. When the sunspot cycle is at its lowest point, few solar flares are detected. They occur more frequently as the sun approaches the high point of its cycle.

Space Weather

Solar flares release a tremendous amount of ionizing radiation. What is ionizing radiation? It carries enough energy to remove electrons from their orbits in atoms to produce ions.

The intense burst of radiation produced during a solar flare can have a direct effect on Earth's magnetic field and atmosphere. The connection between processes on the sun and changes in Earth's environment is often called space weather.

- **Auroras** One obvious consequence of a solar flare is increased aurora activity. Auroras, better known as Northern and Southern Lights, occur when charged particles collide with magnetized gases that surround Earth. Most of the particles are diverted around Earth, but some leak through, especially at the magnetic poles. The result can sometimes be seen as colorful displays of light.
- **Communication on Earth** Ionizing radiation from the sun causes Earth's upper atmosphere to become ionized and expand. The change in the ionosphere can interfere with long distance radio signals. Shortwave radio communications use the ionosphere to transfer signals. Changes can cause radio signals from the ground to be reflected as if from a mirror. Broadcasts that use shortwave radio, such as many police and fire channels, can be disrupted.
- **Power Grids** Ionized particles in the atmosphere can induce electric currents in power lines and cause power surges. These surges can overload a power grid and cause blackouts. Similar currents can be produced in long, uninterrupted oil and gas pipelines. Electricity in pipelines can cause the pipes to corrode at a faster rate.
- **Satellites** Many satellites are in geosynchronous orbits around Earth. In addition, the ionized atmosphere means that objects traveling through it experience greater drag. The orbits of satellites can therefore be disturbed. Scientists estimate that the orbits of more than 1500 satellites slowed or lost altitude due to the solar radiation released during March of 1989.
- **Air Travel** The increase in radiation on airline passengers due to a solar flare is generally not large, but does exist. The level of radiation is about the same as a week's worth of natural background radiation. There are some flares that release higher amounts of energy. However, these types of flares are expected only a few times during the 11-year solar cycle. The radiation exposure is similar to receiving an x-ray.
- **Space Travel** Unlike airline passengers, astronauts are more vulnerable to solar radiation. Although astronauts in the International Space Station are shielded from solar radiation, an astronaut walking on the surface of the moon would be unprotected. Unlike Earth, the moon does not have an atmosphere or magnetic field to protect it. The astronaut would suffer radiation sickness, which includes fatigue, low blood count, and vomiting. Scientists estimate that if astronauts had been on the moon during the solar flare that occurred in 1972, they might have even been killed.

While scientists cannot accurately predict solar flares, NASA and other agencies are working to provide warning about possible dangers as early as possible. The NASA Space Weather Initiative is set up to provide warning so that power stations, astronauts, and airplane crews, for example, can take safety precautions or turn off sensitive equipment.

Questions

1. Why do radio and power blackouts from a solar flare occur only during the day?
2. How is the energy released during a solar flare similar to that released by a twisted rubber band? Cite evidence from the text.
3. Why don't people on Earth experience the effects of most solar flares?
4. Why are instruments used to measure solar flares located in space?
5. How can a solar flare change the orbit of a satellite? Cite evidence from the text.

Now You See It, Now You Don't

Imagine playing the ultimate game of hide-and-seek. Rather than finding a secluded hiding spot, you stand out in plain sight. What's the trick? You surround yourself in a shield that makes you completely invisible. This may sound like science fiction, but if some scientists have their way, a cloak of invisibility may soon be reality.

Hiding in plain sight all comes down to refraction. When a beam of light crosses from one material into another at a slant, the light bends. The amount by which the path of light is altered depends on the refractive indices of the materials. The greater the difference between them, the greater the refraction of the light beam. For all natural materials, the refractive index is a positive value.

In 1967, a Soviet physicist named Victor Veselago suggested that a material could have a negative refractive index. He hypothesized that such a material would not violate any laws of physics. Veselago predicted that this material would reveal optical phenomena that had never before been observed. In particular, he calculated that a material with a negative refractive index would refract light toward its source instead of away from it.

Metamaterials

Veselago could not test his hypothesis because he did not have the materials to do so. No natural materials have a negative refractive index. In the last few years, however, scientists have developed new types of materials known as metamaterials.

Metamaterials are made up of three-dimensional arrays of rings, rods, or other shapes. They consist of metals joined together by electrical insulators, such as fiberglass. The atoms and molecules of original materials are replaced with micro- or even nanocircuits. The resulting material has properties that the original materials do not.

Natural materials refract light to the right of the incident beam at different angles and speeds. If you shine a beam of light on a window at an angle, the beam of light bends away from the source inside the glass.

Metamaterials make it possible to refract light at a negative angle so it emerges on the left side of the incident beam. This is why they are also known as left-handed materials. An electromagnetic wave aimed at a metamaterial bends back toward the source.

Metamaterials can be controlled to respond to electromagnetic waves in predictable ways. By adjusting the molecular structure, the properties can be tuned for specific wavelengths of electromagnetic radiation. They can also be developed to bend light in desired paths.

Possible Applications

The ability to bend light in new ways has many potential applications. One is to make optical lenses with greater resolution. Standard-light microscopes cannot resolve objects smaller than the wavelength of light. This is why it takes an electron microscope to study molecules or atoms.

Researchers hope to use metamaterials to make what has been called a superlens. Metamaterials cause electromagnetic waves to be focused on a point instead of diverging outward. While a normal lens has to be curved to focus light, flat metamaterials focus electromagnetic waves with great precision. They have the potential for the development of a flat superlens with the power to see inside a human cell.

In addition, the superlens would recover light that is usually lost in traditional lenses. The material would actually amplify waves of light to form an image with sharp resolution.

Another possible application is camouflage, or making objects invisible. One approach to making objects invisible is to design a cloaking device out of metamaterials that bend light. The metamaterial can be designed so that it bends light in a particular direction. Light and other forms of electromagnetic radiation could be made to flow like a river streaming around an obstacle. The light would then continue past the ob-

ject as if it had never struck the obstacle. The object would not reflect light and it would not form a shadow. An onlooker would seem to look right through the device without being able to see anything inside it.

In another approach, a microscopically thin film of metamaterial might be used to create an invisible zone. Any objects placed close to either side of the film would disappear from view.

Becoming "invisible" is important for much more than children's games. Researchers hope to use cloaking technology to save lives. Rescuers might one day be able to see through the rubble of an earthquake or hurricane to find survivors. Doctors might be able to see through bones to diagnose medical conditions.

The military would also benefit from this new technology. Currently, military planes, boats, and armored vehicles are coated with materials that either absorb or deflect radar signals. Metamaterials would either force signals to pass around them or cancel them, thereby making the vehicle impossible to detect.

Obstacles

Don't look for cloak of invisibility at your local mall any time soon. Although there has been some experimental evidence supporting the technology, most of the predictions about metamaterials are theoretical and mathematical only. Researchers have worked through the mathematical equations proving a cloaking device is possible. Working past the obstacles has yet to be achieved.

For one thing, researchers have yet to build a metamaterial that can be used with visible light. Existing metamaterials operate only in the microwave or far infrared regions of the electromagnetic spectrum. These electromagnetic waves have much longer wavelengths than visible light. The first prototypes will most likely be devices that hide an object from microwave radiation.

Another concern is that cloaking devices might work only at specific wavelengths. Visible light is made up of light with many different wave-

lengths. Researchers must consider the possibility of constructing a metamaterial that can address this situation.

An additional problem is that light cannot enter the cloaking device. A person inside such a device would be unable to see out of it. The cloaking device would most likely be a solid object. It could not be folded or bent.

Another challenge researchers face is the shape of the object being concealed. The electromagnetic waves at one part of the object will need to curve in a different direction than waves in another part. This suggests that the properties of the metamaterials will need to vary as well. Researchers will need to create a three-dimensional structure from many different pieces of metamaterials.

Questions

1. How is the amount of refraction related to the refractive indices of the materials?

2. Why are metamaterials sometimes called left-handed materials? Support your answer with evidence from the text.

3. What are two possible applications of negatively refracting materials? Support your answer with evidence gathered by Internet research.

4. How does the fact that visible light is made up of many different wavelengths pose an obstacle to the applications of metamaterials?

5. Why might researchers have to use many different metamaterials to construct a sphere in which to cloak an object?

Reversal of Fortune

Imagine you are out hiking in the woods when you pull out your compass to check your direction. What you find is that it points to the South Pole instead of the North Pole! This is exactly what may happen in less than 2000 years. It's not science fiction. If the predictions of many scientists are correct, it will be reality.

Earth as a Magnet

To understand why a compass might change direction, it is first important to recognize how a compass works. The needle of a compass is a small magnet that is allowed to pivot in a circle. When it comes to magnets, opposites attract. The north end of a magnet is attracted to a magnetic south pole and the south end of a magnet is attracted to a magnetic north pole.

The magnetized needle of a compass points toward the north because Earth is surrounded by a magnetic field much like that of a bar magnet. Earth's magnetic poles are located near its geographic poles. This makes it possible to use a compass to navigate direction. The source of Earth's magnetism is its core.

Earth's magnetic field extends far out into space, forming what is known as the magnetosphere. The magnetosphere is altered by the solar wind, which is a stream of charged particles ejected by the sun. The solar wind pushes the magnetosphere inward on the sunward side. It stretches the opposite side of the magnetosphere into a tail that is hundreds of thousands of kilometers long.

Earth's Magnetic Record

Much of what scientists know about Earth's magnetic history has come from studies of rock. Iron in lava becomes aligned with Earth's magnetic field. When the lava hardens into rock, the iron becomes permanently locked in position.

In certain locations on the sea floor, new rock is constantly being formed. By studying samples, scientists have discovered that the polarity of Earth's magnetic field has reversed many times throughout Earth's history. In other words, the north and south poles have flipped. Although they are not common by human standards, reversals are pretty common on a geologic timescale. The poles have reversed hundreds of times throughout the planet's history. The length of time between reversals varies from a thousand years to millions of years. The average is about 200,000 years.

A Decline in Earth's Magnetic Field

In addition to data about the direction of Earth's magnetic field, scientists are interested in the strength of the field. The strength of a magnetic field is known as gauss strength after Carl Friedrich Gauss who invented a device to measure magnetism in 1837.

To learn about the strength of Earth's magnetic field before this time, researchers had to find different sources of data. They found one important source in the logs of ships. It was common practice for ship captains in the 17th and 18th centuries to calibrate their ship's compasses relative to true north. In some cases, they would even measure the steepness at which magnetic field lines entered Earth's surface. These ship logs are not enough to determine Earth's magnetic record in great detail. However, they provide enough information to make large-scale estimates dating back to 1590.

In addition, researchers studied materials used by prehistoric civilizations. Much like the iron in lava, iron minerals in clay also record Earth's magnetic field once the clay is fired. Similarly, minerals in bricks used by early human settlements provide additional magnetic evidence.

The combination of these records indicates that Earth's magnetic field has been steadily declining in the recent past. The decline appears to have been minimal until about 1840. Since that time, the decline has accelerated to a rate of roughly 5 percent per century. If the field continues to decline at this rate, it will disappear in about 2000 years.

A Possible Reversal Ahead

Magnetic data indicates that all magnetic reversals are preceded by periods of weakened magnetic fields. The field must weaken and go to zero before it can reverse itself. Based on this, many researchers suggest that the decline indicates an approaching reversal in polarity. Recall that the average time span between magnetic reversals is about 200,000 years. The last major reversal appears to have occurred 780,000 years ago, so Earth is long overdue for a magnetic flip.

The current weakening of Earth's magnetic field does not guarantee that a reversal is coming. Not every period during which a weakened magnetic field existed was followed by a reversal in polarity. Some periods of weakness were followed by strengthening after what some scientists describe as a failed attempt at reversal. This means that the magnetic field might increase in strength once again. Then, perhaps 10,000 years from now, it will again decline and finally reverse.

The current weakening may be due to a weakness in Earth's magnetic field known as the Southern Atlantic Anomaly (SAA). Located off the coast of Brazil, the SAA is a region in which the magnetic field is 30 percent weaker than anywhere else.

The SAA occurs because the center of Earth's magnetic field is offset from its geographic center. This region is therefore farthest from the magnetic poles and as a result experiences the weakest field. Just as the magnetic poles drift in position over time, the SAA slowly drifts toward the west. In addition, it is steadily growing wider toward the southern Indian Ocean.

The SAA is significant to people and spacecraft that travel through the region. Satellites and spacecraft passing through this area are bombarded with charged particles. The extra dose of radiation creates problems with electronic systems and alterations in astronomical data. The Hubble Space Telescope is affected by the radiation as it passes through the SAA several times each day. Controllers on the ground lower the power on high-voltage instruments on board the Hubble before it enters the SAA to protect them from damage. Astronauts passing through the region must be protected by higher shielding.

The Southern Atlantic Anomaly might be a temporary alteration in the magnetic field that will eventually disappear. Another possibility is that it will instead grow into a complete reversal. Only time will tell.

If the magnetic field does indeed reverse polarity, the results may not be as dramatic as in a science fiction movie but they will be felt around the world. There would be additional radiation exposure that would damage satellites and airplanes. It would also pose health dangers to people on Earth. The increased radiation could knock out power grids, interfere with the communications systems on spacecraft, temporarily increase the size of holes in the ozone layer, and produce more aurora activity.

In addition, animals that rely on Earth's magnetic field for navigation would need to adjust. Keep in mind, however, that the reversal is gradual. It would not occur within the lifetime of a single animal, but rather over several generations.

Whatever the case, many generations from now, humans just may witness a complete magnetic reversal. By then, scientists and society will hopefully understand the process and be prepared to deal with the effects.

Questions

1. What is the source of Earth's magnetism?

2. What evidence has led some scientists to predict that Earth may experience a magnetic reversal within the next 2000 years?

3. How might animals be affected by a magnetic reversal? Cite evidence from the text.

4. How might the SAA be related to current observations about Earth's magnetic field?

Tracking a Storm

On September 8, 1900, the deadliest natural disaster in U.S. history struck Galveston, Texas. It was a hurricane that locals simply called "The Storm" because nothing else could compare to it. Over 8000 people lost their lives. The city was devastated.

Unfortunately, weather forecasters at the time had few tools at their disposal for predicting hurricanes. They could look at sea swells and measure wind speeds. They could evaluate stories about stormy seas from sailors and telegraph reports from the Caribbean. However, they had no conclusive scientific data on which to base their forecasts.

What Is a Hurricane?

A hurricane is a severe tropical cyclone. Any region of circulation that forms over tropical oceans is called a tropical cyclone. A tropical cyclone is like a huge fan. It changes energy from the warm tropical ocean into an intense and dangerous storm.

Near the ocean surface, a hurricane rotates in a counterclockwise direction around a center called the eye. The storm consists of bands of thunderstorms that spiral out from the center.

The winds around the core of a hurricane create large waves in the ocean water below them. When the storm reaches land, it produces what is known as a storm surge. It can also produce tornadoes, torrential rains, and floods.

Hurricanes are described by their strength, or intensity. The scale of hurricane intensity is called the Saffir-Simpson Scale. It rates hurricanes from 1 to 5. The determination is based on the potential damage and flooding the hurricane is expected to cause along the coast when it makes landfall. These estimates result primarily from wind speed. Storms with wind speeds between 75–95 mph are Category 1 storms. Storms with wind speeds greater than 155 mph are Category 5 storms. Wind speeds in between determine Category 2, 3, and 4 storms.

The name of the storm varies depending upon where it forms. They are known as hurricanes in the North Atlantic Ocean. In other regions of the world, they are known as typhoons, tropical cyclones, and severe cyclonic storms.

Satellites

Fortunately, modern forecasters have the ability to identify and track hurricanes days before they reach land. One group of relatively new tools that forecasters use is satellites that continuously orbit Earth. The National Aeronautics and Space Administration (NASA) currently has about twenty satellites in orbit to monitor conditions in different parts of Earth.

One such satellite is Aqua, which was launched in 2002. Given the name of the Latin word for water, Aqua has instruments on board that collect information about Earth's water cycle. Data about evaporation from the oceans, water vapor in the atmosphere, clouds, precipitation, ice, and snow are constantly being sent to researchers.

Another satellite is the Tropical Rainfall Measuring Mission (TRMM) satellite. It was launched as part of a joint mission with the Japanese Aerospace Exploration Agency (JAXA) in 1997. TRMM uses microwave and visible infrared sensors to measure tropical rainfall.

Tropical rainfall makes up more than two-thirds of all the rainfall on Earth. This rainfall distributes heat throughout the atmosphere. In order to predict global climate changes, forecasters must therefore understand tropical rainfall and any changes to it.

TRMM marks a first because its precipitation radar is the first ever to be launched into space. The radar system is particularly important because it can see directly into a hurricane. Researchers use data from the satellite to produce images of the precipitation in a storm.

TRMM does not replace computer models and other methods of data collection. Instead, it adds to them. Each method has advantages and disadvantages. For example, ground radars can observe a hurricane for hours at a time. The orbit of TRMM means that the satellite is continuously moving. It cannot hover over one spot for any period of time. Meanwhile, ground radars give less precise height measurements of rainfall than TRMM's radar. Adding TRMM rainfall data to data collected from other sources more than triples the accuracy of forecasts over a 12-hour period.

Hot Towers

One goal of TRMM is to study a specific formation of rain clouds in tropical cyclones. High rain clouds, known as hot towers, reach to the top of the troposphere. This is the lowest layer of the atmosphere. These towers form in areas of the storm where large amounts of heat are released as water vapor condenses. Researchers use the term latent heat to describe this form of heat.

Sometimes, a group of these towers forms in a portion of the eyewall of the hurricane. The eyewall is a region of intense thunderstorms around the eye of a hurricane. The strongest winds and heaviest rains are usually in the eyewall. A group of towers in the eyewall have a striking appearance on satellite and are known as convective bursts. Other towers form in rainbands, which are the arms that spiral out from the eye of a hurricane.

Hot towers give researchers clues about the nature of a hurricane. One hot tower alone does not provide much information. A quick sequence of towers, however, suggests that some change is occurring inside the storm.

Forecasters have found that hot towers indicate that a hurricane is intensifying. In Hurricane Katrina, which struck New Orleans in 2005, several hot towers were observed. One tower near the eyewall was 16 kilometers (about 10 miles) tall. As a comparison, consider that the tallest mountain in the world (Mt. Everest) is roughly 9 kilometers tall. Shortly after this tower formed, Katrina jumped from a Category 3 to a Category 4 storm. Researchers are investigating this phenomenon in an effort to better understand the process through which storms intensity.

Aircraft

Another method of tracking hurricanes is flying aircraft directly into them. NASA, along with the National Oceanic and Atmospheric Administration (NOAA), send teams commonly known as the hurricane hunters into a hurricane. They observe the eyewall and measure wind speeds. In addition to collecting data about hot towers, the team in the aircraft also gathers information about sinking currents of air in the eye of a storm.

Technology on the Horizon

A new project, named the Global Precipitation Measurement Mission (GPM), is scheduled for 2010. This will be another joint project between NASA and JAXA. The goal is to send a main satellite into orbit. It will have several additional satellites, known as constellation satellites. Together, they will provide global coverage of precipitation. By being able to study how precipitation forms and changes on a global scale, forecasters hope to be able to improve the ability to predict climate changes. In addition, the satellites will improve weather forecasts by providing more accurate measurements of the rate at which rain is falling and how the atmosphere is heating.

Questions

1. What is the source of energy for a tropical cyclone? Cite evidence from the text.

2. In what direction does a tropical cyclone spin?

3. How is the TRMM different from other rain radar instruments?

4. What advantage does ground radar have over the TRMM in tracking a hurricane?

5. What does a sequence of tall hot towers indicate about a hurricane?

6. How does the arsenal of modern tools help modern forecasters study hurricanes? Support your answer with Internet research.

It Takes Guts

The Fundamental Forces

The interactions of all particles can be described by four fundamental forces: gravity, electromagnetism, the weak nuclear force, and the strong nuclear force.

The gravitational force is the dominant force in the universe. It is the force of attraction that exists between every pair of masses in the universe. The force of gravity extends over the greatest of distances—trillions of kilometers. The gravitational force keeps planets in their orbits, prevents stars from exploding, and keeps objects on Earth's surface. The gravitational force, however, it is the weakest of the four forces.

The second of the four forces, electromagnetism, is responsible for causing negatively charged electrons to spin around positively charged nuclei in atoms. For a long time, electric and magnetic forces were considered to be separate forces. Beginning in 1864 with the work of James Maxwell, however, scientists came to recognize that they are two aspects of the same force.

Strong and weak forces are nuclear forces of the atom. The weak nuclear force is a powerful force that is responsible for such phenomena as beta decay. The strong nuclear force keeps protons and neutrons bound together despite the repulsive forces among the positive charges in the atomic nucleus.

The Electroweak Force

In the 1960s, some physicists began working on a subfield of physics called quantum electrodynamics (QED). Simply put, QED extends the quantum theory about subatomic particles to fields of force. According to QED, charged particles interact by means of virtual particles. These particles act as carriers of momentum and force, but do not exist outside of the interaction.

QED led to a unification of the weak force and the electromagnetic force. This means that under certain conditions, the two forces behave as one. The new force, known as the electroweak force, occurs at extremely high temperatures such as those that existed at the beginning of the universe or that are created in particle accelerators.

The theory of the electroweak force, known as the electroweak theory, suggests that the particles that act as mediators of weak interactions behave in much the same way as photons that mediate electromagnetic interactions. These particles, called W and Z bosons, are comparable in strength to the photon. However, unlike photons, W and Z bosons are massive. The result is that beta decay from weak interactions occurs at a much slower rate than electromagnetic decays.

The Standard Model

Like the electroweak theory, the Standard Model describes forces in terms of the interactions of messenger particles. However, the Standard Model goes on to include the strong force and to identify basic types of particles. The two most fundamental types of particles are quarks and leptons. These particles are divided into six categories called flavors, corresponding to three generations of matter. The everyday world is of the first generation. It is made up of the up quark, the down quark, the electron, and the neutrino. These particles can form protons, neutrons, atoms, and molecules.

According to this model, particles transmit forces by exchanging bosons. These mediators carry discrete amounts of energy, called quanta, from one particle to another. Each force has its own boson. As you read earlier, the W and Z bosons carry the weak force and the photon carries the electromagnetic force. The gluon mediates the strong force.

Physicists predict that the gravitational force is also mediated by a boson, which they have named the graviton. However, this hypothetical particle has not yet been discovered. Even if it does exist, it would be extremely difficult to detect because it would be considerably weaker than the other three forces at the subatomic level.

This model has been fairly successful at describing the interactions of subatomic particles. However, it has significant weaknesses that urge

physicists to look for a more complete theory. One such weakness is the use of arbitrary constants. In a fundamental theory, the constants should be evident rather than contrived. In addition, it cannot explain why some particles have mass whereas others do not.

Added to that, the model predicts the existence of a particle called the Higgs boson. However, the particle has not been discovered. Many scientists consider this particle to be the missing piece that will complete the Standard Model and explain observations about particles and mass.

Perhaps one of the biggest shortcomings of the Standard Model is that it does not unify all of the forces because gravity is not included. A Grand Unified Theory (GUTs) is being sought to clearly unify the strong force with the electroweak force and to answer questions that remain in the Standard Model.

The Theory of Everything
Even a GUTs would not include spacetime and therefore gravity. It is hypothesized that a Theory of Everything (TOE) would unify all four fundamental forces, matter, and spacetime. This would identify the single force that controlled the universe at the time it was formed. The hope is to develop a theory that underlies all other theories and cannot be explained by any deeper understanding.

One theory that has been posed as a TOE is that of supergravity. This is a complex theory that attempts to unite particle types using a ten-dimensional spacetime. The theory suggests that there are dimensions of time and space that exist at the quantum level of elementary particles, but do not effect normal experiences.

Another suggested TOE is string theory. According to this theory, everything consists of vibrating strands at its most microscopic level. The properties of particles are a reflection of the different ways in which a string can vibrate. Just as strings on a violin have resonant frequencies, so do the loops of string theory. Instead of producing musical notes, the preferred mass and force

charges are determined by the string's vibrating pattern. For example, an electron is a string vibrating one way. An up-quark is a string vibrating another way, and so on.

Einstein's Quest
When Albert Einstein studied physics, the strong and weak forces had not yet been discovered. The two forces known at the time were gravity and electromagnetism. Einstein found it troubling that two distinct forces could exist.

Einstein spent the last thirty years of his career searching for a theory that would show that these two forces were part of a single underlying principle. In fact, this quest caused Einstein to become isolated from mainstream physicists who were more interested in studying quantum mechanics.

Although many believed Einstein to be somewhat crazy for trying to unite the forces, it turns out that Einstein was simply ahead of his time. Over a half century later, the quest for a unified theory has become a basic goal of physics. Only time and experimentation will tell which, if any, of these theories will explain the finer workings of the universe. And there's always the possibility that the forces cannot be combined. Physicists around the globe will continue to work to uncover the secrets the universe holds.

Questions

1. Which were the first two forces combined by physicists? Cite evidence from the text.
2. How does the electroweak theory depend on mediator particles? Support your answer with Internet research.
3. How would a TOE differ from the Standard Model of Matter?
4. What happened to Einstein as a result of his quest to unify the forces? Cite evidence from the text.

Are We There Yet?

A manufacturer advertises that its vehicle can go from 0 to 60 mph in . . . 4 *days*. Are you interested? You might be if you were an astronomer hoping to study distant objects in the solar system. The vehicle is an ion propulsion rocket and it has opened up new possibilities for research in space.

Chemical Propellants

The ion propulsion rocket is very different from conventional spacecraft. Conventional vehicles use solid or liquid chemical fuels, often liquid hydrogen and oxygen. The fuels are placed into a tank and allowed to mix together. When they mix, they react explosively producing extremely hot gases. The force of the gases being ejected from the rocket results in an equal force in the opposite direction. This force, known as thrust, accelerates the rocket forward.

At the speeds attained by chemical propellants, travel to most objects in the solar system is impractical if not impossible on a human timescale. A trip to the nearest star would take centuries if not longer.

In addition to their limited speeds, chemical rockets are extremely inefficient in terms of fuel. They are like huge gas guzzlers that have to carry all of their fuel with them. That doesn't leave room for much else, such as scientific equipment or human crews.

Ion Propulsion

Rocket scientists have long known that they needed to develop a more efficient propulsion system. One solution is an ion propulsion system, which is an electrical system rather than a chemical system. Essentially, this technology involves ionizing an inert gas to propel a spacecraft. The most common gas used is xenon, which is four times heavier than air.

In this system, electrons are emitted into a chamber filled with xenon gas. There the electrons strike the xenon atoms, causing an electron to be knocked away from each xenon atom. The loss of an electron causes a xenon atom to become a positive ion. The longer the electrons and the xenon atoms are held in the chamber, the greater the chances of forming ions. This makes the process very efficient.

A pair of electrically charged metal grids exerts a force on the xenon ions. The grids accelerate the positively charged xenon ions and focus them toward the spacecraft's exhaust. The ions shoot out the back of the engine at a speed of 100,000 km/h (60,000 mph). Much like in a chemical engine, a thrust is produced that pushes the craft forward. Because the craft is much heavier than the ions, it does not move at the same speed in the opposite direction. Nonetheless, it still moves at extremely high speeds exceeding 60,000 km/h.

To keep the total charge of the exhaust beam neutral, an amount of negative charge equal to the positive charge expelled must also be produced. A device called the neutralizer expels the electrons needed to achieve this balance.

Ion systems require a source of energy to ionize the atoms and charge the engine. Solar power and nuclear power are the two primary options. In a solar electric propulsion system, sunlight and solar cells generate power. In a nuclear electric propulsion system, a source of nuclear heat is attached to an electric generator. Solar electric propulsion engines are preferable for missions between the sun and Mars. Nuclear electric propulsion systems would be needed for missions where sunlight is weak or where high speed is required.

Tradeoffs

An ion engine puts out less than one newton of thrust. That is so gentle it is equivalent to the weight of a sheet of paper lying on your hand. So what's the advantage of this system? Ion systems can operate over long periods of time. In contrast to a chemical propulsion system which has high thrust for a short time, an ion system has a low thrust for a long time.

The low thrust produces a small acceleration continuously for the spacecraft. Over a period of time, the small acceleration results in extremely

high speeds. In fact, ion propulsion can push a spacecraft about ten times as fast as chemical propulsion.

The slow, but steady acceleration of an ion propulsion system is more efficient than bursts of energy produced by chemical methods. This means that ion propulsion systems require less propellant than chemical systems. As a result, spacecraft propelled by ion systems can be smaller, which lowers the direct cost as well as the cost of launching. At the same time, the higher speeds reduce flight times, thereby making missions more practical and scientifically productive.

A disadvantage of ion systems is that the low thrust means ion propulsion cannot be used for applications that require rapid acceleration. In addition, rockets powered by electric propulsion systems cannot generate enough thrust to lift their own weight. Instead, they require a chemically powered launch vehicle.

The Future of Propulsion

Ion propulsion is a viable option for missions that require high amounts of energy. Missions already planned for ion engines include obtaining data and samples from comets, Saturn's rings, the moons Titan and Europa, and the planets Neptune and Venus.

Some other technologies being developed take advantage of natural forces to achieve acceleration or deceleration. Several NASA spacecraft have already used one such method known as aerobraking. In this method, the spacecraft uses friction created by the target planet's atmosphere to slow and steer the craft. Aerobraking replaces traditional methods that require using onboard jets and a large amount of propellant to adjust the orbit of a spacecraft.

In a similar way, scientists hope to develop a technique called aerocapture in which a spacecraft uses the atmosphere of a planet to change its flight path. Aerocapture should be able to put a vehicle in orbit after only one pass without using any fuel. The challenge of this method is to protect the spacecraft from burning up. NASA scientists feel that they can design heat shields that make it possible for missions to be safely captured into orbit.

Another technology being developed is the use of a solar sail, which is a large, lightweight panel of highly reflective material. When the sail is opened, it captures light from the sun. As photons of light reflect off the sails, a thrust is produced on the spacecraft. Much like wind pushes the sails of a boat, photons of light would push the equipment attached to the solar sail.

When a spacecraft's distance from the sun is too great, an onboard laser or microwave could provide power instead. Although the solar sails could carry only robotic equipment, they can eventually reach speeds that are ten times faster than the space shuttle because they are so light and have an endless source of energy. They can also be reusable. Despite some problems with early tests, this technology continues to hold great promise for the future.

Most likely, no single technology will fill every need in the future of space research. Instead, a combination of several technologies may prove to be the most economical method of investigating deep space and possibly carrying astronauts to Mars and beyond.

Questions

1. How does the thrust of a chemical engine compare to that of an ion engine?

2. Why doesn't an ion engine spacecraft accelerate at the same rate as the ions that are emitted from its exhaust?

3. Why does an ion engine need to neutralize its exhaust? Cite evidence from the text.

4. What is one important way that reducing the mass of a spacecraft reduces cost?

5. Why are ion propulsion engines an important option for deep space exploration?

Glossary

absolute error: the absolute value of the difference between the measured value and the accepted value; absolute errors are always expressed as positive numbers.

absolute index of refraction: the ratio of the speed of light in a vacuum to the speed of light in a given medium.

acceleration: the time-rate of change of velocity.

accuracy: the agreement of a measured value with the true or accepted value.

alpha decay: the emission of an alpha particle from a nucleus.

alternating current: electric current that varies in magnitude and alternates in direction.

ammeter: a modified galvanometer used to measure larger amounts of current.

ampere: the fundamental unit of electric current, equal to 1 coulomb/second.

amplitude: the maximum disturbance in a wave cycle.

angle of incidence: the angle formed between the incident ray and a line normal (perpendicular) to the surface.

angle of reflection: the angle formed between the normal and reflected ray.

angle of refraction: the angle formed between the normal and the refracted ray.

anode: the positively charged plate.

antinodes: points of constructive interference.

antiparticle: a particle with the same mass but opposite charge of its counterpart subatomic particle.

atom: the basic unit of matter.

atomic number: the number of protons in a nucleus.

average speed: the distance traveled per unit time.

back EMF: the electromotive force that develops in a circuit from the magnetic effects of the induced current and that opposes the electromotive force producing the current.

baryon: a heavier nuclear particle such as the proton, neutron, and hyperon.

battery: a combination of cells.

best-fit line: a straight line drawn so that data points that do not fall on the line are balanced on both sides of the line.

beta decay: the disintegration of a neutron into a proton and an electron; the electron is emitted from the nucleus.

binding energy: the energy lost as nucleons coalesce to form a stable nucleus; it is equivalent to the work needed to pry the nucleons away from one another.

branch: an alternate pathway for current to flow in a parallel circuit.

breeder reactor: a reactor that uses its own nuclear waste as fuel.

cathode: the negatively charged plate.

cathode ray tube: an evacuated tube that contains a source of electrons at one end and a fluorescent screen at the other end.

cell: a chemical source of potential difference.

center of curvature: the point that corresponds to the center of a sphere, with a segment being a concave mirror.

centripetal acceleration: the acceleration experienced by an object in uniform circular motion.

centripetal force: the force that causes centripetal acceleration; the net force on a circling object that acts to change the object's direction.

chain reaction: a self-sustaining series of collisions between neutrons and radioactive nuclei.

circuit: a closed loop formed by a source of electrons and a conductor with no gaps across which electrons cannot travel.

cloud chamber: a device that reveals the path of a charged particle by creating a trail of condensed vapor in its wake.

coefficient of friction: a value derived by dividing the force of friction by the normal force; it represents the nature of the surfaces rubbing together.

coherent light waves: waves of the same frequency produced by sources in phase.

combined resistance: See **equivalent resistance**.

compass: a magnet used to determine direction on Earth.

concave: thinner in the middle than at the edges.

concave mirror: a small segment of a sphere with a reflecting surface on the inside of the sphere.

concurrent forces: two or more forces acting on an object at the same time and at the same point.

conductivity: the reciprocal of a solid's resistivity.

conductor: a substance that allows electrons to flow through it.

conservative force: any force such that the work done by the force is independent of the path taken.

constructive interference: the production of a larger disturbance at the point where waves meet.

contact: a method of charging a neutral object by making contact with a charged object.

controlled experiment: an experiment in which one variable is changed while the others are held constant.

conventional current: theoretical flow of positive charges from the positive to the negative terminal of a battery.

converging lens: a lens that acts to bring light rays together at a point on the other side of the lens.

converging mirror: a mirror that acts on incoming light rays parallel to the principal axis and converges them to a point.

convex: thicker in the middle than at the edges.

convex mirror: a small segment of a sphere with a reflecting surface on the outside.

coulomb: the unit of charge, symbolized by C, equal to the charge carried by 6.25×10^{18} electrons.

critical angle: an angle of incidence for which the corresponding angle of refraction is 90°.

critical mass: the minimum amount of fissionable material necessary to initiate a chain reaction.

cycle: one complete repetition of the pattern in a periodic wave.

cyclotron: a device that accelerates particles by making them circle in a magnetic field and accelerating them once during every round trip when they pass through an electric field; similar to a synchrotron.

de Broglie wave: See **matter wave**.

derived unit: a unit that consists of a combination of fundamental units; the newton (N) is a derived unit equivalent to 1 kilogram meter per second squared $(kg \cdot m/s^2)$.

destructive interference: the production of a smaller disturbance, or none at all, at the point where waves meet.

diffraction: the spreading of light waves behind a barrier.

dispersive medium: a material that allows waves of different frequencies to pass through it at different velocities.

displacement: a change of position in a particular direction.

diverging lens: a lens that acts to separate light rays as they pass through the lens.

diverging mirror: a mirror that acts on incoming light rays parallel to the principal axis and separates them.

dynamic equilibrium: a condition that exists when no net force acts on an object in motion; the object maintains a constant velocity.

dynamics: the study of the relationship between forces and motion.

elastic potential energy: the energy stored in a spring when it is compressed or stretched.

electric current: a flow of charged particles, usually negative.

electric field: the entity around a charged object that affects other charges.

electric field direction: the direction of force exerted by an electric field on a positive test charge.

electric field intensity: the force that an electric field exerts on one coulomb of positive charge at a given point; also referred to as the field's magnitude.

electric motor: a device that makes use of the force exerted by magnets on currents to convert electrical energy into rotational kinetic energy.

electric potential: the total amount of work an electric field can do on one coulomb of positive charge by moving the charge from a given point in the field to infinity.

electric potential energy: the ability of a charged object to do work, due to its position in an electric field.

electric power: the rate at which electric potential energy is converted to another form of energy.

electromagnet: a solenoid whose magnetic field is intensified by the insertion of certain materials.

electromagnetic induction: the process by which a magnetic field generates an electric current and a potential difference.

electromagnetic radiation: the propagation of electric and magnetic fields away from the vicinity of an accelerating charge.

electromagnetic spectrum: a representation of all types of electromagnetic waves in order of decreasing wavelengths and increasing frequency.

electromagnetic waves: a periodic wave of electric and magnetic fields that is radiated outward from the vicinity of an oscillating charge.

electron: a negatively charged subatomic particle.

electron capture: a radioactive process in which a nucleus absorbs one of the atom's innermost electrons; the electron unites with a proton to form a neutron, decreasing the atomic number by one but leaving the mass number unchanged.

electron volt: unit of energy and work equal to the work done on one elementary charge to move it between two points separated by a potential difference of 1 volt.

electroscope: a device that consists of a metal knob attached to two light metallic leaves; used to detect the presence of excess charge on an object.

elementary unit of charge: an amount equal to the positive charge on one proton or the negative charge on one electron; equal to 1.60×10^{-19} coulombs.

ellipse: a closed curve such that the sum of the distances from any point on the curve to two fixed points, called the foci, is constant.

emission spectrum: electromagnetic radiation of only certain wavelengths and frequencies that is unique to each element.

energy: the ability to do work.

entropy: the amount of disorder in a system.

equilibrant: the balancing force that creates equilibrium.

equilibrium: the state of an object when the vector sum of the concurrent forces acting on it is zero.

equivalent resistance: the single resistance that can replace all the resistances of branches in parallel; also known as the combined resistance.

escape velocity: the minimum speed an object must have to escape the influence of a body's gravitational pull.

excitation: an increase in the orbital distance, or energy, of an electron.

excited state: the condition of an electron that jumps to a higher orbit.

extrapolation: predicting values outside the range of experimental data.

ferromagnetic: a material whose insertion causes the magnetic field of a solenoid to become stronger.

field lines: lines used to diagram the direction and intensity of an electric or magnetic field.

fission: the process of splitting a nucleus into smaller fragments.

fluid friction: the force that resists the motion of an object through a fluid such as water or air.

flux density: the number of magnetic field lines per unit area.

flux lines: magnetic field lines.

focal length: the distance between the center of a lens or mirror and the focal point.

focal point: the point where incoming parallel light rays meet; also called the **principal focus**.

force: a push or a pull.

free fall: motion due solely to the attraction of Earth's gravitational pull.

frequency: the number of cycles produced per second by a vibrating source.

friction: the force that opposes the motion of one surface over another.

fusion: the process of uniting light nuclei to become one heavier nucleus.

galvanometer: a device consisting of a coil-shaped wire placed between the opposite poles of a permanent magnet that is used to measure small amounts of current.

gamma rays: high energy, short wavelength photons.

geiger counter: a device that detects nuclear particles emitted from atoms.

generator: a device that creates electric current by moving a conducting wire across magnetic field lines.

geosynchronous orbit: an orbital period of a satellite equal to the period of Earth's rotation.

gravitational field: the entity around a mass that acts on other masses.

gravitational force: the universal attraction of one piece of matter for another.

gravitational potential energy: the energy an object has due to its position.

grounded: connected to an object so large that it can either accept or give up a significant number of electrons without becoming noticeably charged.

ground state: an electron in the lowest possible orbit with the least amount of energy.

half-life: the time it takes for half the number of nuclei present in a radioactive sample to decay.

heat energy: energy that is transferred from warm objects to cooler ones due to the temperature difference between them.

hertz: a derived unit for frequency that stands for cycles per second.

hyperons: subatomic particles with masses greater than neutrons.

impulse: the product of the net force acting on an object and the time during which the force acts.

in phase: describes points on a periodic wave that are identically displaced from equilibrium and are moving in the same direction from the equilibrium position.

induced current: a current created either by moving a conducting wire across magnetic field lines or by changing the intensity of the magnetic field.

induced EMF: a potential difference created either by moving a conducting wire across magnetic field lines or by changing the intensity of the magnetic field.

induced potential difference: the amount of work in joules done on every coulomb of charge between the ends of a wire segment that is moving in a magnetic field.

induced voltage: See **induced potential difference**.

induction: a method of charging a neutral object by attaching it to a ground and then bringing a charged object near it.

inertia: the property of matter that resists change in motion.

instantaneous velocity: the slope of a line tangent to a displacement-time graph at any given point.

insulator: a material that strongly resists the flow of electrons through it.

intensity: the magnitude, or strength, of a field.

interference pattern: alternating lines of constructive and destructive interference.

internal energy: the total kinetic and potential energy associated with the molecules of an object, apart from any kinetic or potential energy the object as a whole may possess.

interpolation: predicting values within the range of experimental data.

ionization potential: the amount of energy needed to remove an electron from an atom.

isotopes: atoms with the identical atomic numbers but different mass numbers.

joule: the work done on an object when a one-newton force displaces the object one meter.

kinematics: the study of motion.

kinetic energy: the energy an object has due to its motion.

kinetic friction: the force opposing the motion of an object sliding over a surface.

laser: an acronym for Light Amplification by Stimulated Emission of Radiation; a device that emits coherent, monochromatic, very intense light.

linear accelerator: a device that accelerates particles continuously in an electric field as they travel in a straight line.

longitudinal waves: waves in which the disturbances are parallel to the direction of wave motion.

magnet: a long, slender piece of ore that aligns itself, when free to do so, with one end pointing north and the other pointing south.

magnetic field: the entity around a moving charge or magnet that exerts a force on other moving charges or magnets.

magnetic field direction: the direction in which the N-pole of a test magnet is made to point by a magnetic field.

magnetic field intensity: the force that a magnetic field exerts on a one-meter-long wire in the field carrying one ampere of current.

magnetic force: the force that moving charges exert on other moving charges.

magnification: the ratio of image size to object size.

mass defect: the difference between the mass of a nucleus and the sum of the masses of its nucleons.

main line: the part of a parallel circuit through which the full amount of current flows.

mass number: the number of nucleons in a nucleus.

mass spectrometer: a device that separates atoms of the same element with different masses.

matter wave: the wave associated with a moving particle.

mechanics: the study of forces and their effects in producing and changing motion.

medium: a body of matter through which a disturbance travels.

meson: a subatomic particle with a mass between that of an electron and a proton; made up of a quark-antiquark pair.

moderators: materials used in nuclear reactors to slow down neutrons.

momentum: the product of a moving object's mass and velocity.

motor: a device that converts electrical energy to rotational kinetic energy.

net force: the resultant of concurrent forces acting on an object.

newton: the force applied to a one-kilogram mass to accelerate it one meter per second per second.

neutrino: a subatomic particle with no charge and much less mass than an electron.

nodes: points of destructive interference.

nonconservative force: any force such that the work done by the force depends on the path taken.

nondispersive medium: a medium in which the velocity of a wave does not depend on its frequency.

normal: a line perpendicular to a surface.

normal force: the force that presses two surfaces together.

nuclear energy: the potential energy that exists in a nucleus because of the work done on particles by the nuclear force.

nuclear force: an attractive force that acts between masses but is only effective across extremely small distances; it binds nucleons together into a tightly packed and cohesive unit.

nuclear fuel: small pellets of fissionable material.

nuclear fusion: See **fusion**.

nucleon: a particle in the nucleus of an atom.

nucleus: the tiny, heavy core where most of an atom's mass and all of its positive charge are located.

nuclides: nuclei with the same number of protons and neutrons.

observed frequency: the number of wave cycles passing by a given point per second.

ohm: the unit of electrical resistance; defined as the resistance of a material that allows one ampere of current when a potential difference of one volt exists between the ends of the material.

parallel circuit: an electrical circuit in which the current flows through more than one branch.

particle accelerator: a device used to accelerate charged particles to speeds approaching that of light.

period: the time for a complete wave cycle to be produced or to pass a given point.

periodic wave: a regularly repeating series of pulses, also called a wave train.

photoelectric effect: a phenomenon in which electrons are ejected from certain materials and escape into space when electromagnetic radiation strikes the materials.

photoelectrons: electrons emitted as a result of the photoelectric effect.

photoemissive: applied to materials that demonstrate the photoelectric effect.

photon: the fundamental particle of all forms of electromagnetic radiation.

point charges: two or more charged objects that are much smaller than the distance between them.

polarize: to cause light waves to vibrate in a particular plane.

positron: an antiparticle that has the same mass as an electron but a positive charge; also called an anti-electron.

potential difference: the work required to move a test charge of one coulomb from one point to another in an electric field.

potential energy: the energy an object has due to its position or condition.

power: the amount of work done per unit time.

precision: the degree of refinement with which a measurement is made.

pressure: the result of gas molecules colliding with the walls of the container; the force per unit surface area.

primary coil: the coil of a transformer that is connected to an ac source.

principal axis (lens): a line drawn perpendicular to the plane of a lens and through its center.

principal axis (mirror): a line that connects the center of curvature to the geometric center of the mirror.

principal focus: See **focal point**.

projectile: any object that is launched by some force and continues to move by its own inertia.

pulse: a single vibratory disturbance.

quanta: discrete amounts of energy.

quantum theory: the theory that electromagnetic radiation is absorbed and emitted in discrete amounts.

quarks: the proposed constituent particles of baryons and intermediate mass mesons; quarks carry a charge of either one-third or two-thirds of an elementary unit of charge.

radioactive decay: the transmutation of one element into another.

radioactivity: naturally occurring transmutations.

radius of curvature: the distance between the center of curvature and a concave mirror.

random error: an error in measurement due to fluctuations in the environment.

range: the distance traveled by a projectile.

rarefaction: a pocket of expanded air.

real current: the flow of electrons in a conductor.

real image: an image created by the actual convergence of rays of light; such images are always inverted.

refraction: the bending of light as it passes at an angle from one medium to another medium with a different index of refraction.

relative index of refraction: the ratio of the velocity of light in the first medium to the velocity of light in the second medium (v_1/v_2)

resistance: the opposition of a material to the flow of electrons through it.

resistivity: the resistance, in ohms, of a uniform rod of unit-length and unit cross-sectional area.

resistor: any conductor with a measurable resistance.

resonance: a phenomenon produced when an object is disturbed by a wave whose frequency is the same as the object's natural vibration frequency; the amplitude of vibration of the object continues to increase.

resultant: a vector that represents the sum of two or more other vectors.

rolling friction: the force opposing the motion of one object rolling over another; it is usually weaker than sliding friction.

satellite: any body that revolves around a larger body.

scalar quantity: a measurement that has magnitude but no direction.

scientific notation: the expression of a number in the form $A \times 10^n$ where A is any number equal to or greater than 1 but less than 10, and the exponent n is an integer.

scintillation counter: a device that detects nuclear particles emitted from atoms by converting their energy into photons of light.

secondary coil: the coil of a transformer that is not connected to any source of current.

series circuit: an electrical circuit in which only one path exists for the current.

significant figures: those digits in a measurement that are obtained properly and directly from an instrument, including the final estimated digit.

solenoid: a coil-shaped, current-carrying wire.

source: a device that creates an electric field such that a potential difference exists between the ends of a conductor.

sources in phase: wave sources that vibrate at the same frequency and produce waves of equal amplitude.

speed: a scalar quantity equal to the magnitude of the velocity vector.

split-ring commutator: a device in which each end of a wire loop is connected to a conducting material in the shape of a half-ring; used in motors and generators.

standing waves: created by two interfering waves that repeatedly pass through each other in such a way that certain points always show constructive interference while other points always show destructive interference.

static electricity: electricity associated with charges at rest.

static equilibrium: a condition that exists when no net force acts on an object at rest; the object remains at rest.

static friction: the frictional force that must be overcome to start an object moving over a surface; sometimes called starting friction.

statics: the study of the effect of forces on stationary objects.

stationary state: an electron that is circling the nucleus in an allowed orbit and is neither losing nor gaining energy.

step-down transformer: a transformer in which the voltage in the secondary coil is less than that in the primary coil.

step-up transformer: a transformer in which the voltage in the secondary coil is greater than that in the primary coil.

strong nuclear force: the strong, but short-range, force that holds the particles in the nucleus.

superconductor: a material that offers virtually no resistance to the passage of electrons through it; no energy is lost to heat when current flows through such a material.

systematic error: an error in measurement due to flaws in the instruments used.

temperature: how hot or cold an object is with respect to a chosen standard.

test charge: a known charge used to test for the presence of an electric field.

thermal equilibrium: the point at which materials in contact reach the same temperature and heat exchange ceases.

thermionic emission: the emission of electrons from metallic substances when they are heated to incandescence.

thermodynamics: the study of heat and its relationship to other forms of energy and to work.

threshold frequency: a frequency below which no photoelectrons will be emitted no matter how intense the radiation.

transformer: a device inserted into a circuit carrying alternating current to change the voltage to some higher or lower value.

transmitted frequency: the number of cycles produced by a source per second.

transmutation: the change of an unstable nucleus of one element into another as nuclear particles are emitted.

transverse waves: waves in which the disturbances are perpendicular to the direction of the wave motion.

uniform acceleration: a change in the velocity of a moving object by a fixed amount per second.

uniform circular motion: movement of an object along a circular path at a constant speed.

uniform velocity: movement of an object at a constant speed and direction.

universal mass unit (u): a quantity equal to 9.31×10^2 MeV or 1.66×10^{-27} kg.

Van de Graaff generator: a device that accelerates particles by driving them parallel to an intense electric field.

vector: a geometric representation of a vector quantity as an arrow whose length represents the magnitude and whose direction is the same as that of the vector quantity.

vector quantity: a measurement that has magnitude and direction.

velocity: the change in displacement per unit time.

virtual focal point: the point where incident rays parallel to the principal axis appear to be coming from after they are refracted by a diverging lens or reflected by a diverging mirror.

virtual image: an image not produced by waves of light actually coming to a point, but merely appearing to be coming from a point; such images are always upright.

volt: the unit of potential difference, equal to one joule/coulomb.

voltage: another term for the potential difference between two points in an electric field.

voltage drop: the potential difference between the entry and exit points of a particular resistor in a circuit.

voltmeter: a modified galvanometer used to measure potential difference.

watt: the derived unit of power, equal to one joule/second.

wavelength: the length of one complete wave cycle.

wave speed: the distance traveled per unit time for any part of a wave.

weak nuclear force: the interaction between electrons and protons or neutrons needed to explain certain forms of nuclear decay.

weight: the amount of gravitational force an object experiences.

work: the product of the force on an object and the resultant displacement.

work function: the minimum amount of energy an electron needs to escape from a photo-emissive material.

Index

Reference Tables for Physical Setting/PHYSICS
2006 Edition

List of Physical Constants

Name	Symbol	Value
Universal gravitational constant	G	6.67×10^{-11} N•m^2/kg^2
Acceleration due to gravity	g	9.81 m/s^2
Speed of light in a vacuum	c	3.00×10^8 m/s
Speed of sound in air at STP		3.31×10^2 m/s
Mass of Earth		5.98×10^{24} kg
Mass of the Moon		7.35×10^{22} kg
Mean radius of Earth		6.37×10^6 m
Mean radius of the Moon		1.74×10^6 m
Mean distance—Earth to the Moon		3.84×10^8 m
Mean distance—Earth to the Sun		1.50×10^{11} m
Electrostatic constant	k	8.99×10^9 N•m^2/C^2
1 elementary charge	e	1.60×10^{-19} C
1 coulomb (C)		6.25×10^{18} elementary charges
1 electronvolt (eV)		1.60×10^{-19} J
Planck's constant	h	6.63×10^{-34} J•s
1 universal mass unit (u)		9.31×10^2 MeV
Rest mass of the electron	m_e	9.11×10^{-31} kg
Rest mass of the proton	m_p	1.67×10^{-27} kg
Rest mass of the neutron	m_n	1.67×10^{-27} kg

Prefixes for Powers of 10

Prefix	Symbol	Notation
tera	T	10^{12}
giga	G	10^9
mega	M	10^6
kilo	k	10^3
deci	d	10^{-1}
centi	c	10^{-2}
milli	m	10^{-3}
micro	μ	10^{-6}
nano	n	10^{-9}
pico	p	10^{-12}

Approximate Coefficients of Friction

	Kinetic	Static
Rubber on concrete (dry)	0.68	0.90
Rubber on concrete (wet)	0.58	
Rubber on asphalt (dry)	0.67	0.85
Rubber on asphalt (wet)	0.53	
Rubber on ice	0.15	
Waxed ski on snow	0.05	0.14
Wood on wood	0.30	0.42
Steel on steel	0.57	0.74
Copper on steel	0.36	0.53
Teflon on Teflon	0.04	

239

The Electromagnetic Spectrum

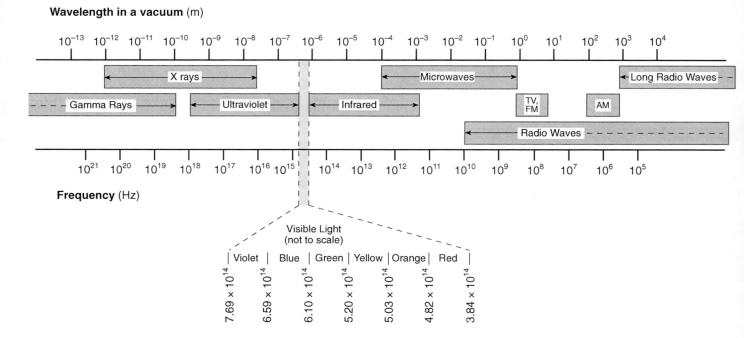

Energy Level Diagrams

Hydrogen

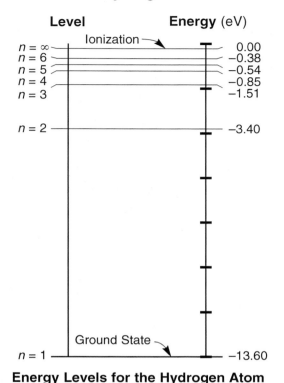

Energy Levels for the Hydrogen Atom

Mercury

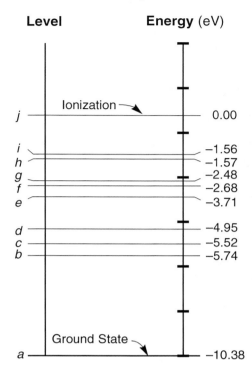

A Few Energy Levels for the Mercury Atom

Classification of Matter

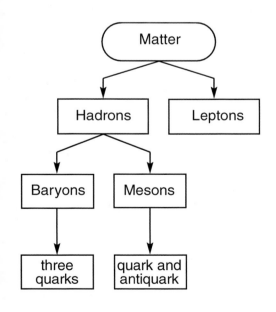

Particles of the Standard Model

Quarks

Name	up	charm	top
Symbol	u	c	t
Charge	$+\frac{2}{3}$ e	$+\frac{2}{3}$ e	$+\frac{2}{3}$ e

	down	strange	bottom
	d	s	b
	$-\frac{1}{3}$ e	$-\frac{1}{3}$ e	$-\frac{1}{3}$ e

Leptons

electron	muon	tau
e	μ	τ
-1e	-1e	-1e

electron neutrino	muon neutrino	tau neutrino
ν_e	ν_μ	ν_τ
0	0	0

Note: For each particle, there is a corresponding antiparticle with a charge opposite that of its associated particle.

Electricity

$$F_e = \frac{kq_1q_2}{r^2}$$

$$E = \frac{F_e}{q}$$

$$V = \frac{W}{q}$$

$$I = \frac{\Delta q}{t}$$

$$R = \frac{V}{I}$$

$$R = \frac{\rho L}{A}$$

$$P = VI = I^2R = \frac{V^2}{R}$$

$$W = Pt = VIt = I^2Rt = \frac{V^2t}{R}$$

A = cross-sectional area

E = electric field strength

F_e = electrostatic force

I = current

k = electrostatic constant

L = length of conductor

P = electrical power

q = charge

R = resistance

R_{eq} = equivalent resistance

r = distance between centers

t = time

V = potential difference

W = work (electrical energy)

Δ = change

ρ = resistivity

Series Circuits

$I = I_1 = I_2 = I_3 = \ldots$

$V = V_1 + V_2 + V_3 + \ldots$

$R_{eq} = R_1 + R_2 + R_3 + \ldots$

Parallel Circuits

$I = I_1 + I_2 + I_3 + \ldots$

$V = V_1 = V_2 = V_3 = \ldots$

$$\frac{1}{R_{eq}} = \frac{1}{R_1} + \frac{1}{R_2} + \frac{1}{R_3} + \ldots$$

Circuit Symbols

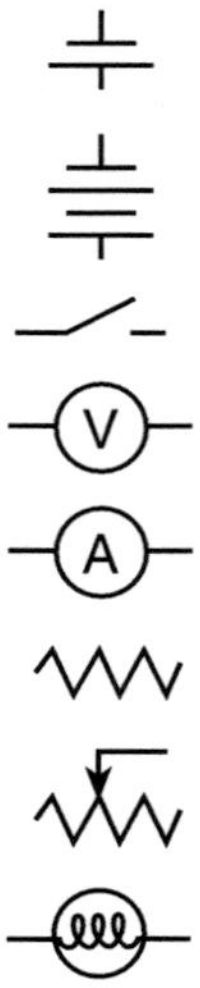

cell

battery

switch

voltmeter

ammeter

resistor

variable resistor

lamp

Resistivities at 20°C	
Material	**Resistivity ($\Omega \cdot$m)**
Aluminum	2.82×10^{-8}
Copper	1.72×10^{-8}
Gold	2.44×10^{-8}
Nichrome	$150. \times 10^{-8}$
Silver	1.59×10^{-8}
Tungsten	5.60×10^{-8}

Waves

$$v = f\lambda$$

$$T = \frac{1}{f}$$

$$\theta_i = \theta_r$$

$$n = \frac{c}{v}$$

$$n_1 \sin \theta_1 = n_2 \sin \theta_2$$

$$\frac{n_2}{n_1} = \frac{v_1}{v_2} = \frac{\lambda_1}{\lambda_2}$$

c = speed of light in a vacuum

f = frequency

n = absolute index of refraction

T = period

v = velocity or speed

λ = wavelength

θ = angle

θ_i = angle of incidence

θ_r = angle of reflection

Modern Physics

$$E_{photon} = hf = \frac{hc}{\lambda}$$

$$E_{photon} = E_i - E_f$$

$$E = mc^2$$

c = speed of light in a vacuum

E = energy

f = frequency

h = Planck's constant

m = mass

λ = wavelength

Geometry and Trigonometry

Rectangle

$$A = bh$$

Triangle

$$A = \frac{1}{2}bh$$

Circle

$$A = \pi r^2$$

$$C = 2\pi r$$

Right Triangle

$$c^2 = a^2 + b^2$$

$$\sin \theta = \frac{a}{c}$$

$$\cos \theta = \frac{b}{c}$$

$$\tan \theta = \frac{a}{b}$$

A = area

b = base

C = circumference

h = height

r = radius

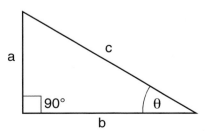

Mechanics

$$\bar{v} = \frac{d}{t}$$

$$a = \frac{\Delta v}{t}$$

$$v_f = v_i + at$$

$$d = v_i t + \frac{1}{2}at^2$$

$$v_f^2 = v_i^2 + 2ad$$

$$A_y = A \sin \theta$$

$$A_x = A \cos \theta$$

$$a = \frac{F_{net}}{m}$$

$$F_f = \mu F_N$$

$$F_g = \frac{Gm_1m_2}{r^2}$$

$$g = \frac{F_g}{m}$$

$$p = mv$$

$$p_{before} = p_{after}$$

$$J = F_{net}t = \Delta p$$

$$F_s = kx$$

$$PE_s = \frac{1}{2}kx^2$$

$$F_c = ma_c$$

$$a_c = \frac{v^2}{r}$$

$$\Delta PE = mg\Delta h$$

$$KE = \frac{1}{2}mv^2$$

$$W = Fd = \Delta E_T$$

$$E_T = PE + KE + Q$$

$$P = \frac{W}{t} = \frac{Fd}{t} = F\bar{v}$$

a = acceleration

a_c = centripetal acceleration

A = any vector quantity

d = displacement or distance

E_T = total energy

F = force

F_c = centripetal force

F_f = force of friction

F_g = weight or force due to gravity

F_N = normal force

F_{net} = net force

F_s = force on a spring

g = acceleration due to gravity or gravitational field strength

G = universal gravitational constant

h = height

J = impulse

k = spring constant

KE = kinetic energy

m = mass

p = momentum

P = power

PE = potential energy

PE_s = potential energy stored in a spring

Q = internal energy

r = radius or distance between centers

t = time interval

v = velocity or speed

$\bar{v}$ = average velocity or average speed

W = work

x = change in spring length from the equilibrium position

Δ = change

θ = angle

μ = coefficient of friction

Sample Examinations

Physical Setting/PHYSICS
June 2016

Part A

Answer all questions in this part.

Directions (1–35): For *each* statement or question, choose the word or expression that, of those given, best completes the statement or answers the question. Some questions may require the use of the *2006 Edition Reference Tables for Physical Setting/Physics.* Record your answers on your separate answer sheet.

1 Which quantity is a vector?

(1) power (3) speed
(2) kinetic energy (4) weight

2 A 65.0-kilogram astronaut weighs 638 newtons at the surface of Earth. What is the mass of the astronaut at the surface of the Moon, where the acceleration due to gravity is 1.62 meters per second squared?

(1) 10.7 kg (3) 105 N
(2) 65.0 kg (4) 638 N

3 When the sum of all the forces acting on a block on an inclined plane is zero, the block

(1) must be at rest
(2) must be accelerating
(3) may be slowing down
(4) may be moving at constant speed

4 The greatest increase in the inertia of an object would be produced by increasing the

(1) mass of the object from 1.0 kg to 2.0 kg
(2) net force applied to the object from 1.0 N to 2.0 N
(3) time that a net force is applied to the object from 1.0 s to 2.0 s
(4) speed of the object from 1.0 m/s to 2.0 m/s

5 A 100.-kilogram cart accelerates at 0.50 meter per second squared west as a horse exerts a force of 60. newtons west on the cart. What is the magnitude of the force that the cart exerts on the horse?

(1) 10. N (3) 60. N
(2) 50. N (4) 110 N

6 Sound waves are described as

(1) mechanical and transverse
(2) mechanical and longitudinal
(3) electromagnetic and transverse
(4) electromagnetic and longitudinal

7 An electrical force of 8.0×10^{-5} newton exists between two point charges, q_1 and q_2. If the distance between the charges is doubled, the new electrical force between the charges will be

(1) 1.6×10^{-4} N (3) 3.2×10^{-4} N
(2) 2.0×10^{-5} N (4) 4.0×10^{-5} N

8 A blue lab cart is traveling west on a track when it collides with and sticks to a red lab cart traveling east. The magnitude of the momentum of the blue cart before the collision is 2.0 kilogram • meters per second, and the magnitude of the momentum of the red cart before the collision is 3.0 kilogram • meters per second. The magnitude of the total momentum of the two carts after the collision is

(1) 1.0 kg • m/s (3) 3.0 kg • m/s
(2) 2.0 kg • m/s (4) 5.0 kg • m/s

9 The diagram below represents the path of a thrown ball through the air.

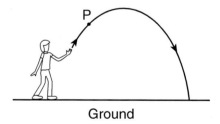

Ground

Which arrow best represents the direction in which friction acts on the ball at point *P*?

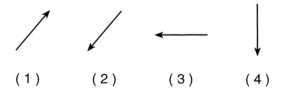

(1) (2) (3) (4)

10 A magnetic field would be produced by a beam of

(1) x rays (3) protons
(2) gamma rays (4) neutrons

11 The diagram below represents the electric field in the region of two small charged spheres, *A* and *B*.

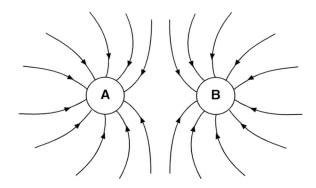

What is the sign of the net charge on *A* and *B*?

(1) *A* is positive and *B* is positive.
(2) *A* is positive and *B* is negative.
(3) *A* is negative and *B* is negative.
(4) *A* is negative and *B* is positive.

12 A horizontal force of 20 newtons eastward causes a 10-kilogram box to have a displacement of 5 meters eastward. The total work done on the box by the 20-newton force is

(1) 40 J   (3) 200 J
(2) 100 J   (4) 1000 J

13 A block initially at rest on a horizontal, frictionless surface is accelerated by a constant horizontal force of 5.0 newtons. If 15 joules of work is done on the block by this force while accelerating it, the kinetic energy of the block increases by

(1) 3.0 J   (3) 20. J
(2) 15 J   (4) 75 J

14 Two objects, *A* and *B*, are held one meter above the horizontal ground. The mass of *B* is twice as great as the mass of *A*. If *PE* is the gravitational potential energy of *A* relative to the ground, then the gravitational potential energy of *B* relative to the ground is

(1) *PE*   (3) $\dfrac{PE}{2}$

(2) 2*PE*   (4) 4*PE*

15 What is the kinetic energy of a 55-kilogram skier traveling at 9.0 meters per second?

(1) 2.5×10^2 J   (3) 2.2×10^3 J
(2) 5.0×10^2 J   (4) 4.9×10^3 J

16 A 5.09×10^{14}-hertz electromagnetic wave is traveling through a transparent medium. The main factor that determines the speed of this wave is the

(1) nature of the medium
(2) amplitude of the wave
(3) phase of the wave
(4) distance traveled through the medium

17 A motor does a total of 480 joules of work in 5.0 seconds to lift a 12-kilogram block to the top of a ramp. The average power developed by the motor is

(1) 8.0 W   (3) 96 W
(2) 40. W   (4) 2400 W

18 A 5.8×10^4-watt elevator motor can lift a total weight of 2.1×10^4 newtons with a maximum constant speed of

(1) 0.28 m/s   (3) 2.8 m/s
(2) 0.36 m/s   (4) 3.6 m/s

19 A stationary police officer directs radio waves emitted by a radar gun at a vehicle moving toward the officer. Compared to the emitted radio waves, the radio waves reflected from the vehicle and received by the radar gun have a

(1) longer wavelength   (3) longer period
(2) higher speed   (4) higher frequency

20 A light wave strikes the Moon and reflects toward Earth. As the light wave travels from the Moon toward Earth, the wave carries

(1) energy, only
(2) matter, only
(3) both energy and matter
(4) neither energy nor matter

21 The time required to produce one cycle of a wave is known as the wave's

(1) amplitude   (3) period
(2) frequency   (4) wavelength

22 A magnetic compass is placed near an insulated copper wire. When the wire is connected to a battery and a current is created, the compass needle moves and changes its position. Which is the best explanation for the production of a force that causes the needle to move?

(1) The copper wire magnetizes the compass needle and exerts the force on the compass needle.
(2) The compass needle magnetizes the copper wire and exerts the force on the compass needle.
(3) The insulation on the wire becomes charged, which exerts the force on the compass needle.
(4) The current in the wire produces a magnetic field that exerts the force on the compass needle.

23 A beam of monochromatic light ($f = 5.09 \times 10^{14}$ Hz) has a wavelength of 589 nanometers in air. What is the wavelength of this light in Lucite?

(1) 150 nm   (3) 589 nm
(2) 393 nm   (4) 884 nm

24 If the amplitude of a sound wave is increased, there is an increase in the sound's

(1) loudness   (3) velocity
(2) pitch   (4) wavelength

25 In the diagram below, point P is located in the electric field between two oppositely charged parallel plates.

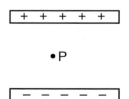

Compared to the magnitude and direction of the electrostatic force on an electron placed at point P, the electrostatic force on a proton placed at point P has

(1) the same magnitude and the same direction
(2) the same magnitude, but the opposite direction
(3) a greater magnitude, but the same direction
(4) a greater magnitude and the opposite direction

26 The effect produced when two or more sound waves pass through the same point simultaneously is called

(1) interference   (3) refraction
(2) diffraction   (4) resonance

27 A gamma ray photon and a microwave photon are traveling in a vacuum. Compared to the wavelength and energy of the gamma ray photon, the microwave photon has a

(1) shorter wavelength and less energy
(2) shorter wavelength and more energy
(3) longer wavelength and less energy
(4) longer wavelength and more energy

28 According to the Standard Model of Particle Physics, a neutrino is a type of

(1) lepton   (3) meson
(2) photon   (4) baryon

29 Which combination of quarks produces a neutral baryon?

(1) cts   (3) uds
(2) dsb   (4) uct

30 When 2.0×10^{-16} kilogram of matter is converted into energy, how much energy is released?

(1) 1.8×10^{-1} J   (3) 6.0×10^{-32} J
(2) 1.8×10^{1} J   (4) 6.0×10^{-8} J

31 A ball is hit straight up with an initial speed of 28 meters per second. What is the speed of the ball 2.2 seconds after it is hit? [Neglect friction.]

(1) 4.3 m/s   (3) 22 m/s
(2) 6.4 m/s   (4) 28 m/s

32 A particle with a charge of 3.00 elementary charges moves through a potential difference of 4.50 volts. What is the change in electrical potential energy of the particle?

(1) 1.07×10^{-19} eV   (3) 1.50 eV
(2) 2.16×10^{-18} eV   (4) 13.5 eV

33 Which circuit has the largest equivalent resistance?

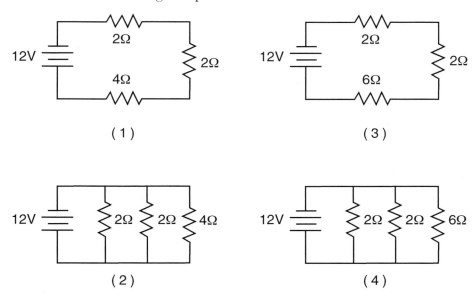

(1)   (3)

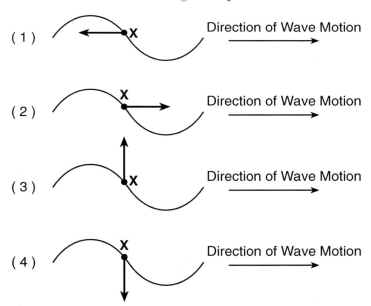

(2)   (4)

34 A transverse wave is moving toward the right in a uniform medium. Point *X* represents a particle of the uniform medium. Which diagram represents the direction of the motion of particle *X* at the instant shown?

(1)   Direction of Wave Motion

(2)   Direction of Wave Motion

(3)   Direction of Wave Motion

(4)   Direction of Wave Motion

35 Which diagram represents magnetic field lines between two north magnetic poles?

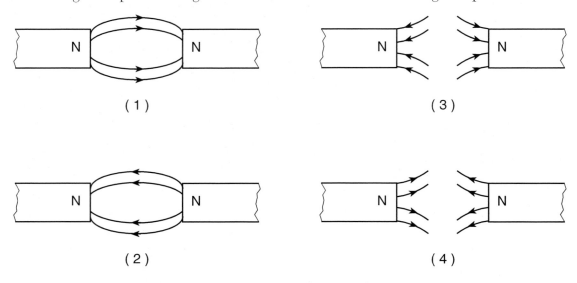

Part B–1

Answer all questions in this part.

Directions (36–50): For *each* statement or question, choose the word or expression that, of those given, best completes the statement or answers the question. Some questions may require the use of the *2006 Edition Reference Tables for Physical Setting/Physics.* Record your answers on your separate answer sheet.

36 Which measurement is closest to 1×10^{-2} meter?

(1) diameter of an atom
(2) width of a student's finger
(3) length of a football field
(4) height of a schoolteacher

37 Which graph represents the relationship between the speed of a freely falling object and the time of fall of the object near Earth's surface?

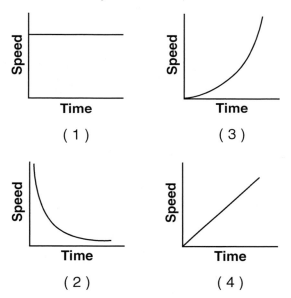

(1)   (3)

(2)   (4)

38 A hair dryer with a resistance of 9.6 ohms operates at 120 volts for 2.5 minutes. The total electrical energy used by the dryer during this time interval is

(1) 2.9×10^3 J   (3) 1.7×10^5 J
(2) 3.8×10^3 J   (4) 2.3×10^5 J

39 A box weighing 46 newtons rests on an incline that makes an angle of 25° with the horizontal. What is the magnitude of the component of the box's weight perpendicular to the incline?

(1) 19 N   (3) 42 N
(2) 21 N   (4) 46 N

40 Which graph represents the motion of an object traveling with a positive velocity and a negative acceleration?

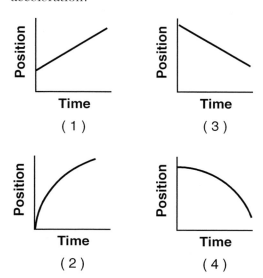

(1)   (3)

(2)   (4)

41 Car *A*, moving in a straight line at a constant speed of 20. meters per second, is initially 200 meters behind car *B*, moving in the same straight line at a constant speed of 15 meters per second. How far must car *A* travel from this initial position before it catches up with car *B*?

(1) 200 m   (3) 800 m
(2) 400 m   (4) 1000 m

42 A 2700-ohm resistor in an electric circuit draws a current of 2.4 milliamperes. The total charge that passes through the resistor in 15 seconds is

(1) 1.6×10^{-4} C   (3) 1.6×10^{-1} C
(2) 3.6×10^{-2} C   (4) 3.6×10^{1} C

43 A 1000.–kilogram car traveling 20.0 meters per second east experiences an impulse of 2000. newton • seconds west. What is the final velocity of the car after the impulse has been applied?

(1) 18.0 m/s east   (3) 20.5 m/s west
(2) 19.5 m/s east   (4) 22.0 m/s west

44 Which graph represents the relationship between the potential difference applied to a copper wire and the resulting current in the wire at constant temperature?

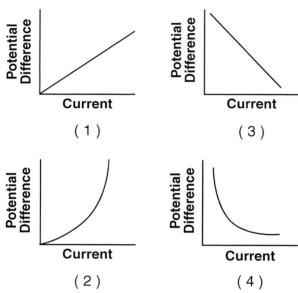

45 A tungsten wire has resistance R at 20°C. A second tungsten wire at 20°C has twice the length and half the cross-sectional area of the first wire. In terms of R, the resistance of the second wire is

(1) $\frac{R}{2}$

(3) $2R$

(2) R

(4) $4R$

46 After an incandescent lamp is turned on, the temperature of its filament rapidly increases from room temperature to its operating temperature. As the temperature of the filament increases, what happens to the resistance of the filament and the current through the filament?

(1) The resistance increases and the current decreases.

(2) The resistance increases and the current increases.

(3) The resistance decreases and the current decreases.

(4) The resistance decreases and the current increases.

47 Parallel wave fronts are incident on an opening in a barrier. Which diagram shows the configuration of wave fronts and barrier opening that will result in the greatest diffraction of the waves passing through the opening? [Assume all diagrams are drawn to the same scale.]

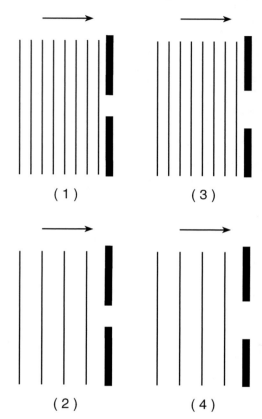

48 A singer demonstrated that she could shatter a crystal glass by singing a note with a wavelength of 0.320 meter in air at STP. What was the natural frequency of the glass?

(1) 9.67×10^{-4} Hz

(3) 1.03×10^{3} Hz

(2) 1.05×10^{2} Hz

(4) 9.38×10^{8} Hz

49 The diagram below represents a standing wave in a string.

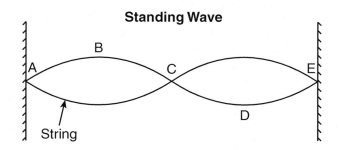

Standing Wave

Maximum constructive interference occurs at the

(1) antinodes A, C, and E
(2) nodes A, C, and E

(3) antinodes B and D
(4) nodes B and D

50 Which circuit diagram represents voltmeter V connected correctly to measure the potential difference across resistor R_2?

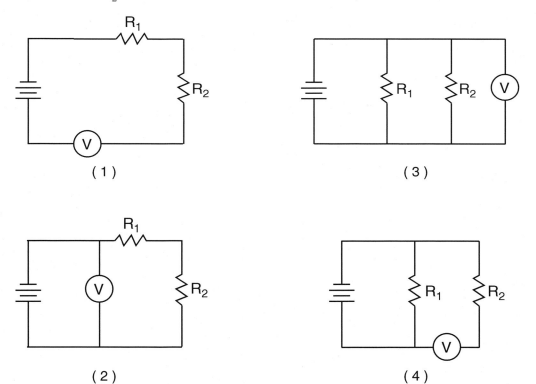

(1)

(3)

(2)

(4)

Part B–2

Answer all questions in this part.

Directions (51–65): Record your answers in the spaces provided in your answer booklet. Some questions may require the use of the *2006 Edition Reference Tables for Physical Setting/Physics.*

Base your answers to questions 51 through 53 on the information and diagram below and on your knowledge of physics.

As represented in the diagram below, a constant 15-newton force, *F*, is applied to a 2.5-kilogram box, accelerating the box to the right at 2.0 meters per second squared across a rough horizontal surface.

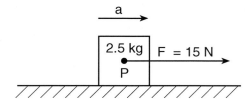

51–52 Calculate the magnitude of the net force acting on the box. [Show all work, including the equation and substitution with units.]   [2]

53 Determine the magnitude of the force of friction on the box.   [1]

Base your answers to questions 54 and 55 on the information and diagram below and on your knowledge of physics.

A ray of light $(f = 5.09 \times 10^{14}$ Hz) is traveling through a mineral sample that is submerged in water. The ray refracts as it enters the water, as shown in the diagram below.

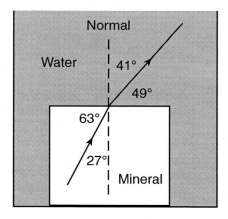

54–55 Calculate the absolute index of refraction of the mineral. [Show all work, including the equation and substitution with units.]   [2]

Base your answers to questions 56 through 58 on the information below and on your knowledge of physics.

A ball is rolled twice across the same level laboratory table and allowed to roll off the table and strike the floor. In each trial, the time it takes the ball to travel from the edge of the table to the floor is accurately measured. [Neglect friction.]

56–57 In trial A, the ball is traveling at 2.50 meters per second when it reaches the edge of the table. The ball strikes the floor 0.391 second after rolling off the edge of the table. Calculate the height of the table. [Show all work, including the equation and substitution with units.] [2]

58 In trial B, the ball is traveling at 5.00 meters per second when it reaches the edge of the table. Compare the time it took the ball to reach the floor in trial B to the time it took the ball to reach the floor in trial A. [1]

_____

Base your answers to questions 59 through 61 on the information and diagram below and on your knowledge of physics.

A toy airplane flies clockwise at a constant speed in a horizontal circle of radius 8.0 meters. The magnitude of the acceleration of the airplane is 25 meters per second squared. The diagram shows the path of the airplane as it travels around the circle.

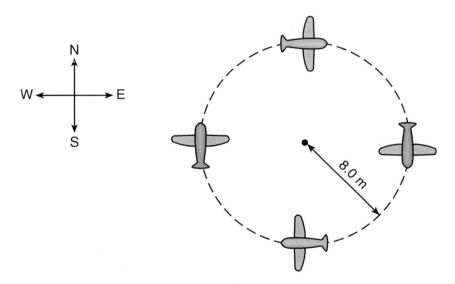

59–60 Calculate the speed of the airplane. [Show all work, including the equation and substitution with units.] [2]

61 State the direction of the velocity of the airplane at the instant the acceleration of the airplane is southward. [1]

_____

Base your answers to questions 62 through 64 on the information and graph below and on your knowledge of physics.

The graph below represents the speed of a marble rolling down a straight incline as a function of time.

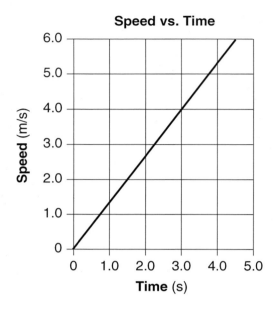

Speed vs. Time

62 What quantity is represented by the slope of the graph?   [1]

63–64 Calculate the distance the marble travels during the first 3.0 seconds. [Show all work, including the equation and substitution with units.]   [2]

_____

65 The graph below represents the relationship between weight and mass for objects on the surface of planet *X*.

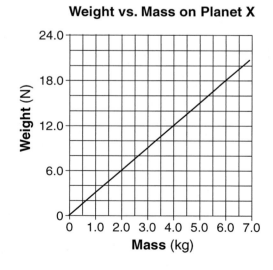

Weight vs. Mass on Planet X

Determine the acceleration due to gravity on the surface of planet *X*.   [1]

_____

Part C

Answer all questions in this part.

Directions (66–85): Record your answers in the spaces provided in your answer booklet. Some questions may require the use of the *2006 Edition Reference Tables for Physical Setting/Physics.*

Base your answers to questions 66 through 69 on the information and vector diagram below and on your knowledge of physics.

A hiker starts at point *P* and walks 2.0 kilometers due east and then 1.4 kilometers due north. The vectors in the diagram below represent these two displacements.

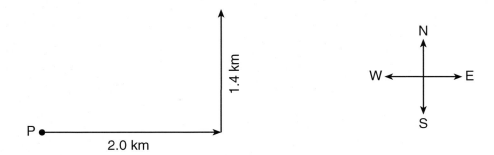

66 Using a metric ruler, determine the scale used in the vector diagram. [1]

67 On the diagram *in your answer booklet,* use a ruler to construct the vector representing the hiker's resultant displacement. [1]

68 Determine the magnitude of the hiker's resultant displacement. [1]

69 Using a protractor, determine the angle between east and the hiker's resultant displacement. [1]

Base your answers to questions 70 through 74 on the information and diagram below and on your knowledge of physics.

>A jack-in-the-box is a toy in which a figure in an open box is pushed down, compressing a spring. The lid of the box is then closed. When the box is opened, the figure is pushed up by the spring. The spring in the toy is compressed 0.070 meter by using a downward force of 12.0 newtons.

70–71 Calculate the spring constant of the spring. [Show all work, including the equation and substitution with units.] [2]

72–73 Calculate the total amount of elastic potential energy stored in the spring when it is compressed. [Show all work, including the equation and substitution with units.] [2]

74 Identify *one* form of energy to which the elastic potential energy of the spring is converted when the figure is pushed up by the spring. [1]

Base your answers to questions 75 through 80 on the information below and on your knowledge of physics.

A 12-volt battery causes 0.60 ampere to flow through a circuit that contains a lamp and a resistor connected in parallel. The lamp is operating at 6.0 watts.

75 Using the circuit symbols shown on the *Reference Tables for Physical Setting/Physics*, draw a diagram of the circuit in the space provided *in your answer booklet.*   [1]

76–77 Calculate the current through the lamp. [Show all work, including the equation and substitution with units.]   [2]

78 Determine the current in the resistor.   [1]

79–80 Calculate the resistance of the resistor. [Show all work, including the equation and substitution with units.]   [2]

Base your answers to questions 81 through 85 on the information below and on your knowledge of physics.

The Great Nebula in the constellation Orion consists primarily of excited hydrogen gas. The electrons in the atoms of excited hydrogen have been raised to higher energy levels. When these atoms release energy, a frequent electron transition is from the excited $n = 3$ energy level to the $n = 2$ energy level, which gives the nebula one of its characteristic colors.

81 Determine the energy, in electronvolts, of an emitted photon when an electron transition from $n = 3$ to $n = 2$ occurs.   [1]

82 Determine the energy of this emitted photon in joules.   [1]

83–84 Calculate the frequency of the emitted photon. [Show all work, including the equation and substitution with units.]   [2]

85 Identify the color of light associated with this photon.   [1]

Physical Setting/PHYSICS
June 2016

ANSWER BOOKLET

Student ....

Teacher ....

School .... Grade ....

Answer all questions in the examination.
Record your answers in this booklet.

Performance Test Score
(Maximum Score: 16)

Part	Maximum Score	Student's Score
A	35	
B-1	15	
B-2	15	
C	20	

Total Written Test Score
(Maximum Raw Score: 85)

Final Score
(from conversion chart)

Rater's Initials:

Rater 1 .... Rater 2 ....

Part A

1	10..............	19	28..............
2	11..............	20	29..............
3	12..............	21	30..............
4	13..............	22	31..............
5	14..............	23	32..............
6	15..............	24	33..............
7	16..............	25	34..............
8	17..............	26	35..............
9	18..............	27	

Part B

36	40..............	44	48..............
37	41..............	45	49..............
38	42..............	46	50..............
39	43..............	47	

☐ Male

Student . . . . Sex: ☐ Female

Teacher . . . .

School . . . . Grade . . . .

Record your answers for Part B–2 and Part C in this booklet.

Part B–2

51–52

53 _____ N

54–55

56–57

58 _____

_____

59–60

61 _____

62 _____

63–64

65 _____ m/s^2

Part C

66 1.0 cm =_____ **km**

67

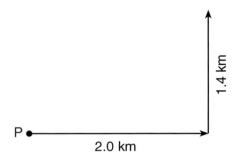

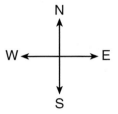

68 _____ **km**

69 _____ °

70–71

72–73

74 _____

75

76–77

78 _____ **A**

79–80

81 _____ eV

82 _____ J

83–84

85 _____

Physical Setting/PHYSICS
June 2017

Part A

Answer all questions in this part.

Directions (1–35): For *each* statement or question, choose the word or expression that, of those given, best completes the statement or answers the question. Some questions may require the use of the *2006 Edition Reference Tables for Physical Setting/Physics*. Record your answers on your separate answer sheet.

1 A unit used for a vector quantity is
  (1) watt
  (2) newton
  (3) kilogram
  (4) second

2 A displacement vector with a magnitude of 20. meters could have perpendicular components with magnitudes of
  (1) 10. m and 10. m
  (2) 12 m and 8.0 m
  (3) 12 m and 16 m
  (4) 16 m and 8.0 m

3 A hiker travels 1.0 kilometer south, turns and travels 3.0 kilometers west, and then turns and travels 3.0 kilometers north. What is the total distance traveled by the hiker?
  (1) 3.2 km
  (2) 3.6 km
  (3) 5.0 km
  (4) 7.0 km

4 A car with an initial velocity of 16.0 meters per second east slows uniformly to 6.0 meters per second east in 4.0 seconds. What is the acceleration of the car during this 4.0-second interval?
  (1) 2.5 m/s^2 west
  (2) 2.5 m/s^2 east
  (3) 4.0 m/s^2 west
  (4) 4.0 m/s^2 east

5 On the surface of planet X, a body with a mass of 10. kilograms weighs 40. newtons. The magnitude of the acceleration due to gravity on the surface of planet X is
  (1) 4.0×10^3 m/s^2
  (2) 4.0×10^2 m/s^2
  (3) 9.8 m/s^2
  (4) 4.0 m/s^2

6 A car traveling in a straight line at an initial speed of 8.0 meters per second accelerates uniformly to a speed of 14 meters per second over a distance of 44 meters. What is the magnitude of the acceleration of the car?
  (1) 0.41 m/s^2
  (2) 1.5 m/s^2
  (3) 3.0 m/s^2
  (4) 2.2 m/s^2

7 An object starts from rest and falls freely for 40. meters near the surface of planet P. If the time of fall is 4.0 seconds, what is the magnitude of the acceleration due to gravity on planet P?
  (1) 0 m/s^2
  (2) 1.3 m/s^2
  (3) 5.0 m/s^2
  (4) 10. m/s^2

8 If a block is in equilibrium, the magnitude of the block's acceleration is
  (1) zero
  (2) decreasing
  (3) increasing
  (4) constant, but not zero

9 The diagram below shows a light ray striking a plane mirror.

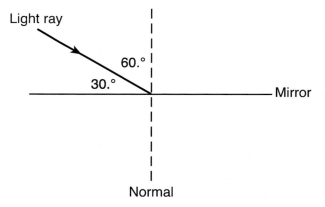

What is the angle of reflection?
  (1) 30.°
  (2) 60.°
  (3) 90.°
  (4) 120.°

10 An electric field exerts an electrostatic force of magnitude 1.5×10^{-14} newton on an electron within the field. What is the magnitude of the electric field strength at the location of the electron?
  (1) 2.4×10^{-33} N/C
  (2) 1.1×10^{-5} N/C
  (3) 9.4×10^4 N/C
  (4) 1.6×10^{16} N/C

11 A 7.0-kilogram cart, *A*, and a 3.0-kilogram cart, *B*, are initially held together at rest on a horizontal, frictionless surface. When a compressed spring attached to one of the carts is released, the carts are pushed apart. After the spring is released, the speed of cart *B* is 6.0 meters per second, as represented in the diagram below.

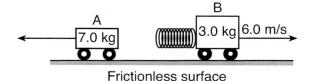

Frictionless surface

What is the speed of cart *A* after the spring is released?

(1) 14 m/s   (3) 3.0 m/s
(2) 6.0 m/s   (4) 2.6 m/s

12 An electron in a magnetic field travels at constant speed in the circular path represented in the diagram below.

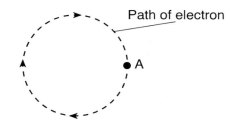

Path of electron

Which arrow represents the direction of the net force acting on the electron when the electron is at position *A*?

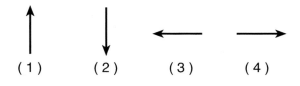

13 The potential difference between two points, *A* and *B*, in an electric field is 2.00 volts. The energy required to move a charge of 8.00×10^{-19} coulomb from point *A* to point *B* is

(1) 4.00×10^{-19} J   (3) 6.25×10^{17} J
(2) 1.60×10^{-18} J   (4) 2.50×10^{18} J

14 Which statement describes the gravitational force and the electrostatic force between two charged particles?

(1) The gravitational force may be either attractive or repulsive, whereas the electrostatic force must be attractive.
(2) The gravitational force must be attractive, whereas the electrostatic force may be either attractive or repulsive.
(3) Both forces may be either attractive or repulsive.
(4) Both forces must be attractive.

15 An electrostatic force exists between two $+3.20 \times 10^{-19}$-coulomb point charges separated by a distance of 0.030 meter. As the distance between the two point charges is *decreased*, the electrostatic force of

(1) attraction between the two charges decreases
(2) attraction between the two charges increases
(3) repulsion between the two charges decreases
(4) repulsion between the two charges increases

16 What is the energy of the photon emitted when an electron in a mercury atom drops from energy level *f* to energy level *b*?

(1) 8.42 eV   (3) 3.06 eV
(2) 5.74 eV   (4) 2.68 eV

17 An observer counts 4 complete water waves passing by the end of a dock every 10. seconds. What is the frequency of the waves?

(1) 0.40 Hz   (3) 40. Hz
(2) 2.5 Hz   (4) 4.0 Hz

18 Copper is a metal commonly used for electrical wiring in houses. Which metal conducts electricity better than copper at 20°C?

(1) aluminum   (3) nichrome
(2) gold   (4) silver

19 A motor does 20. joules of work on a block, accelerating the block vertically upward. Neglecting friction, if the gravitational potential energy of the block increases by 15 joules, its kinetic energy

(1) decreases by 5 J   (3) decreases by 35 J
(2) increases by 5 J   (4) increases by 35 J

20 When only one lightbulb blows out, an entire string of decorative lights goes out. The lights in this string must be connected in

(1) parallel with one current pathway
(2) parallel with multiple current pathways
(3) series with one current pathway
(4) series with multiple current pathways

21 An electric toaster is rated 1200 watts at 120 volts. What is the total electrical energy used to operate the toaster for 30. seconds?

(1) 1.8×10^3 J   (3) 1.8×10^4 J
(2) 3.6×10^3 J   (4) 3.6×10^4 J

22 What is the rate at which work is done in lifting a 35-kilogram object vertically at a constant speed of 5.0 meters per second?

(1) 1700 W   (3) 180 W
(2) 340 W   (4) 7.0 W

23 When a wave travels through a medium, the wave transfers

(1) mass, only
(2) energy, only
(3) both mass and energy
(4) neither mass nor energy

24 Glass may shatter when exposed to sound of a particular frequency. This phenomenon is an example of

(1) refraction   (3) resonance
(2) diffraction   (4) the Doppler effect

25 Which waves require a material medium for transmission?

(1) light waves   (3) sound waves
(2) radio waves   (4) microwaves

26 Which type of oscillation would most likely produce an electromagnetic wave?

(1) a vibrating tuning fork
(2) a washing machine agitator at work
(3) a swinging pendulum
(4) an electron traveling back and forth in a wire

27 If monochromatic light passes from water into air with an angle of incidence of 35°, which characteristic of the light will remain the same?

(1) frequency   (3) speed
(2) wavelength   (4) direction

28 The absolute index of refraction of medium Y is twice as great as the absolute index of refraction of medium X. As a light ray travels from medium X into medium Y, the speed of the light ray is

(1) halved   (3) quartered
(2) doubled   (4) quadrupled

29 The diagram below shows a transverse wave moving toward the right along a rope.

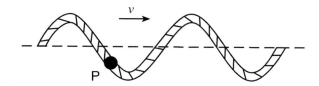

At the instant shown, point *P* on the rope is moving toward the

(1) bottom of the page   (3) left
(2) top of the page   (4) right

30 When an isolated conductor is placed in the vicinity of a positive charge, the conductor is attracted to the charge. The charge of the conductor

(1) must be positive
(2) must be negative
(3) could be neutral or positive
(4) could be neutral or negative

31 The quarks that compose a baryon may have charges of

(1) $+\frac{2}{3}e, +\frac{2}{3}e,$ and $-\frac{1}{3}e$

(2) $+\frac{1}{3}e, -\frac{1}{3}e,$ and $+\frac{2}{3}e$

(3) $-1e, -1e,$ and 0

(4) $+\frac{2}{3}e, +\frac{2}{3}e,$ and 0

32 A rubber block weighing 60. newtons is resting on a horizontal surface of dry asphalt. What is the magnitude of the minimum force needed to start the rubber block moving across the dry asphalt?

(1) 32 N   (3) 51 N
(2) 40. N   (4) 60. N

33 The data table below lists the mass and speed of four different objects.

Object	Mass (kg)	Speed (m/s)
A	2.0	6.0
B	4.0	5.0
C	6.0	4.0
D	8.0	2.0

Which object has the greatest inertia?

(1) *A*   (3) *C*
(2) *B*   (4) *D*

34 The electroscope shown in the diagram below is made completely of metal and consists of a knob, a stem, and leaves. A positively charged rod is brought near the knob of the electroscope and then removed.

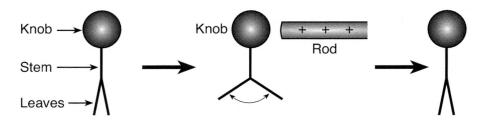

The motion of the leaves results from electrons moving from the

(1) leaves to the knob, only
(2) knob to the leaves, only
(3) leaves to the knob and then back to the leaves
(4) knob to the leaves and then back to the knob

35 Which circuit diagram represents the correct way to measure the current in a resistor?

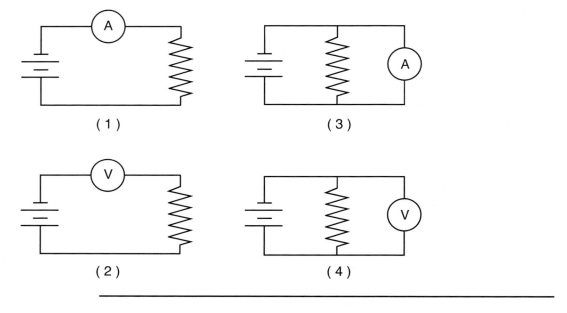

(1)

(3)

(2)

(4)

Part B–1

Answer all questions in this part.

Directions (36–50): For *each* statement or question, choose the word or expression that, of those given, best completes the statement or answers the question. Some questions may require the use of the *2006 Edition Reference Tables for Physical Setting/Physics*. Record your answers on your separate answer sheet.

36 The height of a typical kitchen table is approximately

(1) 10^{-2} m   (3) 10^1 m
(2) 10^0 m   (4) 10^2 m

37 A ball is thrown with a velocity of 35 meters per second at an angle of 30.° above the horizontal. Which quantity has a magnitude of zero when the ball is at the highest point in its trajectory?

(1) the acceleration of the ball
(2) the momentum of the ball
(3) the horizontal component of the ball's velocity
(4) the vertical component of the ball's velocity

38 The graph below represents the relationship between velocity and time of travel for a toy car moving in a straight line.

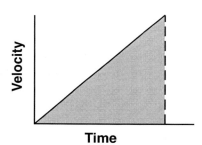

The shaded area under the line represents the toy car's

(1) displacement   (3) acceleration
(2) momentum   (4) speed

39 A spring stores 10. joules of elastic potential energy when it is compressed 0.20 meter. What is the spring constant of the spring?

(1) 5.0×10^1 N/m   (3) 2.5×10^2 N/m
(2) 1.0×10^2 N/m   (4) 5.0×10^2 N/m

Base your answers to questions 40 and 41 on the information below and on your knowledge of physics.

A cannonball with a mass of 1.0 kilogram is fired horizontally from a 500.-kilogram cannon, initially at rest, on a horizontal, frictionless surface. The cannonball is acted on by an average force of 8.0×10^3 newtons for 1.0×10^{-1} second.

40 What is the magnitude of the change in momentum of the cannonball during firing?

(1) 0 kg•m/s   (3) 8.0×10^3 kg•m/s
(2) 8.0×10^2 kg•m/s   (4) 8.0×10^4 kg•m/s

41 What is the magnitude of the average net force acting on the cannon?

(1) 1.6 N   (3) 8.0×10^3 N
(2) 16 N   (4) 4.0×10^6 N

42 A metal sphere, X, has an initial net charge of -6×10^{-6} coulomb and an identical sphere, Y, has an initial net charge of $+2 \times 10^{-6}$ coulomb. The spheres touch each other and then separate. What is the net charge on sphere X after the spheres have separated?

(1) 0 C   (3) -4×10^{-6} C
(2) -2×10^{-6} C   (4) -6×10^{-6} C

43 A constant eastward horizontal force of 70. newtons is applied to a 20.-kilogram crate moving toward the east on a level floor. If the frictional force on the crate has a magnitude of 10. newtons, what is the magnitude of the crate's acceleration?

(1) 0.50 m/s^2 (3) 3.0 m/s^2
(2) 3.5 m/s^2 (4) 4.0 m/s^2

44 Which graph represents the relationship between the energy of photons and the wavelengths of photons in a vacuum?

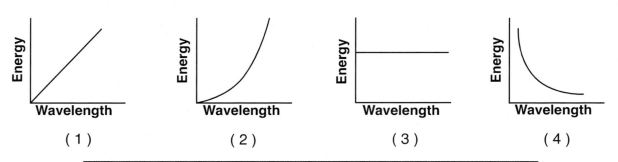

(1)   (2)   (3)   (4)

Base your answers to questions 45 and 46 on the information and diagram below and on your knowledge of physics.

One end of a long spring is attached to a wall. A student vibrates the other end of the spring vertically, creating a wave that moves to the wall and reflects back toward the student, resulting in a standing wave in the spring, as represented below.

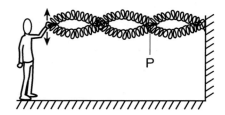

45 What is the phase difference between the incident wave and the reflected wave at point *P*?

(1) 0°   (3) 180°
(2) 90°   (4) 270°

46 What is the total number of antinodes on the standing wave in the diagram?

(1) 6   (3) 3
(2) 2   (4) 4

47 The diagrams below represent four pieces of copper wire at 20.°C. For each piece of wire, ℓ represents a unit of length and A represents a unit of cross-sectional area.

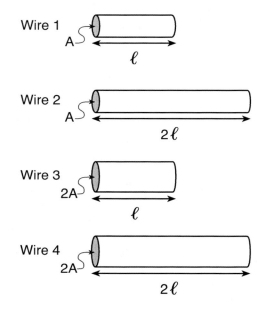

The piece of wire that has the greatest resistance is

(1) wire 1   (3) wire 3
(2) wire 2   (4) wire 4

Base your answers to questions 48 and 49 on the diagram below, which represents two charged, identical metal spheres, and on your knowledge of physics.

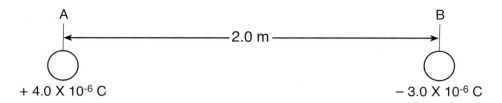

48   The number of excess elementary charges on sphere *A* is

(1) 6.4×10^{-25}

(2) 6.4×10^{-19}

(3) 2.5×10^{13}

(4) 5.0×10^{13}

49   What is the magnitude of the electric force between the two spheres?

(1) 3.0×10^{-12} N

(2) 1.0×10^{-6} N

(3) 2.7×10^{-2} N

(4) 5.4×10^{-2} N

50   The diagram below represents the wave fronts produced by a point source moving to the right in a uniform medium. Observers are located at points *A* and *B*.

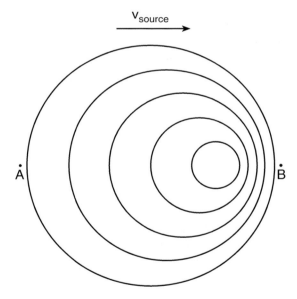

Compared to the wave frequency and wavelength observed at point *A*, the wave observed at point *B* has a

(1) higher frequency and a shorter wavelength

(2) higher frequency and a longer wavelength

(3) lower frequency and a shorter wavelength

(4) lower frequency and a longer wavelength

Part B–2

Answer all questions in this part.

Directions (51–65): Record your answers in the spaces provided in your answer booklet. Some questions may require the use of the *2006 Edition Reference Tables for Physical Setting/Physics.*

51 On the diagram *in your answer booklet*, sketch *at least four* magnetic field lines of force around a bar magnet. [Include arrows to show the direction of each field line.] [1]

Base your answers to questions 52 through 54 on the information below and on your knowledge of physics.

Tritium is a radioactive form of the element hydrogen. A tritium nucleus is composed of one proton and two neutrons. When a tritium nucleus decays, it emits a beta particle (an electron) and an antineutrino to create a stable form of helium. During beta decay, a neutron is spontaneously transformed into a proton, an electron, and an antineutrino.

52 What is the total number of quarks in a tritium nucleus? [1]

53 What is the total charge, in elementary charges, of a proton, an electron, and an antineutrino? [1]

54 What fundamental interaction is responsible for binding together the protons and neutrons in a helium nucleus? [1]

55 The diagram below represents a ball projected horizontally from a cliff at a speed of 10. meters per second. The ball travels the path shown and lands at time *t* and distance *d* from the base of the cliff. [Neglect friction.]

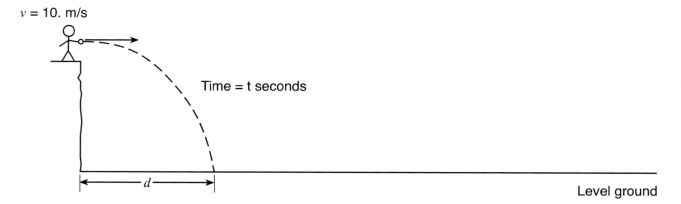

A second, identical ball is projected horizontally from the cliff at 20. meters per second. Determine the distance the second ball lands from the base of the cliff in terms of *d*. [1]

56–57 An operating television set draws 0.71 ampere of current when connected to a 120-volt outlet. Calculate the time it takes the television to consume 3.0×10^5 joules of electric energy. [Show all work, including the equation and substitution with units.]   [2]

58–59 On the centimeter grid *in your booklet*, draw *at least one* cycle of a periodic transverse wave with an amplitude of 2.0 centimeters and a wavelength of 6.0 centimeters.   [2]

60 The diagram below represents a 35-newton block hanging from a vertical spring, causing the spring to elongate from its original length.

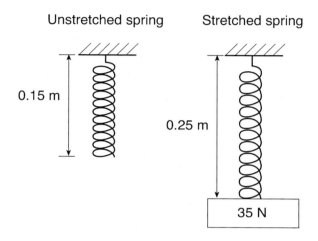

Determine the spring constant of the spring.   [1]

61 Determine the amount of matter, in kilograms, that must be converted to energy to yield 1.0 gigajoule.   [1]

62 Thunder results from the expansion of air as lightning passes through it. The distance between an observer and a lightning strike may be determined if the time that elapses between the observer seeing the lightning and hearing the thunder is known. Explain why the lightning strike is seen before the thunder is heard.   [1]

63–64 A bolt of lightning transfers 28 coulombs of charge through an electric potential difference of 3.2×10^7 volts between a cloud and the ground in 1.5×10^{-3} second. Calculate the average electric current between the cloud and the ground during this transfer of charge. [Show all work, including the equation and substitution with units.]   [2]

65 The diagram below represents two pulses traveling toward each other in a uniform medium.

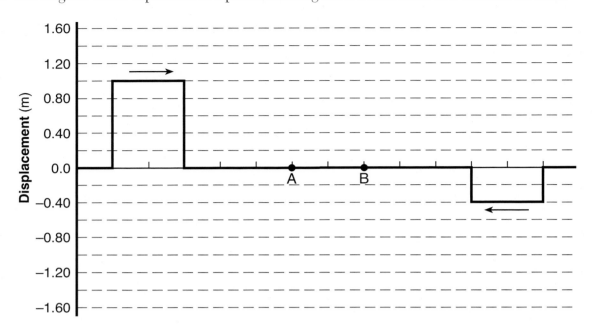

On the grid *in your answer booklet*, draw the resultant displacement of the medium when both pulses are located between points *A* and *B*.   [1]

Part C

Answer all questions in this part.

Directions (66–85): Record your answers in the spaces provided in your answer booklet. Some questions may require the use of the *2006 Edition Reference Tables for Physical Setting/Physics*.

Base your answers to questions 66 through 70 on the information and diagram below and on your knowledge of physics.

As represented in the diagram, a ski area rope-tow pulls a 72.0-kilogram skier from the bottom to the top of a 40.0-meter-high hill. The rope-tow exerts a force of magnitude 158 newtons to move the skier a total distance of 230. meters up the side of the hill at constant speed.

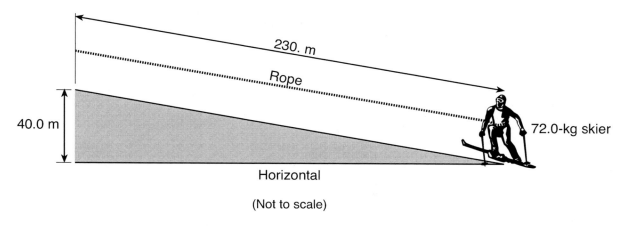

(Not to scale)

66 Determine the total amount of work done by the rope on the skier. [1]

67–68 Calculate the total amount of gravitational potential energy gained by the skier while moving up the hill. [Show all work, including the equation and substitution with units] [2]

69 Describe what happens to the internal energy of the skier-hill system as the skier is pulled up the hill. [1]

70 Describe what happens to the total mechanical energy of the skier-hill system as the skier is pulled up the hill. [1]

Base your answers to questions 71 through 76 on the diagram and information below and on your knowledge of physics.

A 15-ohm resistor, 30.-ohm resistor, and an ammeter are connected as shown with a 60.-volt battery.

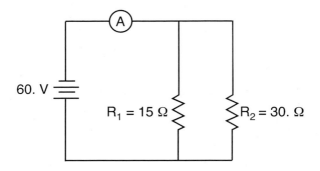

71–72 Calculate the equivalent resistance of R_1 and R_2. [Show all work, including the equation and substitution with units.]   [2]

73 Determine the current measured by the ammeter.   [1]

74–75 Calculate the rate at which the battery supplies energy to the circuit. [Show all work, including the equation and substitution with units.]   [2]

76 If another resistor were added in parallel to the original circuit, what effect would this have on the current through resistor R_1?   [1]

_____

Base your answers to questions 77 through 80 on the information below and on your knowledge of physics.

A gas-powered model airplane has a mass of 2.50 kilograms. A student exerts a force on a cord to keep the airplane flying around her at a constant speed of 18.0 meters per second in a horizontal, circular path with a radius of 25.0 meters.

77–78 Calculate the kinetic energy of the moving airplane. [Show all work, including the equation and substitution with units.]   [2]

79–80 Calculate the magnitude of the centripetal force exerted on the airplane to keep it moving in this circular path. [Show all work, including the equation and substitution with units.]   [2]

_____

Base your answers to questions 81 through 85 on the information and diagram below and on your knowledge of physics.

A ray of light with a frequency of 5.09×10^{14} hertz traveling in medium X is refracted at point P. The angle of refraction is 90.°, as represented in the diagram.

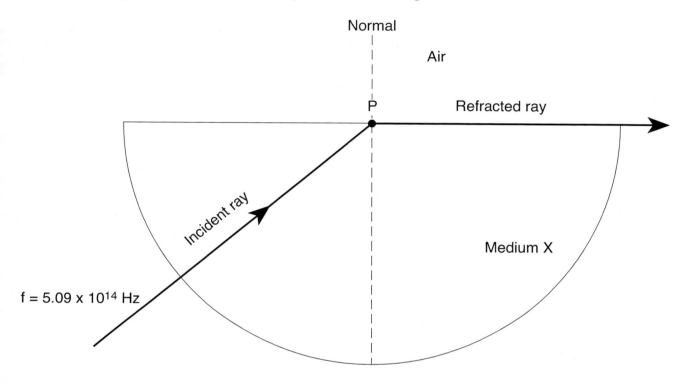

81–82 Calculate the wavelength of the light ray in air. [Show all work, including the equation and substitution with units.] [2]

83 Measure the angle of incidence for the light ray incident at point P and record the value *in your answer booklet.* [1]

84–85 Calculate the absolute index of refraction for medium X. [Show all work, including the equation and substitution with units.] [2]

Physical Setting/PHYSICS
June 2017

ANSWER BOOKLET

Student ....

Teacher ....

School .... Grade ....

Answer all questions in the examination.
Record your answers in this booklet.

Part A

1	10	19	28 ....
2	11	20	29 ....
3	12	21	30 ....
4	13	22	31 ....
5	14	23	32 ....
6	15	24	33 ....
7	16	25	34 ....
8	17	26	35 ....
9	18	27	

Part B

36	40	44	48 ....
37	41	45	49 ....
38	42	46	50 ....
39	43	47	

☐ Male

Student . . . .   Sex: ☐ Female

Teacher . . . .

School . . . .   Grade . . . .

Record your answers for Part B–2 and Part C in this booklet.

Part B–2

51

52 _____ **quarks**

53 _____ **e**

54 _____

55 _____

56–57

58-59

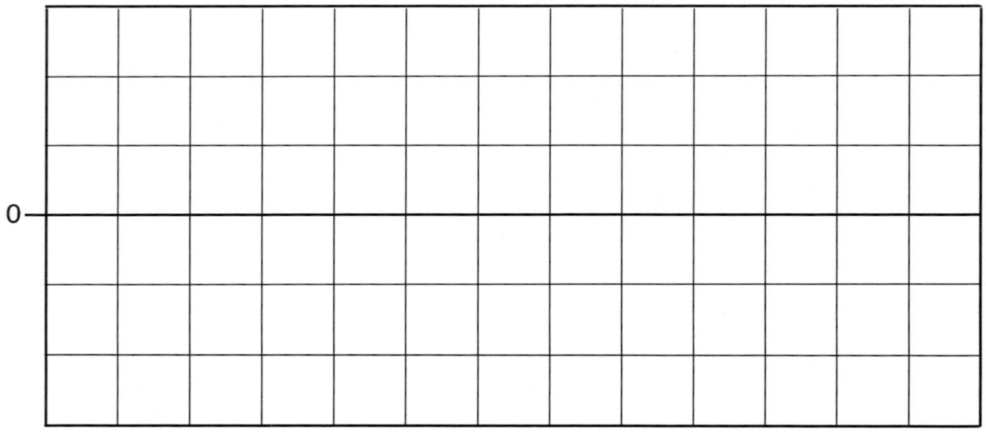

60 _____**N/m**

61 _____ **kg**

62 _____

_____

63–64

65

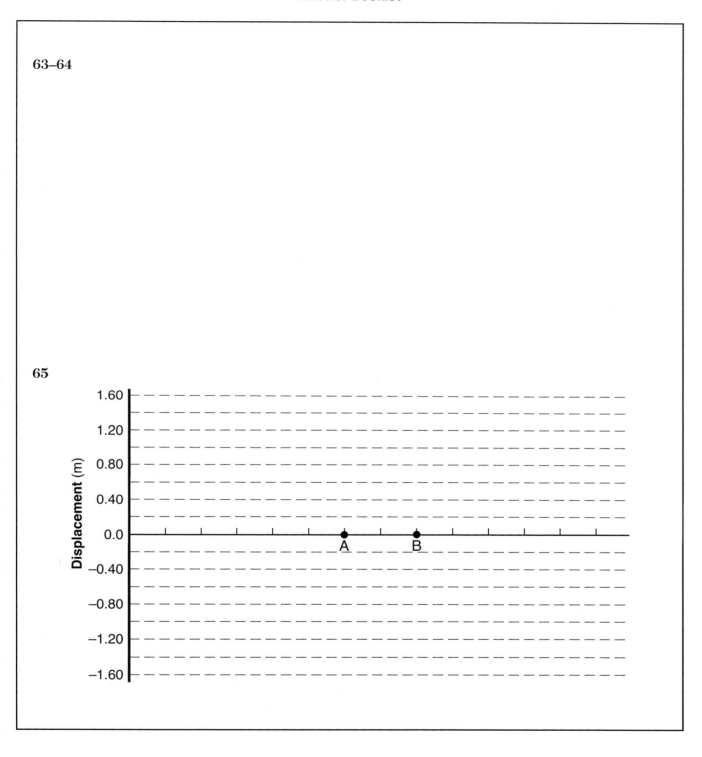

Part C

66 _____ **J**

67–68

69 _____

70 _____

71–72

73 _____ **A**

74–75

76 _____

77–78

79–80

81–82

83 _____ °

84–85